Light Alloys and High-Temperature Alloys

Light Alloys and High-Temperature Alloys

Editor

Lijun Zhang

MDPI • Basel • Beijing • Wuhan • Barcelona • Belgrade • Manchester • Tokyo • Cluj • Tianjin

Editor
Lijun Zhang
State Key Laboratory of
Powder Metallurgy
Central South University
Changsha
China

Editorial Office
MDPI
St. Alban-Anlage 66
4052 Basel, Switzerland

This is a reprint of articles from the Special Issue published online in the open access journal *Materials* (ISSN 1996-1944) (available at: www.mdpi.com/journal/materials/special_issues/Light_Alloy_High_Temp_Alloy).

For citation purposes, cite each article independently as indicated on the article page online and as indicated below:

LastName, A.A.; LastName, B.B.; LastName, C.C. Article Title. *Journal Name* **Year**, *Volume Number*, Page Range.

ISBN 978-3-0365-5660-4 (Hbk)
ISBN 978-3-0365-5659-8 (PDF)

Contents

About the Editor

Lijun Zhang

Lijun Zhang, male, married, born on August 1983 in Jiangxi province, China. He has held a full professor position at the State Key Laboratory of Powder Metallurgy, Central South University, China, since 2013. He serves as the deputy secretary general of the Computational Materials Branch of the Chinese Materials Research Society, the vice chairman of the Youth Science and Technology Association of Central South University, as well as an associate editor/an editorial board member/a youth editorial board member of 10 international and domestic journals.

Prof. Zhang has long been engaged in computational thermodynamics, computational kinetics, and the driven design of high-performance materials as well as their preparation/manufacturing. He has served as the PI of more than 25 projects from the Chinese government and from many companies. Thus far, he has published 1 book, 3 book chapters, as well as more than 120 papers as the first author/corresponding author in international journals such as *Nature Communications, npj Computational Materials, Acta Materialia*, etc. He has organized 5 international/domestic conferences and given more than 30 international plenary/invited presentations.

He was awarded the Alexander von Humboldt Research Fellowship after his Ph.D degree in 2010; a Shenghua Distinguished Professorship at Central South University in 2013; and the following awards: Huxiang Young Talents of Hunan province in 2017, 121 Innovative Talents of Hunan province in 2019, and Natural Science Foundation of Hunan province for Distinguished Young Scholars in 2021. Moreover, he has also won first prize for the Hunan Province Natural Science Award, the Hunan Province Outstanding Doctoral Dissertation Award, and the Annual Best Paper Award of the international journal *CALPHAD*.

Preface to "Light Alloys and High-Temperature Alloys"

Light alloys and high-temperature alloys are widely used as key engineering materials in both the civil and military industries due to their excellent comprehensive properties and performance. To meet the growing demand on the properties/performance of these materials, there is a perpetual need to explore novel light and high-temperature alloys. Over recent decades, huge amounts of theoretical and/or experimental efforts have been devoted to this field and great achievements have been made. Consequently, the purpose of the book titled "Light Alloys and High-Temperature Alloys" is to provide scientists and graduate students in the field of structural materials from around the world with a snapshot of the state-of-the-art research on light alloys and high-temperature alloys in different aspects.

In this collection, 14 research papers contributed by 85 authors at 27 universities/institutes/companies from 9 countries, including China, the USA, the UK, Germany, Spain, Australia, Ukraine, Poland, and Romania are compiled. The topics cover different types of light alloys, including Al-, Mg-, and Ti-based ones (also Ti-based metal matrix composites), and high-temperature alloys, including Ni-, Fe-, Nb-, and Ta-based ones. Two new types of alloys, i.e., complex concentrated alloys (CCAs) and phase change materials, are also included. Moreover, in this book, a variety of multi-scale theoretical methods, ranging from first-principles calculations, first-principles molecular dynamic simulations, and Calculation of Phase Diagram (CALPHAD) modeling to crystal plasticity finite element simulations coupled with knowledge graph, as well as experimental techniques, e.g., casting, powder metallurgy, additive manufacturing, etc. are discussed. Accordingly, the diverse topics and state-of-the-art theoretical/experimental techniques will attract broad interest from materials researchers worldwide.

Great thanks should be given to all authors, reviewers, and academic editors who contributed. Special thanks are also devoted to Ms. Yulia Zhao, the managing editor of the journal *Materials* for her efficient communication and assistance. Without their help, this book would not have been possible.

Lijun Zhang
Editor

materials

MDPI

Article

X-ray Thermo-Diffraction Study of the Aluminum-Based Multicomponent Alloy $Al_{58}Zn_{28}Si_8Mg_6$

Yoana Bilbao [1,*], Juan José Trujillo [2], Iban Vicario [3], Gurutze Arruebarrena [2], Iñaki Hurtado [2] and Teresa Guraya [1]

[1] Department of Mining and Metallurgical Engineering and Materials Science, Faculty of Engineering of Bilbao, University of the Basque Country (UPV/EHU), 48013 Bilbao, Spain; teresa.guraya@ehu.eus
[2] Mechanical and Manufacturing Department, Faculty of Engineering, Mondragon Unibertsitatea, 20500 Arrasate/Mondragon, Spain; jjtrujillo@mondragon.edu (J.J.T.); garruebarrena@mondragon.edu (G.A.); ihurtado@mondragon.edu (I.H.)
[3] Manufacturing Processes and Materials Department, Tecnalia, Basque Research and Technology Alliance (BRTA), 48160 Derio, Spain; iban.vicario@tecnalia.com
* Correspondence: yoana.bilbao@ehu.eus

Abstract: Newly designed multicomponent light alloys are giving rise to non-conventional microstructures that need to be thoroughly studied before determining their potential applications. In this study, the novel $Al_{58}Zn_{28}Si_8Mg_6$ alloy, previously studied with CALPHAD methods, was cast and heat-treated under several conditions. An analysis of the phase evolution was carried out with in situ X-ray diffraction supported by differential scanning calorimetry and electron microscopy. A total of eight phases were identified in the alloy in the temperature range from 30 to 380 °C: α-Al, α′-Al, Zn, Si, Mg_2Si, $MgZn_2$, Mg_2Zn_{11}, and $SrZn_{13}$. Several thermal transitions below 360 °C were determined, and the natural precipitation of the Zn phase was confirmed after nine months. The study showed that the thermal history can strongly affect the presence of the $MgZn_2$ and Mg_2Zn_{11} phases. The combination of X-ray thermo-diffraction with CALPHAD methods, differential scanning calorimetry, and electron microscopy offered us a satisfactory understanding of the alloy behavior at different temperatures.

Keywords: lightweight multicomponent alloys; X-ray thermo-diffraction; differential scanning calorimetry; Al–Zn; Zn precipitation; Mg–Zn phases; strontium modification

Citation: Bilbao, Y.; Trujillo, J.J.; Vicario, I.; Arruebarrena, G.; Hurtado, I.; Guraya, T. X-ray Thermo-Diffraction Study of the Aluminum-Based Multicomponent Alloy $Al_{58}Zn_{28}Si_8Mg_6$. *Materials* **2022**, *15*, 5056. https://doi.org/10.3390/ma15145056

Academic Editor: Lijun Zhang

Received: 21 June 2022
Accepted: 19 July 2022
Published: 20 July 2022

1. Introduction

Historically, metallic alloys have been developed by selecting one or two major components and adding several minor ones that confer specific properties, such as corrosion resistance or higher mechanical properties. The multicomponent alloy concept, however, is based on the design of alloys where there are several main components that cover the central areas of phase diagrams [1].

Initial developments in the field focused mainly on steel-like alloys for industrial applications. They were based on equiatomic and near-equiatomic compositions of Co, Cr, Cu, Fe, Mn, or Ni, sometimes adding Al, Ti, or Zr, that resulted in single or dual phase microstructures [2–7]. Yeh et al. suggested that the prevalence of solid solutions over intermetallic phases could be explained by the high mixing entropy generated by the multiple components in the alloy, hence the term "high-entropy alloys" (HEAs) [8]. The conditions that the alloys should satisfy to be considered HEAs may be found in [9].

Over the past decade, research has been extended to other alloy classifications that have evolved from the original HEA concept: "medium-entropy alloys" (MEAs) [9], "non-equiatomic HEAs", or "multi-phase HEAs" that may contain bulky secondary phases [10]. In fact, the idea of intentionally having secondary phases in this type of alloy was first suggested by Miracle et al. [11]. Other alloy families have also been explored, leading to refractory metal HEAs for high temperature structural applications based on Cr, Hf, Mo, Nb, Ta, Ti, V, and W or lightweight multicomponent alloys in the aeronautical field involving Al, Li, Mg, or Zn. Several studies may be found in the literature as proof of this tendency [12–17].

Among the lightweight multicomponent alloys, Yang et al. explored the Al–Li–Mg–(Zn, Cu, Sn) system, obtaining structures dominated by intermetallic compounds. The aluminum face-centered cubic (FCC) structure predominated only in selected alloy compositions [18]. In fact, Sanchez et al. highlighted the difficulty of forming solid solutions in medium entropy alloys based on aluminum (65–70 at. %) with elements such as Cu, Mg, Cr, Fe, Si, Ni, Zn, or Zr. The magnitude of the negative mixing enthalpy of aluminum with transition metals gave rise to intermetallic phases [19]. The presence of intermetallics was also reported by Tun et al. [20]. Only the most recent research suggests that rapid solidification processes may enhance single phase microstructures in this type of alloy [21]. However, in a previous work by Nagase et al. on Al–Mg–Li–Ca equiatomic and non-equiatomic alloys, a single solid solution could not be obtained, even with rapid solidification [22].

Asadikiya et al. considered that the application of the entropy concept in aluminum alloys may be the answer to the challenge of developing novel Al alloys with improved properties [10]. Therefore, multicomponent lightweight alloys continue to be researched for their potential applications.

In the present study, the objective was to characterize the novel $Al_{58}Zn_{28}Si_8Mg_6$ cast alloy. It was designed to obtain as much solid solution of aluminum and zinc as possible, reinforced with intermetallics based on Zn, Mg, and Si. On the one hand, zinc is highly soluble in aluminum, enhancing the obtention of a solid solution matrix. On the other hand, Mg–Zn phases are the usual precipitates in 7xx.x aluminum cast alloys, while Mg–Si phases are common in 3xx.x alloys. In addition, Al–Zn-based alloys have attracted the interest of researchers beyond their usual use as coatings. In fact, Al–Zn cast alloys have potential applications where tribological and damping properties are required [23–25]. In terms of entropy, our multicomponent alloy would be classified as a multi-phase MEA.

The approach was explored by the CALPHAD (calculation of phase diagrams) method. This technique requires databases that are valid in composition ranges that may not be found in conventional alloys, thus demanding further experimental verification [26]. However, it is considered the most direct method for compositional design [27] and has already been used in the design of lightweight multicomponent alloys with differing degrees of success [28–30].

The study is focused on identifying and evaluating the effect of the temperature on the phases that are generated at different initial thermal conditions. Differential scanning calorimetry (DSC), electron microscopy, and X-ray thermo-diffraction are the techniques used in this evaluation. X-ray thermo-diffraction, also known as "high temperature X-ray diffraction" (HT-XRD), enables the in situ study of the solution and precipitation phenomena in the alloys [31–33].

2. Materials and Methods

2.1. Material Manufacturing

Aluminum was melted at 750 °C in a resistance furnace with forced convection; silicon and zinc were subsequently added. Magnesium followed, and once all the elements were melted, strontium was added as a silicon modifier. Aluminum, magnesium, and silicon were of commercial purity, whereas zinc was incorporated by adding a Zamak Zn4Al1Cu alloy so that the final alloy composition contained some residual copper. Samples were obtained to determine the chemical composition by inductively coupled plasma mass spectrometry (ICP) (Table 1).

The metal was gravity cast into a graphite mold, and samples were obtained that were 50 mm long, 22.5 mm wide, and 4 mm thick.

Table 1. Chemical composition of the alloy analyzed by ICP.

	Al	Zn	Mg	Si	Cu	Fe	Sr
wt. %	41.15	48.40	4.11	5.83	0.43	0.05	0.03
at. %	57.56	27.93	6.38	7.83	0.26	0.03	0.01

2.2. Selection of Sample Thermal Treatments and Study Temperatures

In order to select the temperatures of interest, DSC tests were performed on as-cast samples (Figure 1). A Netzsch STA 449 Fe Jupiter calorimeter was used, and measurements were made under argon atmosphere in a temperature range between 25 and 675 °C with a heating rate of 10 °C/min. Samples were then cooled down back to room temperature under these same conditions.

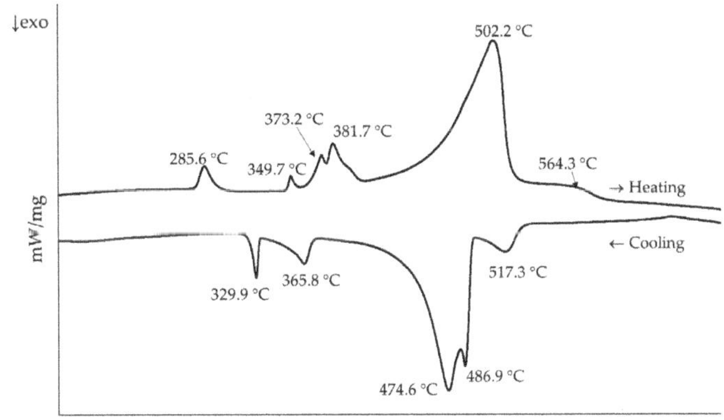

Figure 1. DSC curves for the as-cast sample. Peak temperatures during heating were considered for the thermal treatment selection of the samples.

We decided to subject the samples to seven different thermal conditions to try to separate and simplify the identification of the different phases appearing and disappearing during the heating process (Table 2 and Figure 2).

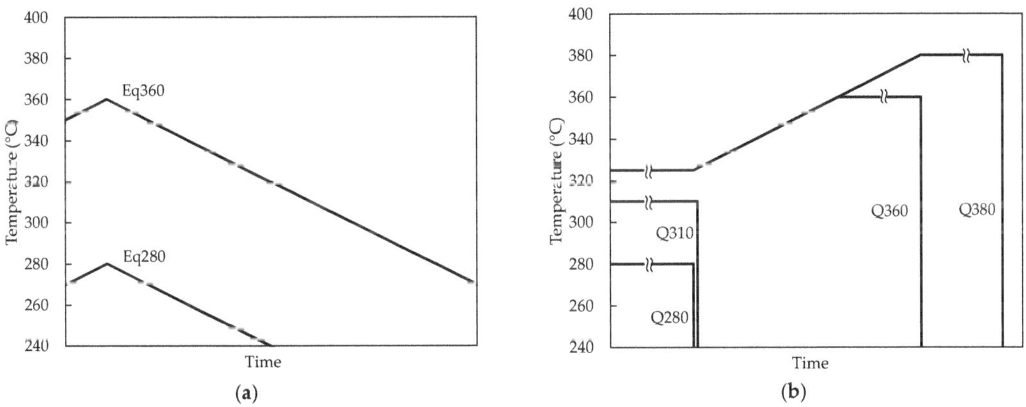

Figure 2. Heating and cooling sequences for the samples, in addition to the as-cast sample. (**a**) Slowly cooled (10 °C/min) samples (Eq280 and Eq360). (**b**) Quenched samples (Q280, Q310, Q360, and Q380).

Table 2. Heat treatments for each sample condition. The two-step solution treatment in Q360 and Q380 was applied to prevent any partial melting during one-step solution treatment at the solution temperature.

Sample Condition	Heat Treatment
as-cast	None.
Eq280	Heated to 280 °C and immediately slowly cooled (10 °C/min) to room temperature.
Eq360	Heated to 360 °C and immediately slowly cooled (10 °C/min) to room temperature.
Q280	Solution treated for 24 h at 280 °C, then water quenched to room temperature.
Q310	Solution treated for 24 h at 310 °C, then water quenched to room temperature.
Q360	Solution treated for 24 h at 325 °C, then heated and kept at 360 °C for 24 h and water quenched.
Q380	Solution treated for 24 h at 325 °C, then heated and kept at 380 °C for 24 h and water quenched.

2.3. Thermodynamic Simulations

Equilibrium and Scheil non-equilibrium solidification simulations were carried out with the CALPHAD method for the cast alloy composition with FactSage 7.3 software, along with the FTlite (2021) database. Only the four main elements in the alloy were considered.

2.4. Microstructural Observations

As-cast and Q380 samples were observed with scanning electron microscopy (SEM of Shottky field emission, JEOL JSM-7000F) and energy dispersive X-ray spectroscopy (INCA EDX detector X-sight Serie Si (Li) pentaFET Oxford) at an electron beam voltage of 5.0 kV at room temperature. Specimens had been previously cleaned, ground, and polished to obtain a proper surface finish for the analysis.

2.5. X-ray Thermo-Diffraction Tests

The equipment used for the X-ray thermo-diffraction tests was a Bruker D8 Advance diffractometer that operated at 30 kV and 20 mA for reflection measurements. It was equipped with a copper anode (λ = 1.5418 Å), a Vantec-1 PSD detector, and an Anton Parr HTK2000 high temperature furnace. The sample holder used, on which the test temperature was controlled, was made of platinum.

The seven specimens, which were 10×10 mm^2 with a thickness between 1 and 2 mm, were subjected to a heating cycle from 30 to 360 °C and cooling again to 30 °C in the diffractometer. Diffraction tests were performed at room temperature (30 °C), at three temperatures during heating (260, 320, 360 °C), and at three temperatures during cooling (260, 180, 30 °C), based on the temperatures of interest found in the DSC curves (Figure 3). The measurements were recorded in the range $10° \leq 2\theta \leq 100°$ at increments of 0.033°, with each stage lasting 0.8 s.

Thermo-diffraction tests were performed three months after the samples were prepared. Twelve months after the preparation; that is, nine months after being subjected to the thermal cycle in the thermo-diffractometer, samples were retested in the same conditions as before but only at 30 °C, in order to observe whether natural precipitation had taken place.

The X-ray diffraction patterns were indexed with the PDF-4+ 2021 database from the International Center for Diffraction Data (ICDD). For the search of non-indexed phases, least squares-based Rietveld refinement was carried out in selected patterns with the FullProf software (FullProf.2k Version 7.40, January 2021, J. Rodriguez-Carvajal, ILL, Grenoble, France). The shape of the Bragg peaks was represented by a Pseudo-Voigt function. Conventional R-values, corrected for background, are given in the figures as agreement of the fitting to the observed values [34,35]. The term "intensity" is used to refer to the "integrated intensity".

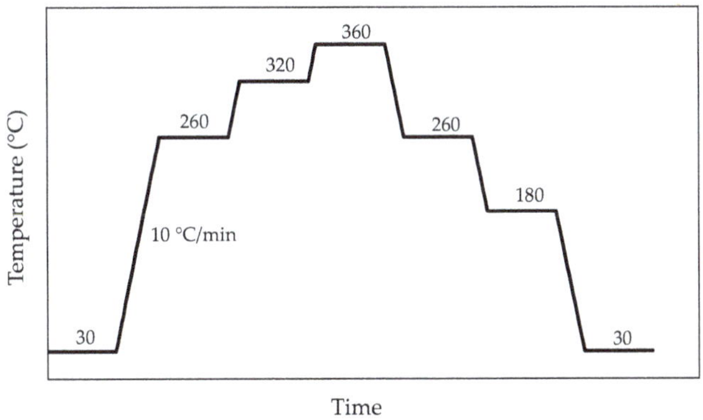

Figure 3. Thermal cycle of the samples in the thermo-diffractometer. Measurements were performed at the temperatures indicated in each step.

3. Results

3.1. Thermodynamic Simulation Results

Thermodynamic simulations performed with FactSage for equilibrium cooling conditions (Figure 4a) predicted a high proportion of the FCC aluminum solid solution at temperatures between 360 and 380 °C, with the Si and Mg_2Si phases being precipitated at these temperatures. As cooling went on, the solid solution decomposed and around 350 °C a second aluminum phase (Al#2) was generated but disappeared soon after. This phase would correspond to the zinc-rich α' aluminum phase of the miscibility gap in the Al–Zn system [36,37]. At about 340 °C, the intermetallic phase Mg_2Zn_{11} was formed, and $MgZn_2$ precipitated from Mg_2Zn_{11} at around 140 °C. The simulation under non-equilibrium conditions (Scheil approximation) predicted the precipitation of Mg_2Zn_{11} and $MgZn_2$ at about 370 °C and that of hexagonal zinc at 350 °C (Figure 4b).

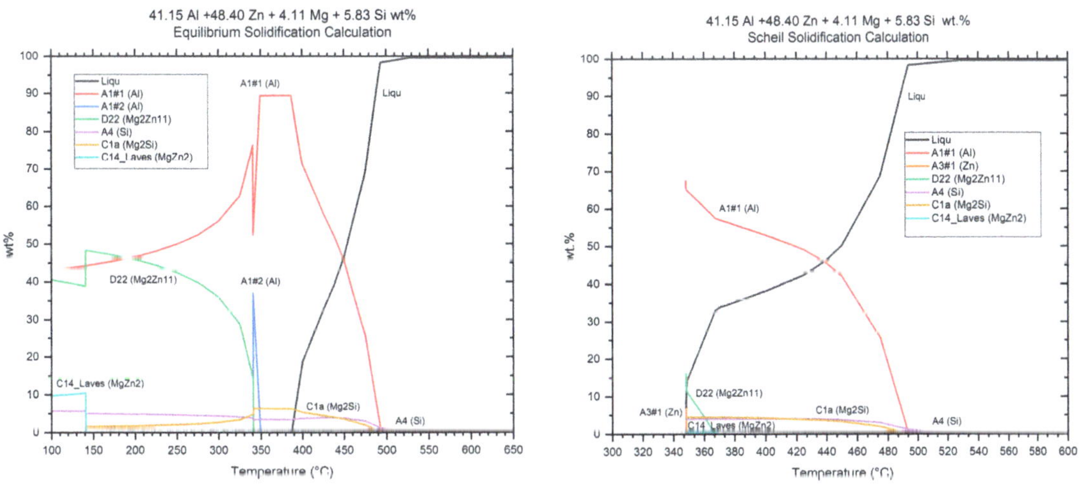

(**a**) (**b**)

Figure 4. Thermodynamic simulations with FactSage for the studied alloy considering (**a**) equilibrium solidification conditions and (**b**) non-equilibrium solidification (Scheil model).

3.2. Microstructure of the Samples Depending on the Initial Thermal Condition

The microstructure resulting from the as-cast state was heterogeneous, with different phases distributed throughout the interdendritic region depending on the solidification rate (Figure 5a). In the Q380 condition (Figure 5b), the globulization and reduction in the size of the phases after the solution treatment were remarkable. The Si, Mg–Si, and Mg–Zn phases were found by EDX measurements. The Si phase solidified in certain areas as eutectic and in other areas as primary silicon. In addition, isolated Al–Fe–Mg–Si phases were detected.

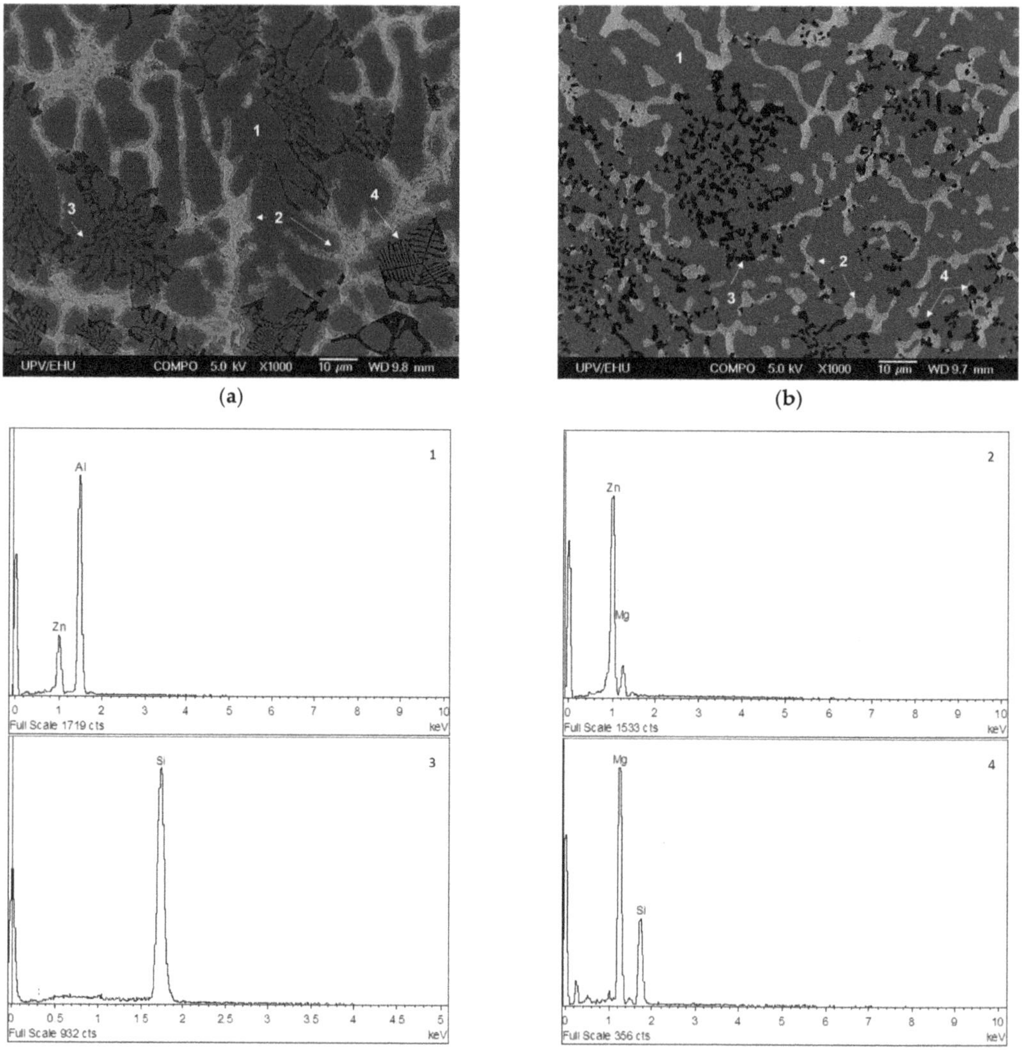

Figure 5. SEM micrographs of the material with $\times 1000$ magnification (**a**) As-cast. (**b**) Q380. Numbers 1 to 4 refer to the EDX results provided below. 1: Al-Zn matrix, 2: Mg-Zn phases, 3: Si phases and 4: Mg-Si phases.

As for the matrix, it showed a two-phase microstructure of aluminum and zinc. Precipitation of the Zn phase was observed in the as-cast material, unlike in sample Q380, where the Zn phase was not detected, indicating that it was dissolved within the matrix (Figure 6).

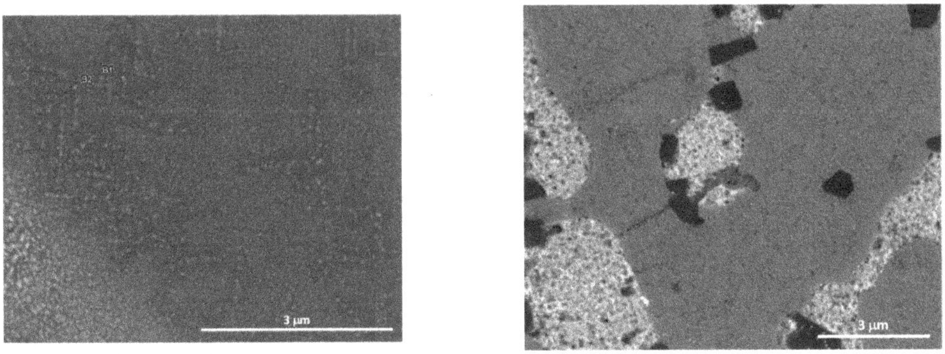

(**a**) (**b**)

Figure 6. SEM micrographs of the Al–Zn matrix. (**a**) As-cast. (**b**) Q380.

The identification of the phases present in each initial thermal condition was performed by room temperature X-ray diffraction (before the heating cycle). As is shown in Figure 7, in addition to the Al phase (PDF: 00-004-0787) and the Pt phase from the sample holder (PDF: 04-013-4766), which are not indicated for clarity, the phases detected were Zn (PDF: 01-078-9363), Si (PDF: 00-027-1402), $MgZn_2$ (PDF: 04-003-2083), Mg_2Zn_{11} (PDF: 04-007-1412), and Mg_2Si (PDF: 01-083-5235).

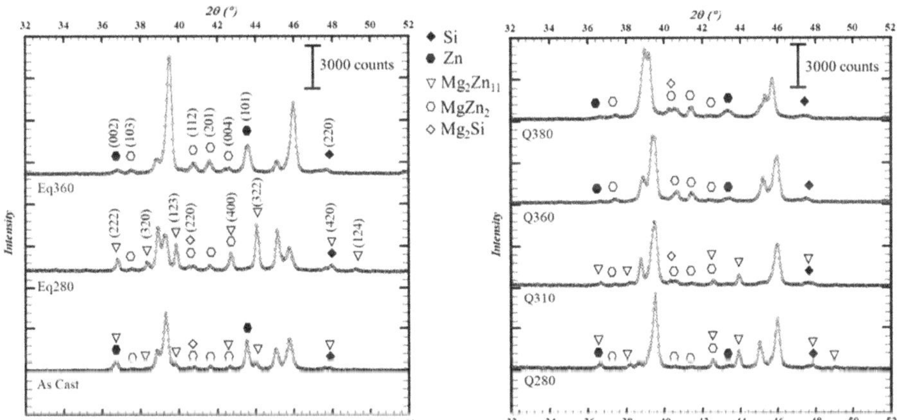

(**a**) (**b**)

Figure 7. Diffraction patterns of the samples in each thermal condition at 30 °C prior to the heating cycle in the thermo-diffractometer. Indexation is shown above the patterns of samples Eq280 (Mg_2Zn_{11} and Mg_2Si phases) and Eq360 (Si, Zn and $MgZn_2$ phases). Peaks corresponding to Al and Pt phases (the latter from the sample holder) are omitted for clarity. (**a**) As-cast samples and those cooled slowly. (**b**) Quenched samples.

However, the microstructure obtained depended on the applied treatment; that is, the temperature at which cooling had started and the cooling rate. In as-cast conditions Zn precipitated, as did both $MgZn_2$ and Mg_2Zn_{11} to a lesser extent. When slowly cooling from 280 °C (Eq280 sample), $MgZn_2$ was obtained again, as in the previous case, but now

Mg_2Zn_{11} precipitated preferentially, while HCP Zn was hardly detected. When the cooling began at 360 °C (Eq360 sample), on the other hand, no precipitation of Mg_2Zn_{11} was observed and zinc was present in the HCP Zn and $MgZn_2$ phases. $MgZn_2$ phases were found in greater quantities than in the as-cast or Eq280 conditions. As for the quenched samples, the Mg_2Zn_{11} phase was dissolved when reaching 360 °C.

Regarding the Fe-bearing quaternary phases observed by SEM, it was not possible to confirm them by X-ray diffraction. The most intense Bragg peak for $Al_8FeMg_3Si_6$ (PDF: 03-065-5936) would overlap with the Al (111) reflection. Given its condition as a minor phase, further peaks could not be detected. Therefore, if other Cu- and Fe-bearing phases found in aluminum alloys containing Zn, Mg, Si, and/or Cu [38,39] were present in very small amounts in this alloy, specific X-ray diffraction conditions and equipment would be required to identify them.

The appearance of the Mg_2Si phase and the dissolution and precipitation of the Zn phase are discussed in the following section.

3.3. *Evolution of the HCP Zn and Intermetallic Phases with Temperature*

The profiles obtained for the as-cast sample are representative of the evolution of the zinc-containing phases with temperature (Figure 8). The description is thus valid for the rest of the samples, while the matrix will be dealt with in the next section. This evolution is summarized below.

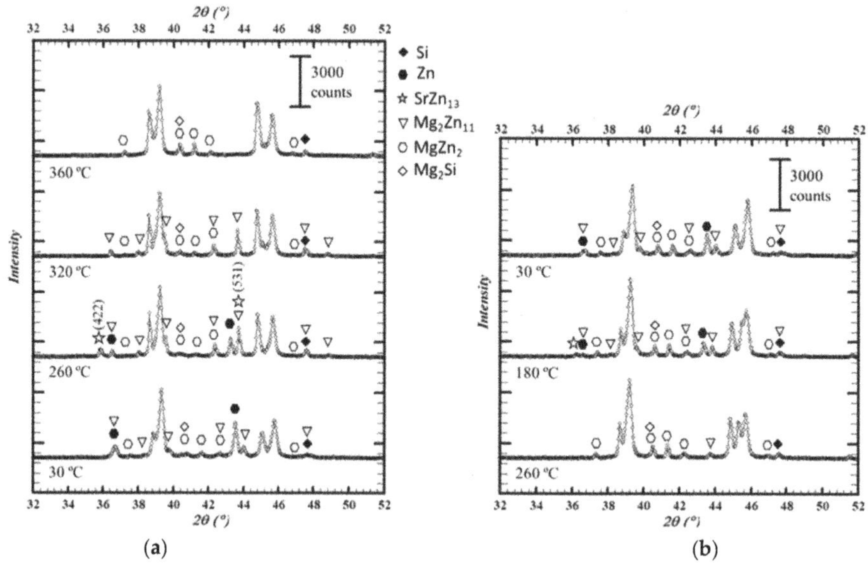

Figure 8. Diffraction patterns of the as-cast sample showing the evolution of the intermetallic phases, Zn, and Si with temperature (**a**) during the heating cycle (from 30 to 360 °C) (**b**) and cooling cycle (from 360 to 30 °C) in the thermo-diffractometer. The main Bragg peaks for the $SrZn_{13}$ phase are identified.

At 30 °C, Zn, $MgZn_2$, and Mg_2Zn_{11} phases were found. At 260 °C, the intensity of Zn peaks decreased while two additional Bragg peaks were detected around 2θ = 35.9° and 2θ = 54.0°. These peaks did not belong to any of the phases already indexed. Assuming they belonged to a new phase, it was clear that it arose at a temperature between 30 and 260 °C and likely dissolved between 260 and 280 °C, since it was absent in the Q280 sample at room temperature and in all the samples at any other temperature during the heating cycle. Indexing was performed considering minor elements present in the alloy, such as Cu, Fe, and Sr, and finally the $SrZn_{13}$ phase was identified (PDF: 04-013-4885).

At 320 °C, both Zn and $SrZn_{13}$ were dissolved. In addition, between 30 and 320 °C the intensity of Mg_2Zn_{11} increased and then became negligible at 360 °C. From the increase in the intensity of the $MgZn_2$ peaks at this temperature, it followed that Mg_2Zn_{11} had not completely dissolved in the matrix and may have become the $MgZn_2$ phase.

Regarding the cooling cycle, the onset of the precipitation of the Mg_2Zn_{11} phase was observed at 260 °C, while that of the Zn phase was not detected until 180 °C. At this temperature, the peaks belonging to $SrZn_{13}$ showed slightly and disappeared again with further cooling. It should be noted that in the final measurement at 30 °C, the distribution of precipitated phases was different from what it had been at the beginning. The proportion of $MgZn_2$ obtained at 360 °C remained stable during cooling and was higher than that found during the initial measurement.

No evolution with temperature was observed for the Mg_2Si phase. There were difficulties with detecting it in some of the measurements (see differences in Figure 7), but this was related to the specific sample (local segregations or inhomogeneities) and not to transformations taking place with temperature.

3.4. Evolution of Aluminum Phases

As was previously mentioned, a two-phase Al–Zn matrix was found. However, a detailed observation of the indexed profiles led to the detection of some peaks whose intensity was higher than expected. These observations were confirmed when performing a Rietveld fitting on one of the profiles (Eq280 sample at 30 °C, before heating). It was verified that some of the peaks could not be fitted with the original model and there was a phase missing (Figure 9a). The addition of a phase with the same spatial group as aluminum ($Fm\bar{3}m$) but a smaller lattice parameter managed to solve the structural model with satisfactory precision (Figure 9b). Due to the smaller atomic size of zinc compared to aluminum, a zinc-rich aluminum phase would show a smaller lattice parameter than α-Al and thus its Bragg peaks would shift to greater angles [32]. Therefore, the new phase observed could be the zinc-rich α′ aluminum metastable phase of the miscibility gap in the Al–Zn system. The samples were retested with room temperature X-ray diffractometry nine months later with the aim of determining whether this was the case, and it was found that precipitation of the Zn phase from the α′ metastable phase had taken place.

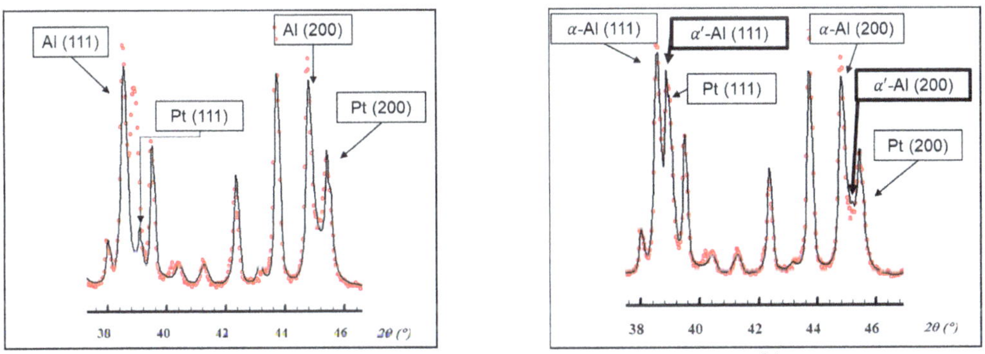

(a) (b)

Figure 9. Rietveld fitting of the Eq280 sample at 30 °C (before heating) with FullProf software. Red dots: experimental data. Black line: Fitting data. (**a**) The following six phases are considered Al (a = 4.0428 Å), Zn, Si, $MgZn_2$, Mg_2Zn_{11}, and Mg_2Si. The Pt phase comes from the sample holder. R-values with background correction: R_p = 36.6 %, R_{wp} = 38.6 %, R_{exp} = 10.72 %, χ^2 = 12.91. (**b**) An α′ phase with a lattice parameter a = 4.0089 Å is added to the previous case. Al phase in (a) is now labeled as α-Al. R-values with background correction: R_p = 20.5 %, R_{wp} = 18.7 %, R_{exp} = 10.54 %, χ^2 = 3.15.

The evolution of the intensity of the Bragg peak corresponding to the (101) plane of the Zn phase was observed (Figure 10). The reason for choosing this peak is simple: it is the one with the maximum intensity of the Zn phase and it does not overlap with signals belonging to any other phase. For these reasons, this reflection is one of those taken as a reference in precipitation studies of Al–Zn alloys [31]. Intensity increased in all cases, although only four of them are shown in the figure.

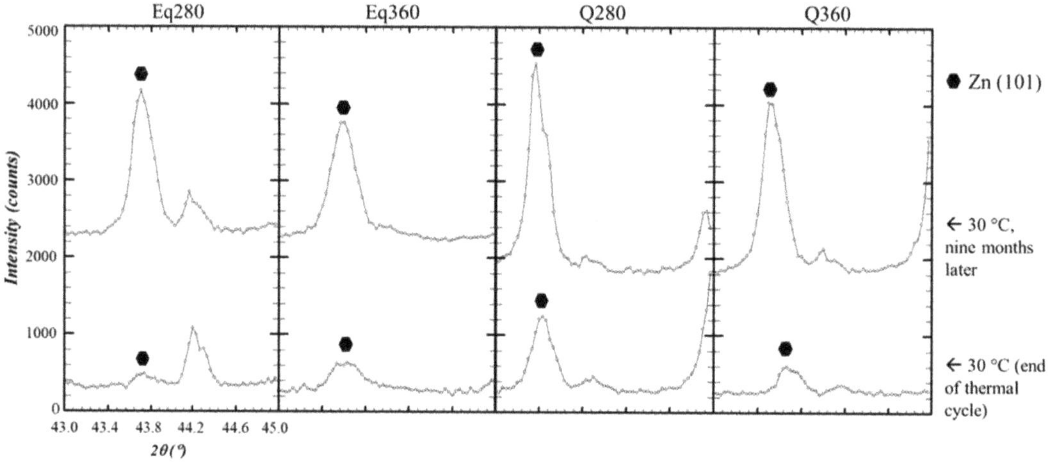

Figure 10. Close-up of the diffraction patterns of samples Eq280, Eq360, Q280, and Q360 showing the Bragg peaks of the (101) plane of the Zn phase. At the bottom, we show the profiles of the last measurements in the thermo-diffractometer at 30 °C at the end of the cooling cycle. At the top, we show the profiles obtained at the same temperature nine months later.

4. Discussion

4.1. Phase Evolution with Temperature

The first transition temperature found in the DSC measurements (Figure 1) was 285 °C, and when compared with the evolution of the phases with the temperature observed in the diffraction patterns (Figure 8) it represents the dissolution of the Zn phase. Tests performed on Al–Zn samples with 24% atomic Zn estimate this reaction at 282 °C [32]. On the other hand, in the Al–Mg–Zn system the reaction would occur at 277 °C together with the partial dissolution of the Mg_2Zn_{11} phase in the matrix [40]. However, it was not possible to determine to what extent this partial dissolution also takes place in the experimental samples at this temperature range.

The second temperature of interest in the DSC curve was about 350 °C. According to the analysis carried out, it corresponded to the complete dissolution of the Mg_2Zn_{11} phase. This agrees with the precipitation temperature range expected for this phase by the equilibrium solidification simulations (Figure 4). In addition, as discussed in the analysis, at 360 °C a higher proportion of $MgZn_2$ was observed so that it is possible that the dissolution of Mg_2Zn_{11} enriched this phase. In fact, the reactions observed in the Al–Mg–Zn system fit this hypothesis, albeit at a lower temperature [40]. The reactions and their experimental and theoretical temperatures are collected in Table 3.

During the cooling cycle in the diffractometer, zinc precipitation was observed at temperatures below 260 °C. However, such a transition was not easily detected in the DSC. According to the studies conducted by Skoko et al. with Al–Zn alloys [32], the precipitation transition occurs over a much larger temperature range than the dissolution one, so that the peak generated when cooling is much smaller. When looking specifically for this peak, it could be the one observed around 195 °C, a temperature consistent with the observations in the diffraction tests (Figure 11).

Table 3. Experimental and theoretical temperatures for the reactions observed in the X-ray thermo-diffraction tests.

Reaction	Experimental T (°C) in $Al_{58}Zn_{28}Si_8Mg_6$ Alloy	T (°C) in Al–Mg–Zn System [40]
(Al) + (Zn) → (Al, Zn)	~285	-
(Al) + (Zn), Mg_2Zn_{11} → (Al, Zn)	-	277
Mg_2Zn_{11} + (Al) → $MgZn_2$ + (Al, Zn)	~350	331

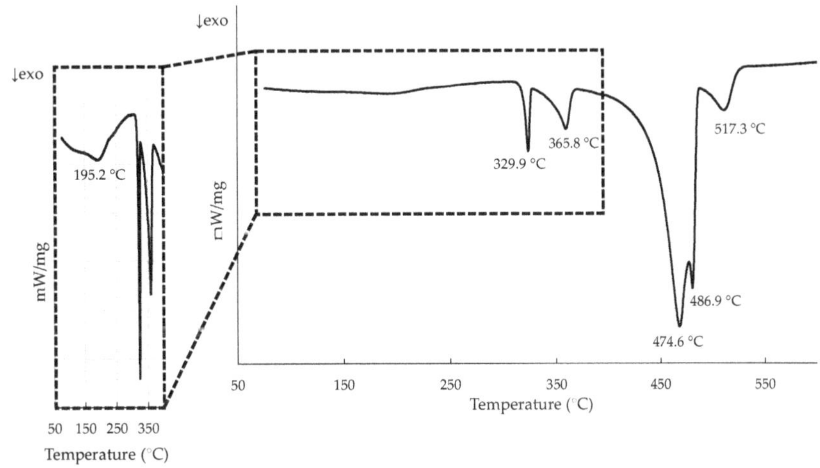

Figure 11. DSC curve for the as-cast sample during cooling. Compressed graph to highlight possible peak showing Zn phase precipitation.

It should be noted that thermodynamic simulations did not agree with the experimental results regarding the $MgZn_2$ and Zn phases.

4.2. Aluminum Phases and Ageing

According to the equilibrium solidification simulations for the alloy under study, phase Al#2, considered the zinc-rich aluminum α′ phase, should have arisen around 350 °C and decomposed shortly thereafter. The simulation with Scheil's approach did not even foresee its appearance. Therefore, the observation of this phase at room temperature was not expected.

However, an additional phase was observed at 30 °C in all samples subjected to the different thermal conditions (Figure 9). After the heating and cooling cycles in the diffractometer, it was still detected. Since this phase would have the same crystal structure as the α-Al phase but a smaller lattice parameter, our first hypothesis was that it was the α′-Al phase.

Analyses carried out nine months later showed that the microstructure of the samples after the diffractometer cycle was metastable, with the Zn phase precipitating during this time (Figure 10). This fact confirmed that the precipitation process usually observed in Al–Zn alloys [31,32,37,41] had occurred, a phenomenon in which the α′ phase intervenes and that has been studied in detail with X ray diffraction [31,32]. Still, it was not possible to determine the evolution of the α′ phase with temperature, so the onset of the solid solution was not detected by diffraction.

4.3. Effect of Strontium

As has been mentioned, the Si phase is observed both as a primary crystal and as an Al-Si eutectic. It can be seen in the literature that, in comparison with Al–Si alloys, the nucleation of primary silicon crystals in Zn–Al–Si alloys is promoted by the presence of zinc. Nevertheless, the addition of strontium promotes the generation of eutectic silicon. Thus, in our microstructure we find both. The presence of strontium in the alloy is expected to modify both primary and eutectic silicon [42,43]. Since strontium is present in all samples in equal proportions, the modification effect could not be evaluated.

However, the test results showed that when heat treating the material between 30 and 280 °C, the strontium ceased to be just a modifier and interacted with zinc to generate $SrZn_{13}$. This is a factor that must be considered since, on the one hand, it can affect the final properties of the alloy, reducing the availability of zinc for the Mg–Zn phases; on the other hand, it can influence subsequent transformation processes that take place in its stability range. Given the interactions that occur, it should be considered whether an alternative modifier, such as sodium or rare earth elements [44], should be used.

5. Conclusions

The design and manufacturing of multicomponent light alloys give rise to non-conventional microstructures. In order to determine their potential applications, it is necessary to analyze their evolution with temperature. In this work, the multicomponent $Al_{58}Zn_{28}Si_8Mg_6$ alloy was studied with CALPHAD methods and then cast and heat-treated under several conditions. Characterization was carried out by X-ray thermo-diffraction, differential scanning calorimetry, and electron microscopy.

As a result, a total of eight phases were identified in the alloy in the 30–380 °C temperature range: α-Al, α′-Al, Zn, Si, Mg_2Si, $MgZn_2$, Mg_2Zn_{11}, and $SrZn_{13}$. The microstructures obtained at room temperature were metastable and the precipitation of Zn from the α′ phase occurred over the course of months.

Moreover, the thermal transitions below 360 °C could be determined; that is, the dissolution and precipitation of Zn and dissolution of Mg_2Zn_{11}. Since the $MgZn_2$ and Mg_2Si phases dissolved above 360 °C, where partial melting may occur, those precipitates are not expected to harden the matrix, as they do in conventional 3xx.x and 7xx.x aluminum alloys with solution and precipitation treatments. However, two remarks should be made. First, a room temperature X-ray diffraction test performed immediately after the Q380 quenching treatment could offer valuable information about the solid solution capability of the alloy and enable a more direct comparison with the simulation results. Second, it was observed that the proportion of $MgZn_2$ and Mg_2Zn_{11} phases was highly dependent on the thermal history, so the microstructure of the alloy is still susceptible to adaptation by heat treatment.

It should be noted that strontium was added to modify silicon phases. Although it was a minor element, as well as Fe and Cu, it interacted with zinc between 30 and 280 °C. This fact should be considered during the postprocessing (thermal and/or thermomechanical treatments) in that temperature range, as it could have unexpected effects.

Finally, it is clear from the experimental results that the used database is not designed to account for high percentages of alloying elements and thus is not able to accurately predict the actual phase evolution in the material; the Al–Zn system turned out to be tricky due to the metastable α′ phase, and the precipitation of $MgZn_2$ in equilibrium conditions was predicted at a temperature that was too low. Nevertheless, the general guidelines given by CALPHAD methods, combined with SEM, DSC, and X-ray thermo-diffraction results, were able to give us a satisfactory understanding of the alloy's behavior at different temperatures.

Author Contributions: Conceptualization, Y.B., J.J.T., I.H. and T.G.; methodology, Y.B., J.J.T., G.A. and T.G.; software, J.J.T.; validation, G.A., I.H. and T.G.; formal analysis, Y.B.; investigation, Y.B. and J.J.T.; resources, J.J.T., I.H. and T.G.; writing—original draft preparation, Y.B.; writing—review and editing, Y.B., J.J.T., I.V., G.A., I.H. and T.G.; visualization, Y.B.; supervision, G.A., I.H. and T.G.; project administration, I.V., I.H. and T.G.; funding acquisition, I.V., I.H. and T.G. All authors have read and agreed to the published version of the manuscript.

Funding: This research was funded by the BASQUE GOVERNMENT through the Elkartek project KK-2020/00047.

Institutional Review Board Statement: Not applicable.

Informed Consent Statement: Not applicable.

Acknowledgments: The authors would like to acknowledge the support from A. Larrañaga (SGIker Advanced Research Facilities, University of the Basque Country (UPV/EHU)) with the X-ray thermo-diffraction analysis.

Conflicts of Interest: The authors declare no conflict of interest.

References

1. Cantor, B. Multicomponent and High Entropy Alloys. *Entropy* **2014**, *16*, 4749–4768. [CrossRef]
2. Cantor, B.; Chang, I.T.H.; Knight, P.; Vincent, A.J.B. Microstructural Development in Equiatomic Multicomponent Alloys. *Mater. Sci. Eng. A* **2004**, *375–377*, 213–218. [CrossRef]
3. Huang, P.-K.; Yeh, J.-W.; Shun, T.-T.; Chen, S.-K. Multi-Principal-Element Alloys with Improved Oxidation and Wear Resistance for Thermal Spray Coating. *Adv. Eng. Mater.* **2004**, *6*, 74–78. [CrossRef]
4. Gao, M.C.; Alman, D.E. Searching for Next Single-Phase High-Entropy Alloy Compositions. *Entropy* **2013**, *15*, 4504–4519. [CrossRef]
5. Tsai, M.-H.; Yeh, J.-W. High-Entropy Alloys: A Critical Review. *Mater. Res. Lett.* **2014**, *2*, 107–123. [CrossRef]
6. Sharma, P.; Dwivedi, V.K.; Dwivedi, S.P. Development of High Entropy Alloys: A Review. *Mater. Today Proc.* **2021**, *43*, 502–509. [CrossRef]
7. Steurer, W. Single-Phase High-Entropy Alloys—A Critical Update. *Mater. Charact.* **2020**, *162*, 110179. [CrossRef]
8. Yeh, J.-W.; Chen, S.-K.; Lin, S.-J.; Gan, J.-Y.; Chin, T.-S.; Shun, T.-T.; Tsau, C.-H.; Chang, S.-Y. Nanostructured High-Entropy Alloys with Multiple Principal Elements: Novel Alloy Design Concepts and Outcomes. *Adv. Eng. Mater.* **2004**, *6*, 299–303. [CrossRef]
9. Yeh, J.-W. Alloy Design Strategies and Future Trends in High-Entropy Alloys. *JOM* **2013**, *65*, 1759–1771. [CrossRef]
10. Asadikiya, M.; Yang, S.; Zhang, Y.; Lemay, C.; Apelian, D.; Zhong, Y. A Review of the Design of High-Entropy Aluminum Alloys: A Pathway for Novel Al Alloys. *J. Mater. Sci.* **2021**, *56*, 12093–12110. [CrossRef]
11. Miracle, D.B.; Miller, J.D.; Senkov, O.N.; Woodward, C.; Uchic, M.D.; Tiley, J. Exploration and Development of High Entropy Alloys for Structural Applications. *Entropy* **2014**, *16*, 494–525. [CrossRef]
12. Senkov, O.N.; Wilks, G.B.; Miracle, D.B.; Chuang, C.P.; Liaw, P.K. Refractory High-Entropy Alloys. *Intermetallics* **2010**, *18*, 1758–1765. [CrossRef]
13. Youssef, K.M.; Zaddach, A.J.; Niu, C.; Irving, D.L.; Koch, C.C. A Novel Low-Density, High-Hardness, High-Entropy Alloy with Close-Packed Single-Phase Nanocrystalline Structures. *Mater. Res. Lett.* **2015**, *3*, 95–99. [CrossRef]
14. Kumar, A.; Gupta, M. An Insight into Evolution of Light Weight High Entropy Alloys: A Review. *Metals* **2016**, *6*, 199. [CrossRef]
15. Tsai, M.-H. Three Strategies for the Design of Advanced High-Entropy Alloys. *Entropy* **2016**, *18*, 252. [CrossRef]
16. Tseng, K.; Yang, Y.; Juan, C.; Chin, T.; Tsai, C.; Yeh, J. A Light-Weight High-Entropy Alloy $Al_{20}Be_{20}Fe_{10}Si_{15}Ti_{35}$. *Sci. China Technol. Sci.* **2018**, *61*, 184–188. [CrossRef]
17. Sanchez, J.M.; Vicario, I.; Albizuri, J.; Guraya, T.; Garcia, J.C. Phase Prediction, Microstructure and High Hardness of Novel Light-Weight High Entropy Alloys. *J. Mater. Res. Technol.* **2019**, *8*, 795–803. [CrossRef]
18. Yang, X.; Chen, S.Y.; Cotton, J.D.; Zhang, Y. Phase Stability of Low-Density, Multiprincipal Component Alloys Containing Aluminum, Magnesium, and Lithium. *JOM* **2014**, *66*, 2009–2020. [CrossRef]
19. Sanchez, J.M.; Vicario, I.; Albizuri, J.; Guraya, T.; Acuña, E.M. Design, Microstructure and Mechanical Properties of Cast Medium Entropy Aluminium Alloys. *Sci. Rep.* **2019**, *9*, 6792. [CrossRef]
20. Tun, K.S.; Murugan, P.; Srivatsan, T.S.; Gupta, M. Synthesis and Characterization of Aluminium Based Multicomponent Alloys. *Mater. Today Proc.* **2021**, *46*, 1210–1214. [CrossRef]
21. Nagase, T.; Todai, M.; Wang, P.; Sun, S.-H.; Nakano, T. Design and Development of (Ti, Zr, Hf)-Al Based Medium Entropy Alloys and High Entropy Alloys. *Mater. Chem. Phys.* **2022**, *276*, 125409. [CrossRef]
22. Nagase, T.; Terayama, A.; Nagaoka, T.; Fuyama, N.; Sakamoto, T. Alloy Design and Fabrication of Ingots of Al–Mg–Li–Ca Light-Weight Medium Entropy Alloys. *Mater. Trans.* **2020**, *61*, 1369–1380. [CrossRef]
23. Krajewski, W.K.; Buraś, J.; Krajewski, P.K.; Greer, A.L.; Faerber, K.; Schumacher, P. New Developments of Al-Zn Cast Alloys. *Mater. Today Proc.* **2015**, *2*, 4978–4983. [CrossRef]
24. Krajewski, P.K.; Greer, A.L.; Krajewski, W.K. Main Directions of Recent Works on Al-Zn-Based Alloys for Foundry Engineering. *J. Mater. Eng. Perform.* **2019**, *28*, 3986–3993. [CrossRef]
25. Sivasankaran, S.; Ramkumar, K.R.; Ammar, H.R.; Al-Mufadi, F.A.; Alaboodi, A.S.; Irfan, O.M. Microstructural Evolutions and Enhanced Mechanical Performance of Novel Al-Zn Die-Casting Alloys Processed by Squeezing and Hot Extrusion. *J. Mater. Process. Technol.* **2021**, *292*, 117063. [CrossRef]
26. Gorsse, S.; Senkov, O.N. About the Reliability of CALPHAD Predictions in Multicomponent Systems. *Entropy* **2018**, *20*, 899. [CrossRef] [PubMed]

27. Li, R.; Xie, L.; Wang, W.Y.; Liaw, P.K.; Zhang, Y. High-Throughput Calculations for High-Entropy Alloys: A Brief Review. *Front. Mater.* **2020**, *7*, 290. [CrossRef]
28. Feng, R.; Gao, M.C.; Lee, C.; Mathes, M.; Zuo, T.; Chen, S.; Hawk, J.A.; Zhang, Y.; Liaw, P.K. Design of Light-Weight High-Entropy Alloys. *Entropy* **2016**, *18*, 333. [CrossRef]
29. Sanchez, J.M.; Vicario, I.; Albizuri, J.; Guraya, T.; Koval, N.E.; Garcia, J.C. Compound Formation and Microstructure of As-Cast High Entropy Aluminums. *Metals* **2018**, *8*, 167. [CrossRef]
30. Asadikiya, M.; Zhang, Y.; Wang, L.; Apelian, D.; Zhong, Y. Design of Ternary High-Entropy Aluminum Alloys (HEAls). *J. Alloy Compd.* **2022**, *891*, 161836. [CrossRef]
31. Popović, S.; Gržeta, B. Precipitation and Dissolution Phenomena in Al-Zn Alloys. *Croat. Chem. Acta* **1999**, *72*, 621–643.
32. Skoko, Ž.; Popović, S.; Štefanić, G. Microstructure of Al-Zn and Zn-Al Alloys. *Croat. Chem. Acta* **2009**, *82*, 405–420.
33. Hayoune, A.; Hamana, D. A Dilatometric and High-Temperature X-ray Diffraction Study of Cold Deformation Effect on the Interaction between Precipitation, Recovery and Recrystallization Reactions in Al–12 wt.% Mg Alloy. *Mater. Sci. Eng. A* **2010**, *527*, 7261–7264. [CrossRef]
34. McCusker, L.B.; Von Dreele, R.B.; Cox, D.E.; Louër, D.; Scardi, P. Rietveld Refinement Guidelines. *J. Appl. Cryst.* **1999**, *32*, 36–50. [CrossRef]
35. Toby, B.H. R Factors in Rietveld Analysis: How Good Is Good Enough? *Powder Diffr.* **2006**, *21*, 67–70. [CrossRef]
36. Mondolfo, L.F. *Metallography of Aluminum Alloys*; John Wiley & Sons: New York, NY, USA; Chapman & Hall: London, UK, 1943.
37. Murray, J.L. The Al−Zn (Aluminum-Zinc) System. *Bull. Alloy Phase Diagr.* **1983**, *4*, 55–73. [CrossRef]
38. Safyari, M.; Moshtaghi, M.; Kuramoto, S. On the Role of Traps in the Microstructural Control of Environmental Hydrogen Embrittlement of a 7xxx Series Aluminum Alloy *J. Alloy Compd.* **2021**, *855*, 157300. [CrossRef]
39. Mao, G.; Liu, S.; Gao, W.; Wang, J.; Liu, D. Effect of Zn or Zn + Cu Addition on the Precipitation in Al–Mg–Si Alloys: A Review. *Trans. Indian Inst. Met.* **2021**, *74*, 2925–2938. [CrossRef]
40. Petrov, D.; Watson, A.; Gröbner, J.; Rogl, P.; Tedenac, J.; Bulanova, M.; Turkevich, V.; Lukas, H.L. Al-Mg-Zn (aluminium-magnesium-zinc): Light metal ternary systems: Phase diagrams, crystallographic and thermodynamic data. In *Light Metal Systems. Part 3*; Effenberg, G., Ilyenko, S., Eds.; Landolt-Börnstein—Group IV Physical, Chemistry; Springer: Berlin/Heidelberg, Germany, 2005; Volume 11A3, pp. 1–19; ISBN 978-3-540-25013-5.
41. Jacobs, M.H. The Nucleation and Growth of Precipitates in Aluminium Alloys. Ph.D. Thesis, University of Warwick, Coventry, UK, 1969.
42. Mao, F.; Chen, F.; Han, Q.; Han, J.; Cao, Z.; Wang, T.; Li, T. Real Time Observation on the Solidification of Strontium-Modified Zinc–Aluminum–Silicon Alloys by Synchrotron Microradiography. *J. Alloy Compd.* **2014**, *608*, 343–351. [CrossRef]
43. Mao, F.; Chen, F.; Yan, G.; Wang, T.; Cao, Z. Effect of Strontium Addition on Silicon Phase and Mechanical Properties of Zn–27Al–3Si Alloy. *J. Alloy Compd.* **2015**, *622*, 871–879. [CrossRef]
44. Mao, F.; Wei, S.; Chen, C.; Zhang, C.; Wang, X.; Cao, Z. Modification of the Silicon Phase and Mechanical Properties in Al-40Zn-6Si Alloy with Eu Addition. *Mater. Des.* **2020**, *186*, 108268. [CrossRef]

materials MDPI

Article

A First-Principles Study of the Cu-Containing β″ Precipitates in Al-Mg-Si-Cu Alloy

Shaozhi He [1], Jiong Wang [1,*], Donglan Zhang [1], Qing Wu [2], Yi Kong [1] and Yong Du [1]

[1] State Key Laboratory of Powder Metallurgy, Central South University, Changsha 410083, China; heshaozhi@csu.edu.cn (S.H.); 203301046@csu.edu.cn (D.Z.); yikong@csu.edu.cn (Y.K.); yong-du@csu.edu.cn (Y.D.)
[2] Information and Network Center, Central South University, Changsha 410083, China; wuqing@csu.edu.cn
* Correspondence: wangjionga@csu.edu.cn

Abstract: The nanostructured β″ precipitates are critical for the strength of Al-Mg-Si-(Cu) aluminum alloys. However, there are still controversial reports about the composition of Cu-containing β″ phases. In this work, first-principles calculations based on density functional theory were used to investigate the composition, mechanical properties, and electronic structure of Cu-containing β″ phases. The results predict that the Cu-containing β″ precipitates with a stoichiometry of $Mg_{4+x}Al_{2-x}CuSi_4$ (x = 0, 1) are energetically favorable. As the concentration of Cu atoms increases, Cu-containing β″ phases with different compositions will appear, such as $Mg_4AlCu_2Si_4$ and $Mg_4Cu_3Si_4$. The replacement order of Cu atoms in β″ phases can be summarized as one Si3/Al site → two Si3/Al sites → two Si3/Al sites and one Mg1 site. The calculated elastic constants of the considered β″ phases suggest that they are all mechanically stable, and all β″ phases are ductile. When Cu atoms replace Al atoms at Si3/Al sites in β″ phases, the values of bulk modulus (B), shear modulus (G), and Young's modulus (E) all increase. The calculation of the phonon spectrum shows that $Mg_{4+x}Al_{2-x}CuSi_4$ (x = 0, 1) are also dynamically stable. The electronic structure analysis shows that the bond between the Si atom and the Cu atom has a covalent like property. The incorporation of the Cu atom enhances the electron interaction between the Mg2 and the Si3 atom so that the Mg2 atom also joins the Si network, which may be one of the reasons why Cu atoms increase the structure stability of the β″ phases.

Keywords: Al-Mg-Si-Cu alloys; Cu-containing β″; atomic configuration; mechanical properties; electronic structure

Citation: He, S.; Wang, J.; Zhang, D.; Wu, Q.; Kong, Y.; Du, Y. A First-Principles Study of the Cu-Containing β″ Precipitates in Al-Mg-Si-Cu Alloy. *Materials* **2021**, *14*, 7879. https://doi.org/10.3390/ma14247879

Academic Editors: Elena Pereloma, Antonino Squillace and Lijun Zhang

Received: 14 November 2021
Accepted: 14 December 2021
Published: 19 December 2021

1. Introduction

Heat treatable Al-Mg-Si(-Cu) alloys in the 6xxx series are a common category of structural materials used in the construction and transportation industries. These alloys can be customized to have a desirable combination of properties, such as good formability, high specific strength, and corrosion resistance [1–3]. After proper aging treatment, the strength of the alloy can be greatly improved. This is mainly due to the precipitates that can contribute to the strengthening mechanisms by hindering the dislocation movement [4,5], particle strengthening σ_p [6], and coherency of the particles [7]. The mechanical properties of these alloys can be greatly influenced by the composition, morphology, scale, and distribution of these solute atom nanostructures [8]. The precipitation sequence for Al-Mg-Si alloys is generally considered to be [9,10]:

$$SSSS \rightarrow \text{solute clusters} \rightarrow GP-\text{zones} \rightarrow \beta'' \rightarrow \beta', U1, U2, B' \rightarrow \beta, Si$$

The supersaturated solid solution is denoted by the abbreviation SSSS. The Guinier-Preston zones (GP-zones) were first discovered in the Al-Cu system by Guinier [11] and Preston [12]. The GP-zones mainly refer to the nanoprecipitate phases formed in the early stage of aging, which is characterized by a certain ordered structure and completely coherent with the matrix.

Among the precipitates [4,13–15] formed in the aged Al-Mg-Si alloys, needle-like β″ precipitate is the most effective strengthening phase [14] responsible for the peak-hardening effect [16]. The β″ phase is a metastable precipitate phase, which is semi-coherent with the Al matrix in the needle cross-section, the space group is $C2/m$, a = 15.16 Å, b = 4.05 Å, c = 6.74 Å, and β = 105.3° [17,18]. The monoclinic β″ phase was originally proposed to have the composition of Mg_5Si_6 [17]. However, according to recent experimental and theoretical studies, the composition of β″ would fluctuate around $Mg_5Al_2Si_4$ [19–22]. Furthermore, the most recent density functional theory (DFT) calculations inferred very minor formation enthalpy differences for β″-$Mg_{5+x}Al_{2-x}Si_4$ ($-1 < x < 1$) [21]. These results indicate that the composition of β″ phase in Al matrix may change under certain conditions. For example, the dispersed nano-precipitates can be affected by the addition of Mg and/or Si, as well as other elements like Cu [5,23–27]. The addition of Cu is demonstrated to increase the age-hardening response, and it promotes the generation of higher number density and smaller size precipitates [14,28–32]. Therefore, a certain amount of Cu is usually added into Al-Mg-Si alloys. The addition of Cu increases the complexity of the precipitation sequence [32,33]. The precipitation sequence of Al-Mg-Si-Cu alloys is reported as [34]:

$$\text{SSSS} \rightarrow \text{solute clusters} \rightarrow \text{GP}-\text{zones} \rightarrow \beta'', \text{L/S/C, QP, QC} \rightarrow \beta', \text{Q}' \rightarrow \text{Q, Si}$$

Previous work used various experimental and theoretical methods to study the incorporation of Cu in β″, and analyzed the Cu atoms as foreign solute atoms in the phases [20]. Cu addition could further enhance the positive effect of pre-aging on bake hardening for Al-Mg-Si alloys [35]. It has been demonstrated by high-angle annular dark-field scanning transmission electron microscopy (HAADF-STEM) that Cu is mainly confined to the Si3/Al sites (Si or Al atoms completely occupy) of the β″ structure [26,35–37], which as mentioned was also supported from DFT-based calculations [38]. The β″ precipitates in Al-Mg-Si-Cu alloy were detected with an average composition of 28.6Al-38.7Mg-26.5Si-5.17Cu (at. %) using atom probe tomography (APT) and high-resolution energy-dispersive X-ray (EDX) mapping [36]. Furthermore, the addition of Cu has no effect on the type of β″ precipitate, Cu atoms incorporate in β″ and some of Mg, Si and Al in β″ unit cell are substituted by Cu atoms [39].

As mentioned above, the β″ precipitation behavior in Al-Mg-Si-Cu alloys has been investigated using various characterization methods. However, the detailed structures and stabilities are still unclear of Cu-containing β″ phases in these alloys, and these structural refinements could be supported by first-principles results [40]. In addition, we predict energy-lowering site occupations and stoichiometries of the β″ phases, where experimental information is incomplete. Understanding the structure of Cu-containing β″ precipitates is essential to elucidate the precipitation sequence in heat-treatable Al-Mg-Si (-Cu) alloys.

In the present work, first-principles calculations based on density functional theory (DFT) [41] were used to study the Cu-containing β″ phases. Based on the structural information obtained by experimental methods, first-principles atomistic calculations can provide structural, chemical, and energetic information [40]. A large number of Cu-containing β″ structures were constructed searching for possible stable configurations and structural stability, kinetic stability, and mechanical stability were also considered. Finally, the characteristics of Cu atoms occupying sites were analyzed through the electronic structure.

2. Materials and Methods

2.1. Atomic Model

For the β″ phases, the formation enthalpies and lattice parameters of $Mg_4Al_3Si_4$, $Mg_5Al_2Si_4$, Mg_6AlSi_4, and Mg_5Si_6 were computed for each of the models of the crystal structures available in the literature [17,18,21], allowing a critical assessment of the validity of the models. Figure 1 shows four atomic models of the β″ without Cu. The Wyckoff site information of the energetically most favorable β″-$Mg_5Al_2Si_4$ is shown in Table 1 [18,19].

MDPI

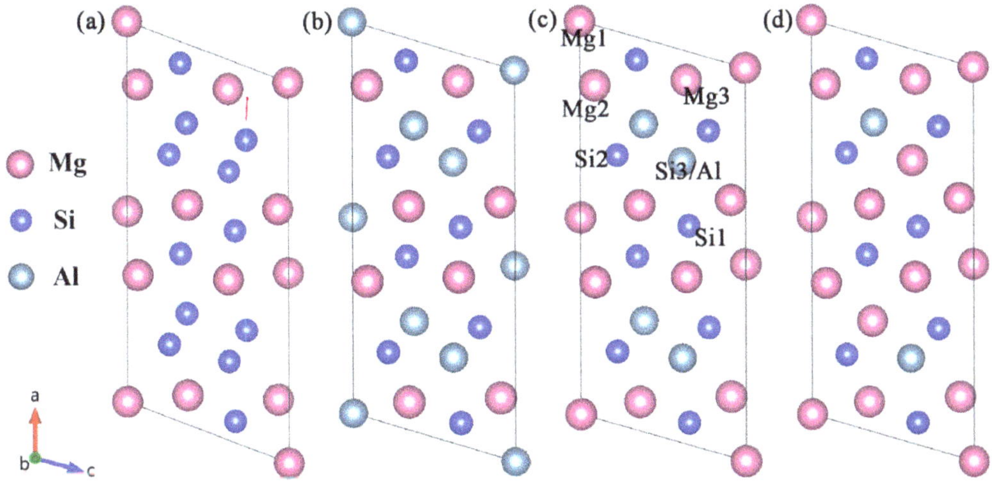

Figure 1. Four atomic models of the β″ available in the literature [17,21]. (**a**) Mg_5Si_6 from Zandbergen [17]; (**b**) $Mg_4Al_3Si_4$ and (**c**) the $Mg_5Al_2Si_4$ from Hasting [19]; (**d**) Mg_6AlSi_4 from Ehlers [21]. The relative location of each site is marked in (**c**).

Table 1. Wyckoff site information (*x*, *y*, *z*) in the β″-$Mg_5Al_2Si_4$ phase [18,19]; atomic configuration is shown schematically in Figure 1c.

Site	Occupation	*x*	*y*	*z*
Mg1	2*a*	0	0	0
Mg2	4*i*	0.3419	0	0.099
Mg3	4*i*	0.4225	0	0.659
Si1	4*i*	0.0501	0	0.678
Si2	4*i*	0.1876	0	0.225
Si3/Al	4*i*	0.2213	0	0.618

2.2. Computational Details

The first-principles calculations were performed utilize the plane wave pseudopotential method, as implemented in the highly efficient Vienna ab initio simulation package (VASP) [42,43], The electron-ion interactions were described through projector augmented wave (PAW) [44,45]. The exchange-correlation function were constructed by the generalized gradient approximation (GGA) of Perdew-Burke-Ernzerhof (PBE) [46]. All structures were fully relaxed with respect to atomic positions as well as all lattice parameters in order to find the lowest-energy structure. The electron wave function was expanded in plane waves up to a cutoff energy of 450 eV. The β″phase was represented by a conventional cell with 22 atoms according to the experimental results, and 3 × 12 × 8 Γ-centered k-point meshes were employed in the Brilluion zone sampling and generated automatically by following the Monkhorst-Pack sampling scheme [47], while the 3 × 3 × 8 Γ-centered k-point meshes and 1 × 4 × 1 supercells were employed for calculation of "replacement energy" (the detailed definition is explained below). Atoms were relaxed until their residual forces converged to 0.01 eV/Å. The phonon spectra were obtained using the Phonopy package [48].

The four-parameter Birch–Murnaghan equation of state with its linear form [49] is employed to estimate the equilibrium total energy (E_0), volume (V_0),

$$E(V) = a + bV^{-2/3} + cV^{-4/3} + dV^{-2} \quad (1)$$

where *a*, *b*, *c*, and *d* are fitting parameters. More details can be found in our previous work [50].

Compared with the energy of solid solution containing a Cu atom, the energy gain of the Cu atoms incorporated in β″ is referred to as "replacement energy". In order to construct the Cu-containing β″ phases, it is necessary to determine the possible occupation sites of Cu in β″ phases. Additionally, computing the replacement energy (see Ref. [51]) can be used as a criterion for the possible occupation sites of solute atoms. There have been previous studies addressing the first-principles calculations for describing replacement energies of different sides. Since the replacement energy of Cu atoms at Mg2 and Mg3 sites were not shown in Saito's work [38], one Cu atom was introduced into a $1 \times 4 \times 1$ supercell and the preference of Cu atoms for each non-equivalent site in β″ was evaluated using the method described by Saito et al. [51], but with higher calculation precision.

To solve the compositional uncertainty preliminarily, the reported $C2/m$ symmetries [18] were deliberately reduced to the level where only pairs of atoms (e.g., the two Cu atoms) were regarded as equivalent. This implies that space group $P2/m$ was used throughout and there are 11 different sites within the unit cell. Besides, no partial occupancies were considered and vacancies were ignored. The replacement energy for Cu incorporation in β″ can be described as follows:

$$\begin{aligned}\Delta H(\beta''_0\colon \mathrm{X} \rightarrow \Xi) = \; & H(\beta''_0 : 3 \times \{\mathrm{Al} \rightarrow \Xi\}; 1 \times \{\mathrm{X} \rightarrow \Xi\}) \\ & + H(\mathrm{fcc\ Al}) - H(\beta''_0 : 4 \times \{\mathrm{Al} \rightarrow \Xi\}) \\ & - H(\mathrm{fcc\ Al} : 1 \times \{\mathrm{X} \rightarrow \mathrm{S}\})\end{aligned} \tag{2}$$

where H are the calculated enthalpy of the system, β''_0 are the Cu-free structure, Ξ are the sides in β''_0, X are the solute atoms incorporated in the precipitates, and S are substitutional sites in the Al matrix. A certain atom X incorporates on site Ξ is referred to as "{X → Ξ}". The formation enthalpy of solid solution (SS), ΔH^{form}_{SS}, was used to find out the most energetically favorable configurations in the atomic models. Since there is no stable fcc structure for Mg and Si, their formation energies in relation to SS were determined as follows:

$$\begin{aligned}\Delta H^{form}_{SS}\,(\mathrm{Mg}_a\mathrm{Al}_b\mathrm{Cu}_c\mathrm{Si}_d) = \; & E(\mathrm{Mg}_a\mathrm{Al}_b\mathrm{Cu}_c\mathrm{Si}_d) - aE^{sub}(\mathrm{Mg}) \\ & - bE(\mathrm{Al}) - cE(\mathrm{Cu}) - dE^{sub}(\mathrm{Si})\end{aligned} \tag{3}$$

where E^{sub} (Mg) and E^{sub} (Si) are the enthalpies of substituting Al atoms by Mg and Si atoms, respectively. E^{sub} (Mg) and E^{sub} (Si) were calculated in a $3 \times 3 \times 3$ Al supercell with one Mg/Si atom and 107 Al atoms with a k-point meshes of $5 \times 5 \times 5$. The enthalpy of substituting a Mg atom was defined as:

$$E^{sub}(\mathrm{Mg}) = E(\mathrm{Al}_{107}\mathrm{Mg}) - 107/108E(\mathrm{Al}) \tag{4}$$

where E (Al) is the enthalpy of a $3 \times 3 \times 3$ Al supercell. The definition of E^{sub} (Mg) was also feasible for E^{sub} (Si).

Finally, in order to compare the structures with different Al content, the formation enthalpy can also be expressed in kJ/mol of solute atoms, instead of kJ/mol [52]. This transformation is achieved as follows: ΔH_{SS} [kJ/mol solute] = ΔH_{SS} [kJ/mol]/(x_{Mg} + x_{Si} + x_{Cu}), where x_{Mg} and x_{Si} and x_{Cu} are the atomic fractions of Mg and Si and Cu in the β″ phases $\mathrm{Mg}_a\mathrm{Al}_b\mathrm{Cu}_c\mathrm{Si}_d$ ($a = x_{Mg}$, $b = x_{Al}$, $c = x_{Cu}$, $d = x_{Si}$). This is a common definition of formation enthalpy in the literature [9,21,52].

The elastic constant can be represented by a 6×6 matrix. Based on the symmetry of the crystal structure, the independent elastic constants of the monoclinic crystal are reduced to 13, as shown in Formula (5):

$$C_{ij} = \begin{pmatrix} C_{11} & C_{12} & C_{13} & 0 & C_{15} & 0 \\ & C_{22} & C_{23} & 0 & C_{25} & 0 \\ & & C_{33} & 0 & C_{35} & 0 \\ & & & C_{44} & 0 & C_{46} \\ & & & & C_{55} & 0 \\ & & & & & C_{66} \end{pmatrix} \tag{5}$$

The stress-strain method based on the generalized Hooke's theorem is used to calculate the elastic constants of each crystal [53]. For more detailed stress-strain method description, please refer to [54]. The relationship between elastic constant C_{ijkl}, stress tensor δ_{kl}, and strain tensor δ_{kl} can be expressed as:

$$\sigma_{ij} = C_{ijkl}\delta_{kl} \tag{6}$$

The Hill model [55] is used to further obtain the bulk modulus (B), shear modulus (G), and Youngs modulus (E) of the crystal through the elastic constant. The Hill model takes into account that the calculation results of the Voigt model and the Reuss model will be high and low, respectively, and take the arithmetic mean of the values of the Voigt model and the Reuss model. For monoclinic crystal structure, the formula for calculating the bulk modulus (B) and shear modulus (G) of monoclinic crystals using Voigt model and Reuss model are [56]:

$$B_V = \frac{1}{9}[C_{11} + C_{22} + C_{33} + 2(C_{12} + C_{13} + C_{23})] \tag{7}$$

$$\begin{aligned} B_R = \; & \Omega[a(C_{11} + C_{22} - 2C_{12}) + b(2C_{12} - 2C_{11} - C_{23}) + c(C_{15} - 2C_{25}) + \\ & d(2C_{12} + 2C_{23} - C_{13} - 2C_{22}) + 2e(C_{25} - C_{15}) + f]^{-1} \end{aligned} \tag{8}$$

$$G_V = (1/15)[C_{11} + C_{22} + C_{33} + 3(C_{44} + C_{55} + C_{66}) - (C_{12} + C_{13} + C_{23})] \tag{9}$$

$$\begin{aligned} G_R = \; & 15\{4[a(C_{11} + C_{22} + C_{12}) + b(C_{11} - C_{12} - C_{23}) + \\ & c(C_{15} + C_{25}) + d(C_{22} - C_{12} - C_{23} - C_{13}) + e(C_{15} - C_{25}) + f]/\Omega + \\ & 3[g/\Omega + (C_{44} + C_{66})/(C_{44}C_{66} - C_{46}^2)]\}^{-1} \end{aligned} \tag{10}$$

wherein:

$$a = C_{33}C_{55} - C_{35}^2 \tag{11}$$

$$b = C_{23}C_{55} - C_{25}C_{35} \tag{12}$$

$$c = C_{13}C_{35} - C_{15}C_{33} \tag{13}$$

$$d = C_{13}C_{55} - C_{15}C_{35} \tag{14}$$

$$e = C_{13}C_{25} - C_{15}C_{23} \tag{15}$$

$$\begin{aligned} f = \; & C_{11}(C_{22}C_{55} - C_{25}^2) - C_{12}(C_{12}C_{55} - C_{15}C_{25}) + \\ & C_{15}(C_{12}C_{25} - C_{15}C_{22}) + C_{25}(C_{23}C_{35} - C_{25}C_{33}) \end{aligned} \tag{16}$$

$$g = C_{11}C_{22}C_{33} - C_{11}C_{23}^2 - C_{22}C_{13}^2 - C_{33}C_{12}^2 + 2C_{12}C_{13}C_{23} \tag{17}$$

$$\begin{aligned} \Omega = \; & 2[C_{15}C_{25}(C_{33}C_{12} - C_{13}C_{23}) + C_{15}C_{35}(C_{22}C_{13} - C_{12}C_{23}) + \\ & C_{25}C_{35}(C_{11}C_{23} - C_{12}C_{13})] - [C_{15}^2(C_{22}C_{33} - C_{23}^2) + C_{25}^2(C_{11}C_{33} - C_{13}^2) + \\ & C_{35}^2(C_{11}C_{22} - C_{12}^2)] + gC_{55} \end{aligned} \tag{18}$$

The formula for calculating the bulk modulus (B), shear modulus (G), and elastic modulus (E) of monoclinic crystal by Hill model [55] is:

$$B_H = \frac{1}{2}(B_V + B_R) \tag{19}$$

$$G_H = \frac{1}{2}(G_V + G_R) \tag{20}$$

$$E = 9BG/(3B + G) \tag{21}$$

3. Results and Discussion

3.1. Structure Stability

The replacement energy is shown in Figure 2 and alternative solute atoms Mg/Si were incorporated for comparison with Cu at different sites. In order to more intuitively express the competitive occupation sites of Cu atoms, the variable Δ is introduced and the

Δ values of Cu atoms at different sites in different β″ configurations are shown in Figure 3. The Δ represents the "competitiveness" between Cu atoms and other solute atoms at each site, it is the difference between the lowest replacement energy of Mg/Si solute atoms and the replacement energy of Cu atoms. The larger the value of Δ, the more likely the Cu atom will occupy the site. Consequently, due to the low Cu occupancy in β″, only three designated Cu sites (Si1, Si3, Mg1, see Figure 1c) were allowed to host Cu atoms according to the relative value of replacement energy (refer to Figure 2). This conclusion is consistent with previous research [38].

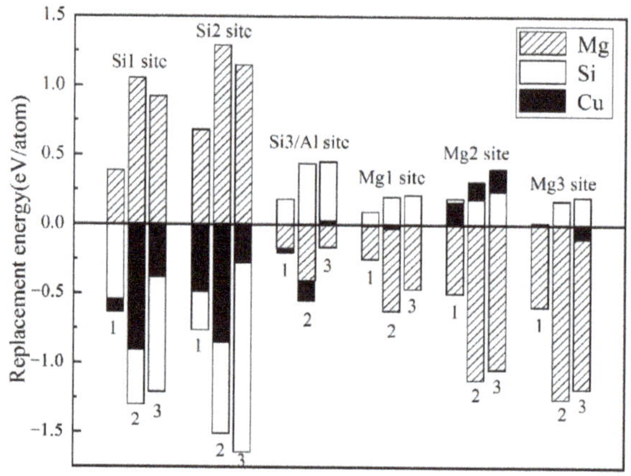

Figure 2. Calculated replacement energies for Cu and alternative solute atoms Mg/Si on the different sites of three different β″ configurations. 1: $Mg_4Al_3Si_4$, 2: $Mg_5Al_2Si_4$, and 3: Mg_6AlSi_4. The position of each column represents a different position in a different configuration. Cu, Mg, and Si replacement energies are labelled with black, shaded, and white bars, respectively.

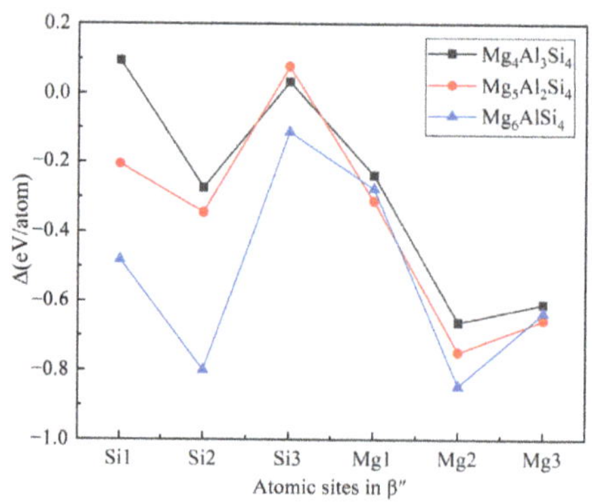

Figure 3. The competitiveness (Δ values) of Cu atoms at different sites in different β″ configurations. The black square, red circle, and blue triangle represent Cu atoms in the $Mg_4Al_3Si_4$, $Mg_5Al_2Si_4$, and Mg_6AlSi_4 configurations, respectively.

For checking the reliability of the calculations, Table 2 displays the structural parameters for selected β″ configuration without Cu atom, along with the results of earlier theoretical and experimental studies of β″. Available calculation results of formation enthalpies are shown in Table 3. The formation enthalpies of the 33 possible unit cells have been plotted in Figure 4, including the configuration without Cu atom. Since the given formation enthalpy of per solute atom (eV/solute atom) essentially presents the solute chemical potentials, the zero-temperature convex hull can be constructed to deduce the precipitation order of the system [57]. It can be seen that Cu occupying one column of each Si3 column pair is found to be the energetically most favorable option for the set of $Mg_4Al_2CuSi_4$ compositions. While the formation enthalpy of $Mg_4Al_2CuSi_4$ is −0.337 eV/solute atom, the formation enthalpy of $Mg_5AlCuSi_4$ is −0.335 eV/solute atom, which is similar to that of $Mg_4Al_2CuSi_4$. This is consistent with the observed in previous experiments that Cu atoms mainly occupy Si3 sites [36]. The energy gained when replacing Mg/Si/Al with at the Wyckoff sites is clearly varying with x. When Cu atoms occupy two sites (that is, $x = 2$), $Mg_4AlCu_2Si_4$ is the energetically most favorable phase, and Cu atoms occupy two Si3 columns. When Cu atoms occupy three sites, $Mg_4Cu_3Si_4$ is the most stable structure, in which Cu atoms occupy one Mg1 site and two Si3 sites, which is consistent with experimental observations [36]. The results show that stoichiometry of Cu-containing β″ phase is suggested as $Mg_4Al_{3-x}Cu_xSi_4$ ($1 \leq x \leq 3$). Since the formation enthalpy of $Mg_5AlCuSi_4$ is very close to that of $Mg_4Al_2CuSi_4$, it can also be taken into account. This result emphasizes the possibility of fluctuations between various compositions as a function of the local alloying element concentration for the physical system during precipitated phases growth. Then the structural parameters of low energy configurations from Figure 4 also have been displayed in Table 2. As discussed above, sole minimization of the β″ phase formation enthalpy supports the well-defined $Mg_{4+x}Al_{2-x}CuSi_4$ (x = 0, 1) unit cell shown in Figure 5.

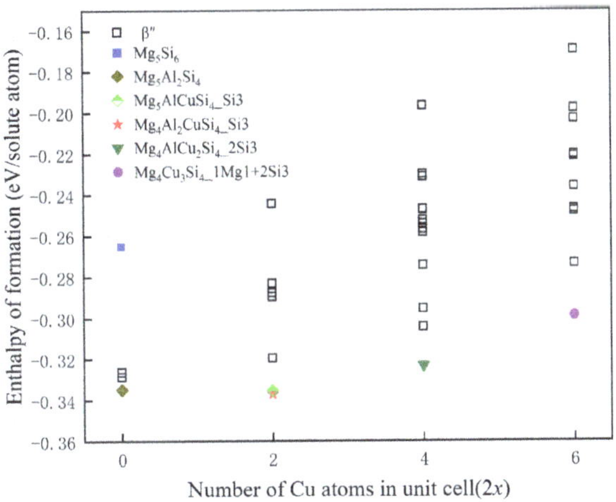

Figure 4. Formation enthalpies of calculated β″ phases. When Cu atoms occupy x sites the number of Cu atoms in unit cell is corresponding to $2x$. The lower energy configuration is marked with a specific shape, and Mg_5Si_6 is also marked as a reference. ■ represents the Mg_5Si_6, ◆ represents the $Mg_5Al_2Si_4$, ◆ represents the $Mg_5AlCuSi_4$ where Cu atoms occupy a Si3 site, ★ represents the $Mg_4Al_2CuSi_4$ where Cu atoms occupy a Si3 site, ▼ represents the $Mg_4AlCu_2Si_4$ where Cu atoms occupy two Si3 sites, and ● represents the $Mg_4Cu_3Si_4$ where Cu atoms occupy a Mg1 site and two Si3 sites.

Table 2. First-principles (VASP-GGA) and experimental lattice parameters of β″ phases of Al-Mg-Si-(Cu) system. For the Cu-containing β″ phases, only the most stable crystal structures under different Cu concentrations are listed.

Configurations	a (Å)	b (Å)	c (Å)	β (°)	Ref.
Mg_5Si_6	15.12	4.04	6.99	110.6	
Mg_5Si_6 (exp.)	15.16 ± 0.02	4.05	6.74 ± 0.02	105.3 ± 0.5	[17]
Mg_5Si_6 (GGA)	15.11	4.080	6.932	110.4	[21]

Table 2. *Cont.*

Configurations	a (Å)	b (Å)	c (Å)	β (°)	Ref.
Mg_5Si_6 (GGA)	15.13	4.05	6.96	110	[58]
Mg_5Si_6 (GGA)	15.12	4.084	6.928	110.5	[59]
Mg_5Si_6 (GGA)	15.14	4.05	6.94	110	[60]
$Mg_4Al_3Si_4$	15.05	4.16	6.59	106.6	
$Mg_4Al_3Si_4$ (GGA)	15.11	4.131	6.615	106.6	[21]
$Mg_5Al_2Si_4$	15.36	4.05	6.79	105.7	
$Mg_5Al_2Si_4$ (GGA)	15.32	4.075	6.778	105.9	[21]
$Mg_5Al_2Si_4$ (GGA)	15.50	4.05	6.74	106	[19]
Mg_6AlSi_4	15.63	4.06	6.82	105.9	
Mg_6AlSi_4 (GGA)	15.59	4.069	6.830	106.1	[21]
$Mg_4Al_2CuSi_4$	14.78	4.02	6.69	107.3	
$Mg_5AlCuSi_4$	15.08	3.95	6.86	106.2	
$Mg_4AlCu_2Si_4$	14.46	4.03	6.68	109.2	
$Mg_4Cu_3Si_4$	14.17	4.11	6.37	107.5	

Table 3. Formation enthalpies of β″ phases with different configurations in this work. The Cu occupied sites and its number are also listed in detail.

Configurations	Cu Occupied Sites	$x_{Mg}/(x_{Mg} + x_{Si})$	x_{Cu}	$\Delta E_{\beta''}$ (eV/Solute Atom)
Mg_5Si_6	-	0.45	0.00	−0.2650
Mg_5Si_6 [21]	-	0.45	0.00	−0.2665
$Mg_4Al_3Si_4$	-	0.50	0.00	−0.3264
$Mg_5Al_2Si_4$	-	0.56	0.00	−0.3348
$Mg_5Al_2Si_4$ [21]	-	0.56	0.00	−0.3456
Mg_6AlSi_4	-	0.60	0.00	−0.3286
Mg_6AlSi_4 [21]	-	0.60	0.00	−0.3380
$Mg_4Al_3CuSi_3$	1 Si1	0.57	0.09	−0.2896
$Mg_4Al_2CuSi_4$	1 Si3/Al	0.50	0.09	−0.3370
$Mg_4Al_2CuSi_4$	1 Mg1	0.50	0.09	−0.3196
$Mg_4Al_3Cu_2Si_2$	2 Si1	0.67	0.18	−0.2525
$Mg_4AlCu_2Si_4$	2 Si3/Al	0.50	0.18	−0.3232
$Mg_4Al_2Cu_2Si_3$	1 Si1 and 1 Si3	0.57	0.18	−0.2567
$Mg_4Al_2Cu_2Si_3$	1 Si1 and 1 Mg1	0.57	0.18	−0.2583
$Mg_4AlCu_2Si_4$	1 Si3/Al and 1 Mg1	0.50	0.18	−0.3043
$Mg_4Al_2Cu_3Si_2$	2 Si1 and 1 Si3/Al	0.67	0.27	−0.2205
$Mg_4AlCu_3Si_3$	1 Si1 and 2 Si3/Al	0.57	0.27	−0.2737
$Mg_4Al_2Cu_3Si_2$	2 Si1/Al and 1 Mg1	0.67	0.27	−0.2215
$Mg_4Cu_3Si_4$	2 Si3/Al and 1 Mg1	0.50	0.27	−0.2988
$Mg_4AlCu_3Si_3$	1 Si1 and 1 Si3/Al and 1 Mg1	0.57	0.27	−0.2482
$Mg_5Al_2CuSi_3$	1 Si1	0.63	0.09	−0.2831
$Mg_5AlCuSi_4$	1 Si3/Al	0.56	0.09	−0.3352
$Mg_5Al_2Cu_2Si_2$	2 Si1	0.71	0.18	−0.2298
$Mg_5Cu_2Si_4$	2 Si3/Al	0.56	0.18	−0.2955
$Mg_5AlCu_2Si_3$	1 Si1 and 1 Si3	0.63	0.18	−0.2542
$Mg_5AlCu_3Si_2$	2 Si1 and 1 Si3	0.71	0.27	−0.2033
$Mg_5Cu_3Si_3$	1 Si1 and 2 Si3	0.63	0.27	−0.2471
$Mg_6AlCuSi_3$	1 Si1	0.67	0.09	−0.2444
Mg_6CuSi_4	1 Si3/Al	0.60	0.09	−0.2876
$Mg_6AlCu_2Si_2$	2 Si1	0.75	0.18	−0.1967
$Mg_6Cu_2Si_3$	1 Si1 and 1 Si3	0.67	0.18	−0.2311
$Mg_5AlCu_2Si_3$	1 Si1 and 1 Mg1	0.63	0.18	−0.2472
$Mg_5Cu_2Si_4$	1 Si3/Al and 1 Mg1	0.56	0.18	−0.2744
$Mg_5AlCu_3Si_2$	2 Si1 and 1 Mg1	0.71	0.27	−0.1981
$Mg_6Cu_3Si_2$	2 Si1 and 1 Si3	0.75	0.27	−0.1696
$Mg_5Cu_3Si_3$	1 Si1 and 1 Si3 and 1 Mg1	0.63	0.27	−0.2361

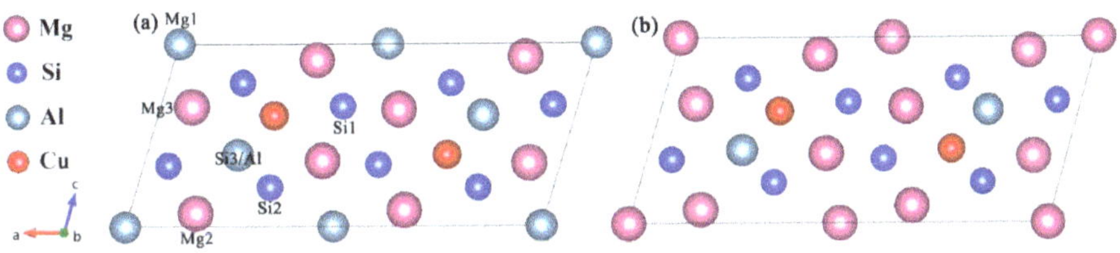

Figure 5. The well-defined $Mg_{4+x}Al_{2-x}CuSi_4$ (x = 0, 1) unit cell that sole minimization of the β″ phase formation enthalpy supports. (**a**) $Mg_4Al_2CuSi_4$; (**b**) $Mg_5AlCuSi_4$. The relative location of each site is marked in (**a**).

3.2. Elastic Properties

Here, we compare the mechanical properties of β″ with or without Cu atoms. The elastic constants of key β″ phases that are most likely to precipitate during aging were calculated by using fully relaxed crystal structures, and the results are listed in Table 4. According to the Born stability criterion [61], the elastic constants of $Mg_4Al_2CuSi_4$ and $Mg_5AlCuSi_4$ all meet the stability criteria of monoclinic crystals. This further supports the stability of $Mg_{4+x}Al_{2-x}CuSi_4$ (x = 0, 1) obtained from the formation enthalpy. The elastic constants C_{11}, C_{22}, and C_{33} are much greater than the other elastic constants in all calculated β″ phases, resulting in an obvious elastic anisotropy. In order to understand the anisotropic characteristics of these precipitation phases, the Young's modulus anisotropies are evaluated by three-dimensional map as shown in Figure 6. Comparing Figure 6a and c, it can be seen that after Cu atoms substituted Al atoms on the Si3/Al sites, the Young's modulus (E) anisotropy increases significantly; similar results are also shown in Figure 6b,d. This phenomenon indicates that the growth rate of the Cu-containing β″ phases may be faster than that of the β″ without Cu. It is consistent with the previous study that Cu can accelerates the age-hardening response [14,28,30].

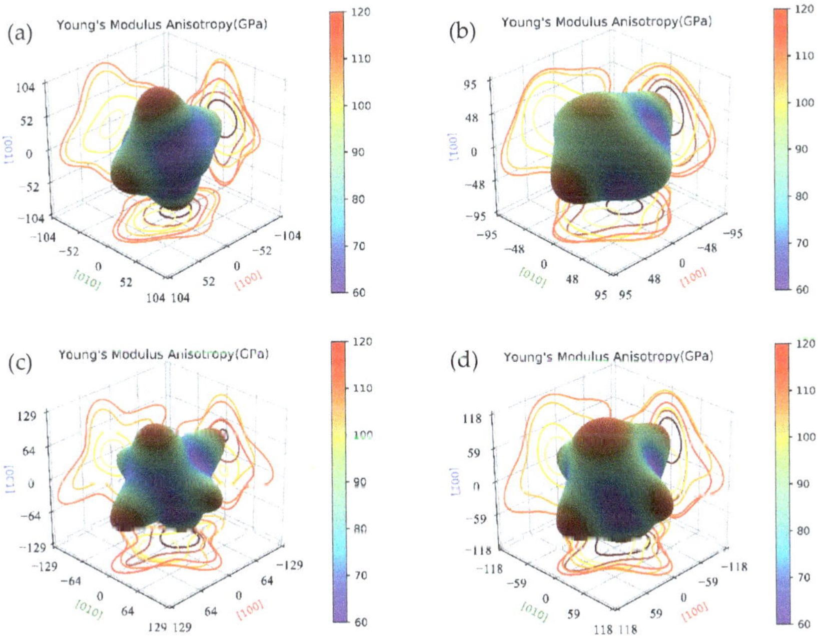

Figure 6. The Young's modulus anisotropies three-dimensional map of β″ phases. (**a**) $Mg_4Al_3Si_4$; (**b**) $Mg_5Al_2Si_4$; (**c**) $Mg_4Al_2CuSi_4$; (**d**) $Mg_5AlCuSi_4$.

Table 4. Calculated single crystal elastic stiffness constants (C_{ij}'s) of the reported β″ phases and energy favorable Cu-containing β″ phases.

Configuration	C_{11}	C_{12}	C_{13}	C_{15}	C_{22}	C_{23}	C_{25}	C_{33}	C_{35}	C_{44}	C_{46}	C_{55}	C_{66}
Mg_5Si_6	110	42	42	−3	103	49	4	94	11	19	5	17	25
Mg_5Si_6 [62]	106	49	50	−11	90	46	6	88	9	17	1	33	30
Mg_5Si_6 [63]	98	50	48	8	84	46	6	88	5.4	22	−10	29	51
$Mg_4Al_3Si_4$	119	52	35	−3	99	47	3	122	10	19	−1	29	20
$Mg_4Al_3Si_4$ [62]	114	46	48	−4	104	49	6	104	7	21	0	34	23
$Mg_4Al_3Si_4$ [63]	107	47	48	9	97	48	6	97	9	26	6	36	46
$Mg_5Al_2Si_4$	111	38	44	−4	102	46	3	106	7	25	4	31	25
$Mg_5Al_2Si_4$ [62]	108	42	48	−3	95	46	5	100	3	23	4	33	27
$Mg_5Al_2Si_4$ [63]	107	40	46	−13	95	43	4	99	12	27	5	36	49
Mg_6AlSi_4	121	28	40	−5	125	28	2	117	6	28	4	35	21
$Mg_4Al_2CuSi_4$	136	44	48	−13	133	43	9	130	14	25	3	35	23
$Mg_5AlCuSi_4$	127	41	46	−9	128	32	5	131	6	31	4	38	22
$Mg_4AlCu_2Si_4$	128	44	70	−3	136	53	6	103	10	28	7	32	22
$Mg_4Cu_3Si_4$	128	47	75	7	153	43	3	115	−6	20	3	51	26

Based on the elastic constants in Table 4, the bulk modulus (*B*), shear modulus (*G*), and Young's modulus (*E*) of polycrystalline are calculated by the Hill model [55], and the results are listed in Table 5. Comparing the values of *E*, *G*, and *B* of $Mg_4Al_3Si_4$ and $Mg_4Al_2CuSi_4$, it can be seen that the values of *E*, *G*, and *B* of β″ with Cu atoms are higher than that of β″ without Cu atoms. This relationship is also shown between $Mg_5Al_2Si_4$ and $Mg_5AlCuSi_4$. In general, the Young's modulus (*E*) can be used to measure the stiffness of the material. The stiffness of the material is greater with the increasing of Young's modulus (*E*) [64]. It is obvious that the stiffness is enhanced after Cu incorporate into Si3 sites. Pugh [65] proposes using the ratio of the bulk and shear modulus, B/G, to predict brittle or ductile behavior of materials. According to the Pugh criterion, if B/G is more than 1.75, ductile behavior is expected; otherwise, the material would be brittle. From Table 4, the B/G values of calculated β″ phases are all larger than 1.75, therefore, all the compounds of β″ phase are ductile with or without Cu atoms and the ductility decreases after Cu atoms incorporate into β″. In addition, Poisson's ratio v has been used to measure the shear stability of the lattice, which usually ranges from −1 to 0.5. The smaller the value, the stronger the ability of the crystal to maintain stability during shear deformation [66]. The value of Poisson's ratio $v > 0.26$ means the ductility of the materials, and the Poisson's ratio of metals is usually $0.25 < v < 0.35$ [67]. As one can see, all β″ configurations show ductility with minor differences. It is consistent with the conclusion based on Pugh criterion.

Table 5. Calculated mechanic properties of the reported β″ phases and energy favorable Cu-containing β″ phases.

Configurations	*B* (GPa)	*G* (GPa)	*E* (GPa)	*B*/*G*	*v*
Mg_5Si_6	62	22	60	2.77	0.34
Mg_5Si_6 [63]	62	-	-	-	-
$Mg_4Al_3Si_4$	67	26	69	2.57	0.33
$Mg_4Al_3Si_4$ [63]	64	-	-	-	-
$Mg_5Al_2Si_4$	63	28	74	2.23	0.30
$Mg_5Al_2Si_4$ [63]	61	-	-	-	-
Mg_6AlSi_4	62	33	84	1.87	0.27
$Mg_4Al_2CuSi_4$	72	32	84	2.26	0.31
$Mg_5AlCuSi_4$	69	34	88	2.01	0.29
$Mg_4AlCu_2Si_4$	76	28	74	2.73	0.34
$Mg_4Cu_3Si_4$	81	32	84	2.54	0.33

3.3. Phonon Spectra

In addition, the dynamic stability is also taken into account. The phonon spectra of $Mg_4Al_2CuSi_4$ and $Mg_5AlCuSi_4$ are shown in Figure 7. From Figure 7, one can see that there is no virtual frequency of configuration $Mg_4Al_2CuSi_4$ and $Mg_5AlCuSi_4$, which is generally considered to be dynamically stable.

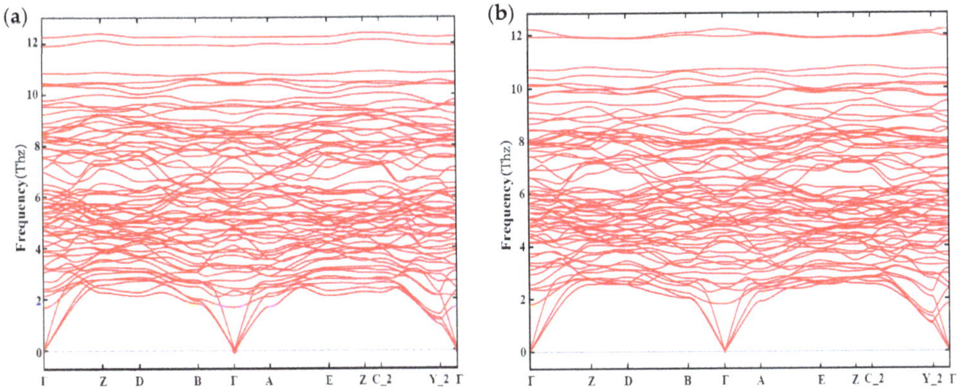

Figure 7. The phonon spectrum along a highly symmetric K-points path of (**a**) $Mg_4Al_2CuSi_4$; (**b**) $Mg_5AlCuSi_4$.

3.4. Electronic Structure

The total and partial electronic density of states (TDOSs and PDOSs) for four types of β″ configurations are calculated to explore the influence mechanism of electronic interaction on structural stability and mechanical properties, as shown in Figure 8, with the Fermi level set to zero. It is evident that incorporating Cu does not change the metallic characteristic of the β″ phase due to the finite DOS at the Fermi level. At the Fermi level, the TDOS for four types of β″ configurations at the Fermi level varies. The greatest n (E_f) is 7.41 states/eV/cell in $Mg_5Al_2Si_4$, followed by 6.40 states/eV/cell in $Mg_4Al_3Si_4$, 5.27 states/eV/cell in $Mg_4Al_2CuSi_4$, and 4.18 states/eV/cell in $Mg_5AlCuSi_4$. This indicates that the Cu-containing β″ phases have a smaller n (E_f). In general, a smaller pseudo gap value n (E_f) corresponds to a more stable structure [68]. This indicates that $Mg_{4+x}Al_{2-x}CuSi_4$ ($x = 0, 1$) are more stable than the β″ phases without Cu. The Si-s (range from around 11 eV to 7 eV) and Si-p states (from around 7 eV to the Fermi level) dominate the TDOS of $Mg_4Al_3Si_4$ and $Mg_5Al_2Si_4$ below the Fermi level. In between (ranging from about −7 eV up to −4 eV) regimes, a mixture of s and p character exists, indicating strong hybridization. Especially from −7 eV to −5 eV, the shapes of Si-s and Si-p are very similar, indicating that there is a strong interaction between Si atoms. This may be the origin for the formation of the Si-network; the Si-network acts as a stable skeleton of these phases [32,69]. One can see that Mg-s/Al-s and Si-p in the range from −7 to −4 eV, originating mainly from the s-p hybridization of Si atoms and Mg/Al atoms. The s-states and p-states of Al, Mg, and Si are strongly hybridized above the Fermi level. From Figure 8, it should be noted that, below the Fermi level, the Cu-d state is formed. The s/p orbitals of Mg, Si, and Al all interact with the Cu-d state, and there is obvious electron transfer. The Si-p orbital and the Cu-d orbital are hybridized to form a covalent like bonding, and more electrons are transferred to the new orbital formed by the p-d hybridization.

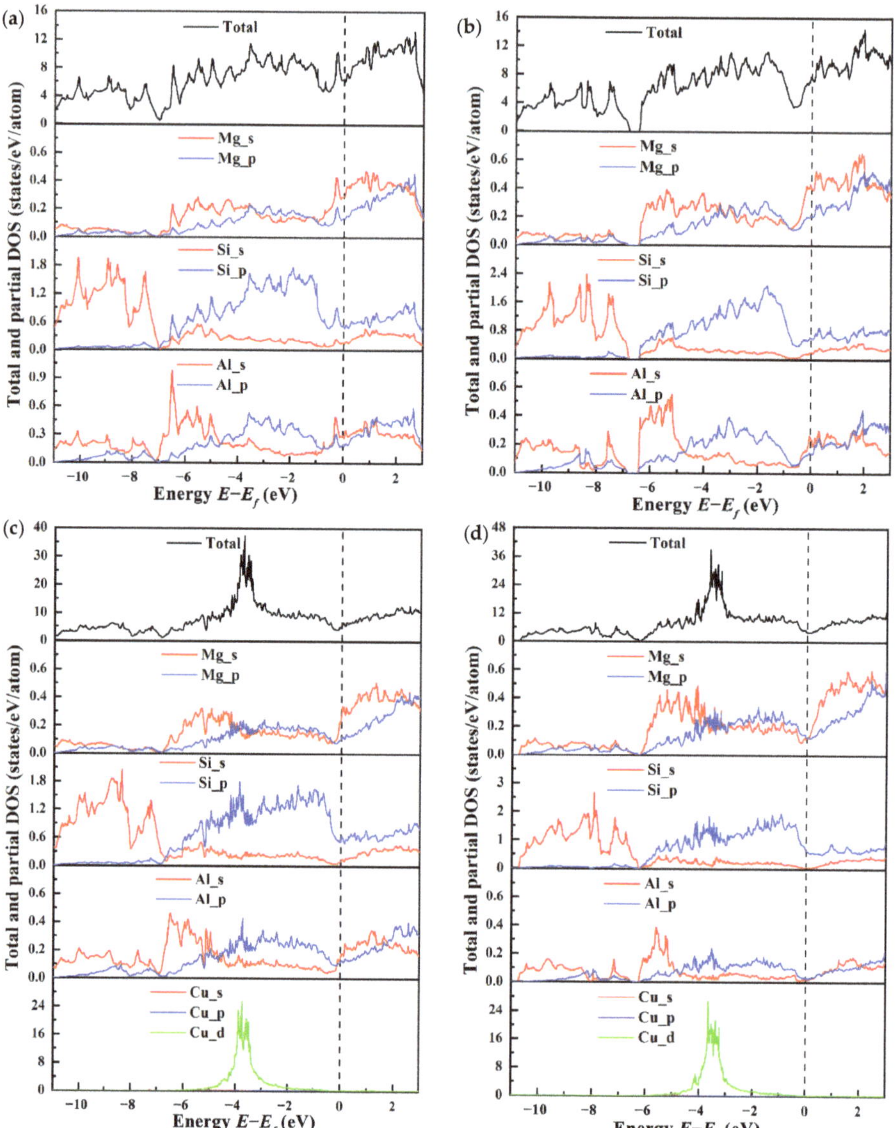

Figure 8. The total and partial electronic density of states (PDOSs and TDOSs) for four β″ type compounds. (**a**) $Mg_4Al_3Si_4$; (**b**) $Mg_5Al_2Si_4$; (**c**) $Mg_4Al_2CuSi_4$; (**d**) $Mg_5AlCuSi_4$.

In order to gain a better understanding of the electronic structure of the studied system, the charge density distributions were used as an additional method. The charge-density difference between the (DFT) converged charge density and the isolated atomic charge densities were employed. Figure 9 shows the charge density difference contour plot for the (010) plane to analyze the interaction between Al, Mg, Si, and Cu atoms for the β″ phases. Here we clearly see that there has indeed been a transfer of charge to all the Si–Si bond regions, it is consistent with the analysis by Derlet et al. [69]. A dominant feature of Figure 9a,b is the concentration of charge between the Si1-S3/Al-Si2-Mg1-Si1 nearest neighbors, and to a lesser extent, between the Si3 and Mg1 nearest neighbors, indicating that covalency plays a role in this system, which was also reported in previous research [70]. Meanwhile, the charge distribution looks like a "charge loop", which can lead to the formation of an "Si network". Strong covalent bonds network can significantly increase the structural stability of β″ phases. Such a charge transfer to the bonding regions originates from the core regions of both atoms on the Mg and Si sites, in addition to the homogeneous interstitial region between the Mg atoms. The depletion of charge from the Mg3 sites indicates that for this system both metallicity and covalency are present in the bonding. Moreover, the charge transfer density between the Si3 and Si2 sites is slightly decreased, and the charge transfer density between the Mg2 and Si2 sites is increased, indicating the bonds between atoms on Mg2 and Si2 sites are covalent. As shown in Figure 9, the charge ionization of all Mg3 sites is strong, and when the Cu atom on the Si3 site charge ionization becomes stronger, it means both Mg and Cu valence electron are delocalized. The difference is that Mg uniformly provides charges to the surroundings to form a metallic environment [69], while the charges of Cu atoms are delocalized toward Si atoms in unit cells, forming a directional covalent like bond.

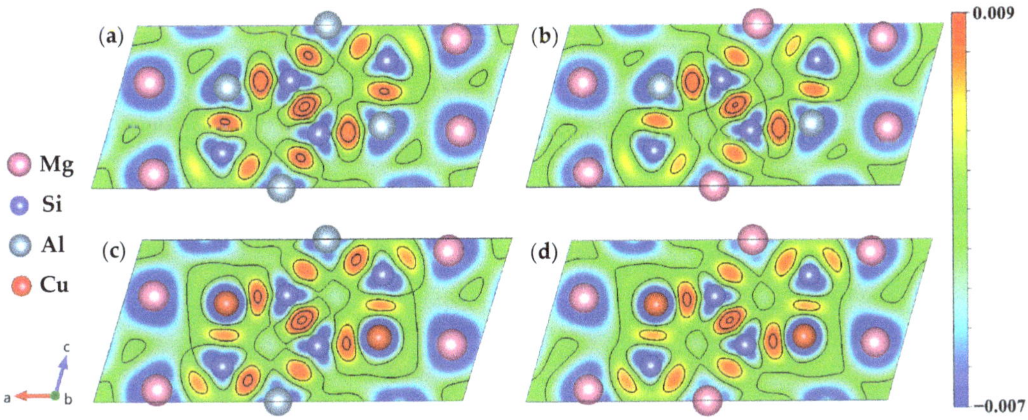

Figure 9. Charge density difference of four calculated β″ phases. (**a**) $Mg_4Al_3Si_4$; (**b**) Mg5Al2Si4; (**c**) $Mg_4Al_2CuSi_4$; (**d**) $Mg_5AlCuSi_4$.

According to the analysis of the thermodynamic results of replacement energies and formation enthalpies in Section 3.1 "Structure stability", the stoichiometry of Cu-containing β″ phases in the precipitation sequence and the sequence of Cu atoms substituting sites in the β″ phases can be inferred. For the stable phases determined from the thermodynamics, the elastic properties of β″ phases with and without Cu were calculated in Section 3.2 "Elastic properties", further supporting the proposed stoichiometry. Besides, the "Phonon spectra" study in Section 3.3 shows that they are also dynamically stable. In summary, the proposed compositions $Mg_4Al_{3-x}Cu_xSi_4$ ($1 \leq x \leq 3$) are reasonable, which is consistent with the results observed in the experiment [36]. In the Section 3.4 "Electronic structure", the origin for the stability of the Cu-containing β″ phases is analyzed from the perspective of electron interaction.

4. Conclusions

(1) The calculation of the formation enthalpies of 33 Cu containing β″ phases shows that the replacement order of Cu atoms in β″ phases can be summarized as one Si3/Al site → two Si3/Al sites → two Si3/Al sites and one Mg1 site.

(2) The Cu atoms strongly favor occupying one of each pair of Si3/Al sites and the most stable Cu-containing β″ phases were expected to have a stoichiometry of $Mg_{4+x}Al_{2-x}CuSi_4$ (x = 0, 1). In addition, taking into account the change of Cu content in β″ phases, the stoichiometry of $Mg_4Al_{3-x}Cu_xSi_4$ ($1 \leq x \leq 3$) may precipitate.

(3) The calculated mechanical properties show that all calculated β″ phases are mechanically stable. The incorporation of Cu atoms improves the values of bulk modulus (*B*), shear modulus (*G*), and Young's modulus (*E*) of β″, respectively, and all β″ phases calculated show ductile behavior. Furthermore, the calculation of the phonon spectra shows that $Mg_{4+x}Al_{2-x}CuSi_4$ ($x = 0, 1$) are dynamically stable.
(4) The electronic structure results shows that the Cu atom will join the Si network, and the bond between the Si atom and the Cu atom has the covalent property. The incorporation of Cu atom increases the electron interaction between the Mg2 and the Si3 atom, which may be one of the reasons why the incorporation of Cu atom increases the stability of the β″ phase structure.

Author Contributions: Conceptualization, S.H. and J.W.; methodology, S.H. and J.W.; investigation, S.H., J.W.; resources, J.W., Q.W. and Y.D.; data curation, S.H. and J.W.; writing—original draft preparation, S.H.; writing—review and editing, J.W., D.Z., Y.K., Q.W. and Y.D.; visualization, S.H. and D.Z.; supervision, J.W. and Y.D.; project administration, J.W. and Y.D.; funding acquisition, J.W. and Y.K. All authors have read and agreed to the published version of the manuscript.

Funding: This research was funded by National Key Research and Development Program of China, grant numbers 2019YFB2006500, and the National Natural Science Foundation of China, grant numbers 52171024, 51771234 and 51601228.

Institutional Review Board Statement: Not applicable.

Informed Consent Statement: Not applicable.

Data Availability Statement: Data are contained within the article.

Acknowledgments: This work was financially supported by National Key Research and Development Program of China (grant numbers: 2019YFB2006500), the National Natural Science Foundation of China (grant numbers 52171024, 51771234 and 51601228) are greatly acknowledged. First-principles calculations were partially carried out at the High Performance Computing of Central South University.

Conflicts of Interest: The authors declare no conflict of interest.

References

1. Li, Y.; Li, H.; Katgerman, L.; Du, Q.; Zhang, J.; Zhuang, L. Recent advances in hot tearing during casting of aluminium alloys. *Prog. Mater. Sci.* **2021**, *117*, 100741. [CrossRef]
2. Abouarkoub, A.; Thompson, G.E.; Zhou, X.; Hashimoto, T.; Scamans, G. The influence of prolonged natural aging on the subsequent artificial aging response of the AA6111 automotive alloy. *Met. Mater. Trans. A* **2015**, *46*, 4380–4393. [CrossRef]
3. Zhang, D.-L.; Wang, J.; Kong, Y.; Zou, Y.; Du, Y. First-principles investigation on stability and electronic structure of Sc-doped θ′/Al interface in Al-Cu alloys. *Trans. Nonferrous Met. Soc. China* **2021**, *31*, 3342–3355. [CrossRef]
4. Chakrabarti, D.; Laughlin, E.D. Phase relations and precipitation in Al–Mg–Si alloys with Cu additions. *Prog. Mater. Sci.* **2004**, *49*, 389–410. [CrossRef]
5. Gupta, A.; Lloyd, D.; Court, S. Precipitation hardening in Al–Mg–Si alloys with and without excess Si. *Mater. Sci. Eng. A* **2001**, *316*, 11–17. [CrossRef]
6. Safyari, M.; Moshtaghi, M.; Hojo, T.; Akiyama, E. Mechanisms of hydrogen embrittlement in high-strength aluminum alloys containing coherent or incoherent dispersoids. *Corros. Sci.* **2021**, *194*, 109895. [CrossRef]
7. Ardell, A.J. Precipitation hardening. *Met. Mater. Trans. A* **1985**, *16*, 2131–2165. [CrossRef]
8. Polmear, I. In Aluminium alloys—A century of age hardening. *Mater. Forum* **2004**, *28*, 1–14.
9. Van Huis, M.; Chen, J.; Sluiter, M.; Zandbergen, H. Phase stability and structural features of matrix-embedded hardening precipitates in Al–Mg–Si alloys in the early stages of evolution. *Acta Mater.* **2007**, *55*, 2183–2199. [CrossRef]
10. Edwards, G.; Stiller, K.; Dunlop, G.; Couper, M. The precipitation sequence in Al–Mg–Si alloys. *Acta Mater.* **1998**, *46*, 3893–3904. [CrossRef]
11. Guinier, A. Structure of age-hardened aluminium-copper alloys. *Nat. Cell Biol.* **1938**, *142*, 569–570. [CrossRef]
12. Preston, G.D. Structure of age-hardened aluminium-copper alloys. *Nat. Cell Biol.* **1938**, *142*, 570. [CrossRef]
13. Matsuda, K.; Ikeno, S.; Terayama, K.; Matsui, H.; Sato, T.; Uetani, Y. Comparison of precipitates between excess Si-type and balanced-type Al-Mg-Si alloys during continuous heating. *Met. Mater. Trans. A* **2005**, *36*, 2007–2012. [CrossRef]
14. Miao, W.F.; Laughlin, D.E. Effects of Cu content and preaging on precipitation characteristics in aluminum alloy 6022. *Met. Mater. Trans. A* **2000**, *31*, 361–371. [CrossRef]

15. Marioara, C.D.; Nordmark, H.; Andersen, S.J.; Holmestad, R. Post-β″ phases and their influence on microstructure and hardness in 6xxx Al-Mg-Si alloys. *J. Mater. Sci.* **2006**, *41*, 471–478. [CrossRef]
16. Esmaeili, S.; Lloyd, D.; Poole, W. A yield strength model for the Al-Mg-Si-Cu alloy AA6111. *Acta Mater.* **2003**, *51*, 2243–2257. [CrossRef]
17. Zandbergen, H.W.; Andersen, S.J.; Jansen, J. Structure determination of Mg_5Si_6 particles in Al by dynamic electron diffraction studies. *Science* **1997**, *277*, 1221–1225. [CrossRef]
18. Andersen, S.J.; Zandbergen, H.; Jansen, J.; Træholt, C.; Tundal, U.; Reiso, O. The crystal structure of the β″ phase in Al–Mg–Si alloys. *Acta Mater.* **1998**, *46*, 3283–3298. [CrossRef]
19. Hasting, H.S.; Frøseth, A.G.; Andersen, S.J.; Vissers, R.; Walmsley, J.C.; Marioara, C.D.; Danoix, F.; Lefebvre, W.; Holmestad, R. Composition of β″ precipitates in Al–Mg–Si alloys by atom probe tomography and first principles calculations. *J. Appl. Phys.* **2009**, *106*, 123527. [CrossRef]
20. Pogatscher, S.; Antrekowitsch, H.; Leitner, H.; Sologubenko, A.; Uggowitzer, P. Influence of the thermal route on the peak-aged microstructures in an Al–Mg–Si aluminum alloy. *Scr. Mater.* **2013**, *68*, 158–161. [CrossRef]
21. Ehlers, F.J. *Ab initio* interface configuration determination for β″ in Al–Mg–Si: Beyond the constraint of a preserved precipitate stoichiometry. *Comput. Mater. Sci.* **2014**, *81*, 617–629. [CrossRef]
22. Ninive, P.H.; Strandlie, A.; Gulbrandsen-Dahl, S.; Lefebvre, W.; Marioara, C.D.; Andersen, S.J.; Friis, J.; Holmestad, R.; Løvvik, O.M. Detailed atomistic insight into the β″ phase in Al–Mg–Si alloys. *Acta Mater.* **2014**, *69*, 126–134. [CrossRef]
23. Xiao, Q.; Liu, H.; Yi, D.; Yin, D.; Chen, Y.; Zhang, Y.; Wang, B. Effect of Cu content on precipitation and age-hardening behavior in Al-Mg-Si-xCu alloys. *J. Alloys Compd.* **2017**, *695*, 1005–1013. [CrossRef]
24. Man, J.; Jing, L.; Jie, S.G. The effects of Cu addition on the microstructure and thermal stability of an Al–Mg–Si alloy. *J. Alloys Compd.* **2007**, *437*, 146–150. [CrossRef]
25. Ding, L.; Jia, Z.; Liu, Y.; Weng, Y.; Liu, Q. The influence of Cu addition and pre-straining on the natural aging and bake hardening response of Al-Mg-Si alloys. *J. Alloys Compd.* **2016**, *688*, 362–367. [CrossRef]
26. Saito, T.; Marioara, C.D.; Andersen, S.J.; Lefebvre, W.; Holmestad, R. Aberration-corrected HAADF-STEM investigations of precipitate structures in Al–Mg–Si alloys with low Cu additions. *Philos. Mag.* **2014**, *94*, 520–531. [CrossRef]
27. Saito, T.; Mørtsell, E.A.; Wenner, S.; Marioara, C.D.; Andersen, S.J.; Friis, J.; Matsuda, K.; Holmestad, R. Atomic structures of precipitates in Al-Mg-Si alloys with small additions of other elements. *Adv. Eng. Mater.* **2018**, *20*, 18. [CrossRef]
28. Murayama, M.; Hono, K.; Miao, W.F.; Laughlin, D.E. The effect of Cu additions on the precipitation kinetics in an Al-Mg-Si alloy with excess Si. *Met. Mater. Trans. A* **2001**, *32*, 239–246. [CrossRef]
29. Jia, Z.; Ding, L.; Cao, L.; Sanders, R.; Li, S.; Liu, Q. The influence of composition on the clustering and precipitation behavior of Al-Mg-Si-Cu alloys. *Met. Mater. Trans. A* **2016**, *48*, 459–473. [CrossRef]
30. Zandbergen, M.; Cerezo, A.; Smith, G. Study of precipitation in Al–Mg–Si Alloys by atom probe tomography II. Influence of Cu additions. *Acta Mater.* **2015**, *101*, 149–158. [CrossRef]
31. Bobel, A.; Kim, K.; Wolverton, C.; Walker, M.; Olson, G.B. Equilibrium composition variation of Q-phase precipitates in aluminum alloys. *Acta Mater.* **2017**, *138*, 150–160. [CrossRef]
32. Ehlers, F.J.H.; Wenner, S.; Andersen, S.J.; Marioara, C.D.; Lefebvre, W.; Boothroyd, C.B.; Holmestad, R. Phase stabilization principle and precipitate-host lattice influences for Al–Mg–Si–Cu alloy precipitates. *J. Mater. Sci.* **2014**, *49*, 6413–6426. [CrossRef]
33. Torsæter, M.; Ehlers, F.; Marioara, C.; Andersen, S.J.; Holmestad, R. Applying precipitate–host lattice coherency for compositional determination of precipitates in Al–Mg–Si–Cu alloys. *Philos. Mag.* **2012**, *92*, 3833–3856. [CrossRef]
34. Marioara, C.D.; Andersen, S.J.; Stene, T.N.; Hasting, H.; Walmsley, J.; van Helvoort, A.T.J.; Holmestad, R. The effect of Cu on precipitation in Al–Mg–Si alloys. *Philos. Mag.* **2007**, *87*, 3385–3413. [CrossRef]
35. Weng, Y.; Jia, Z.; Ding, L.; Liu, M.; Wu, X.; Liu, Q. Combined effect of pre-aging and Ag/Cu addition on the natural aging and bake hardening in Al-Mg-Si alloys. *Prog. Nat. Sci.* **2018**, *28*, 363–370. [CrossRef]
36. Li, K.; Beche, A.; Song, M.; Sha, G.; Liu, X.; Zhang, K.; Du, Y.; Ringer, S.; Schryvers, D. Atomistic structure of Cu-containing β″ precipitates in an Al–Mg–Si–Cu alloy. *Scr. Mater.* **2014**, *75*, 86–89. [CrossRef]
37. Sunde, J.K.; Marioara, C.D.; Holmestad, R. The effect of low Cu additions on precipitate crystal structures in overaged Al-Mg-Si(-Cu) alloys. *Mater. Charact.* **2020**, *160*, 110087. [CrossRef]
38. Saito, T.; Ehlers, F.J.; Lefebvre, W.; Hernandez-Maldonado, D.; Bjørge, R.; Marioara, C.D.; Andersen, S.J.; Mørtsell, E.A.; Holmestad, R. Cu atoms suppress misfit dislocations at the β″/Al interface in Al–Mg–Si alloys. *Scr. Mater.* **2016**, *110*, 6–9. [CrossRef]
39. Dong, L.; Chu, S.; Hu, B.; Zeng, X.; Chen, B. Atomic-scale investigation into precipitated phase thickening in Al-Si-Mg-Cu alloy. *J. Alloys Compd.* **2018**, *766*, 973–978. [CrossRef]
40. Wolverton, C. Crystal structure and stability of complex precipitate phases in Al–Cu–Mg–(Si) and Al–Zn–Mg alloys. *Acta Mater.* **2001**, *49*, 3129–3142. [CrossRef]
41. Kohn, W.; Sham, L.J. Self-consistent equations including exchange and correlation effects. *Phys. Rev.* **1965**, *140*, 1133–1138. [CrossRef]
42. Kresse, G.; Furthmüller, J. Efficiency of ab-initio total energy calculations for metals and semiconductors using a plane-wave basis set. *Comput. Mater. Sci.* **1996**, *6*, 15–50. [CrossRef]
43. Kresse, G.; Furthmüller, J. Efficient iterative schemes for *ab initio* total-energy calculations using a plane-wave basis set. *Phys. Rev. B* **1996**, *54*, 11169–11186. [CrossRef]

44. Kresse, G.; Joubert, D. From ultrasoft pseudopotentials to the projector augmented-wave method. *Phys. Rev. B* **1999**, *59*, 1758. [CrossRef]
45. Blöchl, P.E. Projector augmented-wave method. *Phys. Rev. B* **1994**, *50*, 17953–17979. [CrossRef]
46. Perdew, J.P.; Burke, K.; Ernzerhof, M. Generalized gradient approximation made simple. *Phys. Rev. Lett.* **1996**, *77*, 3865–3868. [CrossRef]
47. Monkhorst, H.J.; Pack, J.D. Special points for Brillouin-zone integrations. *Phys. Rev. B* **1976**, *13*, 5188–5192. [CrossRef]
48. Togo, A.; Tanaka, I. First principles phonon calculations in materials science. *Scr. Mater.* **2015**, *108*, 1–5. [CrossRef]
49. Shang, S.-L.; Wang, Y.; Kim, D.; Liu, Z.-K. First-principles thermodynamics from phonon and Debye model: Application to Ni and Ni_3Al. *Comput. Mater. Sci.* **2010**, *47*, 1040–1048. [CrossRef]
50. Wang, J.; Shang, S.-L.; Wang, Y.; Mei, Z.-G.; Liang, Y.; Du, Y.; Liu, Z.-K. First-principles calculations of binary Al compounds: Enthalpies of formation and elastic properties. *Calphad* **2011**, *35*, 562–573. [CrossRef]
51. Saito, T.; Ehlers, F.; Lefebvre, W.; Hernandez-Maldonado, D.; Bjørge, R.; Marioara, C.D.; Andersen, S.J.; Holmestad, R. HAADF-STEM and DFT investigations of the Zn-containing β'' phase in Al–Mg–Si alloys. *Acta Mater.* **2014**, *78*, 245–253. [CrossRef]
52. Van Huis, M.; Chen, J.; Zandbergen, H.; Sluiter, M. Phase stability and structural relations of nanometer-sized, matrix-embedded precipitate phases in Al–Mg–Si alloys in the late stages of evolution. *Acta Mater.* **2006**, *54*, 2945–2955. [CrossRef]
53. Mehl, M.J.; Osburn, J.E.; Papaconstantopoulos, D.A.; Klein, B.M. Structural properties of ordered high-melting-temperature intermetallic alloys from first-principles total-energy calculations. *Phys. Rev. B* **1990**, *41*, 10311–10323. [CrossRef] [PubMed]
54. Wang, J.; Du, Y.; Tao, X.; Ouyang, Y.; Zhang, L.; Chen, Q.; Engström, A. First-principles generated mechanical property database for multi-component Al alloys: Focusing on Al-rich corner. *J. Min. Met. Sect. B Met.* **2017**, *53*, 1–7. [CrossRef]
55. Hill, R. The elastic behaviour of a crystalline aggregate. *Proc. Phys. Soc. Sect. A* **1952**, *65*, 349–354. [CrossRef]
56. Wu, Z.-J.; Zhao, E.-J.; Xiang, H.-P.; Hao, X.-F.; Liu, X.-J.; Meng, J. Crystal structures and elastic properties of superhard IrN_2 and IrN_3 from first principles. *Phys. Rev. B* **2007**, *76*, 054115. [CrossRef]
57. Wang, D.; Amsler, M.; Hegde, V.I.; Saal, J.E.; Issa, A.; Zhou, B.-C.; Zeng, X.; Wolverton, C. Crystal structure, energetics, and phase stability of strengthening precipitates in Mg alloys: A first-principles study. *Acta Mater.* **2018**, *158*, 65–78. [CrossRef]
58. Ravi, C. First-principles study of crystal structure and stability of Al–Mg–Si–(Cu) precipitates. *Acta Mater.* **2004**, *52*, 4213–4227. [CrossRef]
59. Wang, Y.; Liu, Z.-K.; Chen, L.-Q.; Wolverton, C. First-principles calculations of β''-Mg_5Si_6/α-Al interfaces. *Acta Mater.* **2007**, *55*, 5934–5947. [CrossRef]
60. Zhao, D.; Zhou, L.; Kong, Y.; Wang, A.; Wang, J.; Peng, Y.; Du, Y.; Ouyang, Y.; Zhang, W. Structure and thermodynamics of the key precipitated phases in the Al–Mg–Si alloys from first-principles calculations. *J. Mater. Sci.* **2011**, *46*, 7839–7849. [CrossRef]
61. Grimvall, G.; Magyari-Köpe, B.; Ozolins, V.; Persson, K.A. Lattice instabilities in metallic elements. *Rev. Mod. Phys.* **2012**, *84*, 945–986. [CrossRef]
62. Jain, A.C.P.; Marchand, D.; Glensk, A.; Ceriotti, M.; Curtin, W.A. Machine learning for metallurgy III: A neural network potential for Al-Mg-Si. *Phys. Rev. Mater.* **2021**, *5*, 053805. [CrossRef]
63. Kobayashi, R.; Giofré, D.; Junge, T.; Ceriotti, M.; Curtin, W.A. Neural network potential for Al-Mg-Si alloys. *Phys. Rev. Mater.* **2017**, *1*, 053604. [CrossRef]
64. Guan, Y.; Zhang, H.; Li, W. First-principles study on alloying stability, electronic structure, and mechanical properties of Al-based intermetallics. *Phys. B Condens. Matter* **2011**, *406*, 1149–1153. [CrossRef]
65. Pugh, S. XCII. Relations between the elastic moduli and the plastic properties of polycrystalline pure metals. *Lond. Edinb. Dublin Philos. Mag. J. Sci.* **1954**, *45*, 823–843. [CrossRef]
66. Ravindran, P.; Fast, L.; Korzhavyi, P.; Johansson, B.; Wills, J.; Eriksson, O. Density functional theory for calculation of elastic properties of orthorhombic crystals: Application to $TiSi_2$. *J. Appl. Phys.* **1998**, *84*, 4891–4904. [CrossRef]
67. Greaves, G.N.; Greer, A.L.; Lakes, R.S.; Rouxel, T. Poisson's ratio and modern materials. *Nat. Mater.* **2011**, *10*, 823–837. [CrossRef]
68. Ouyang, Y.; Liu, F.; Lu, T.; Tao, X.; Du, Y.; He, Y. First-principles investigation of the mechanical, electronic and thermophysical properties of Q-phase in Al–Mg–Si–Cu alloys. *Comput. Mater. Sci.* **2013**, *67*, 334–340. [CrossRef]
69. Derlet, P.M.; Andersen, S.J.; Marioara, C.D.; Froseth, A.G. A first-principles study of the β''-phase in Al-Mg-Si alloys. *J. Phys. Condens. Matter* **2002**, *14*, 4011–4024. [CrossRef]
70. Frøseth, A.G.; Høier, R.; Derlet, P.M.; Andersen, S.J.; Marioara, C.D. Bonding in MgSi and Al-Mg-Si compounds relevant to Al-Mg-Si alloys. *Phys. Rev. B* **2003**, *67*, 224106. [CrossRef]

materials

MDPI

Article

High-Throughput Computing Assisted by Knowledge Graph to Study the Correlation between Microstructure and Mechanical Properties of 6XXX Aluminum Alloy

Xiaoyu Zheng [1], Yi Kong [1,*], Tingting Chang [2], Xin Liao [2], Yiwu Ma [3] and Yong Du [1]

1 State Key Laboratory of Powder Metallurgy, Central South University, Changsha 410083, China; xiaoyu_zheng@csu.edu.cn (X.Z.); yong-du@csu.edu.cn (Y.D.)
2 College of Computer Science and Electronic Engineering, Hunan University, Changsha 410082, China; tingtingchang@hnu.edu.cn (T.C.); xinliao@hnu.edu.cn (X.L.)
3 National Supercomputing Center in Changsha, Hunan University, Changsha 410082, China; myw@hnu.edu.cn
* Correspondence: yikong@csu.edu.cn

Abstract: It is of great academic and engineering application to study the evolution of microstructure and properties of age-strengthened aluminum alloys during heat treatment and to establish quantitative prediction models that can be applied to industrial production. The main factors affecting the peak aging state strength of age-strengthened aluminum alloys are the precipitates, solid solution elements, grain size effects, and textures formed during the material processing. In this work, these multi-scale factors are integrated into the framework of the knowledge graph to assist the following crystal plasticity finite elements simulations. The constructed knowledge graph is divided into two parts: static data and dynamic data. Static data contains the basic properties of the material and the most basic property parameters. Dynamic data is designed to improve awareness of static data. High-throughput computing is performed to further obtain clear microstructure-property relationships by varying the parameters of materials properties and the characteristics of the structure models. The constructed knowledge graph can be used to guide material design for 6XXX Al-Mg-Si based alloys. The past experimental values are used to calibrate the phenomenological parameters and test the reliability of the analysis process.

Keywords: knowledge graph; high-throughput computing; microstructure design; crystal plasticity; Al alloys

Citation: Zheng, X.; Kong, Y.; Chang, T.; Liao, X.; Ma, Y.; Du, Y. High-Throughput Computing Assisted by Knowledge Graph to Study the Correlation between Microstructure and Mechanical Properties of 6XXX Aluminum Alloy. *Materials* **2022**, *15*, 5296. https://doi.org/10.3390/ma15155296

Academic Editor: Jae Wung Bae

Received: 10 June 2022
Accepted: 27 July 2022
Published: 1 August 2022

1. Introduction

Aluminum alloy combines the advantages of low density, good electrical conductivity, high corrosion resistance, good heat dissipation, high specific strength, and easy processing and is widely used in transportation, aerospace, and other industries [1–3]. The relationship between "process-structure performance" of aluminum alloy industrial production is quite complex. Specifically, the process includes determining the composition, heat treatment, deformation processing, etc.; the structure involves grain shape and orientation, composition segregation, and second equivalence. The performance includes elasticity, plasticity, fracture toughness, etc.

The concept of Integrated Computational Materials Engineering (ICME) [4] was introduced by the US government in 2008 to integrate the tools of computational materials science into a holistic and systematic materials development process to achieve efficient development, manufacturing, and use of advanced materials by bridging the gap between materials design and manufacturing. ICME is now widely recognized and adopted by industry and academia and will play a significant role in materials development.

The primary strengthening mechanism of 6xxx-series aluminum alloys is the obstruction of dislocation movement by second-phase particles precipitated during aging, and microstructural parameters such as morphology, size, number, and distribution of the second phase determine the strengthening effect [5,6]. The precipitation process of the second phase is mainly influenced by the process parameters such as alloy composition, aging temperature, and aging time. By establishing a quantitative model between process parameters and microstructure and correlating microstructure parameters with alloy properties, the influence of process parameters on alloy properties can be quantitatively studied, which is of great value for the rational design of alloy composition, optimization of heat treatment conditions, and improvement of alloy properties.

In general, the macroscopic mechanical properties of a material depend on the microstructure, spanning several scales from micro to macro. A complete multi-scale simulation starts from first-principles calculations, molecular dynamics, and Monte Carlo simulations to calculate material physical property parameters such as elastic constants, intrinsic strains, interfacial energies, diffusion coefficients, etc. Then around the specific production process parameters, based on the phase diagram thermodynamics and kinetics to summarize the phase transition law, using the phase-field method, the meta-cellular automata method can be obtained microstructure. Finally, the stress-strain behavior of the material is simulated using finite elements. Figure 1 shows a schematic of multi-scale calculations, from component design through performance simulation, and finally improving component design based on performance simulation results.

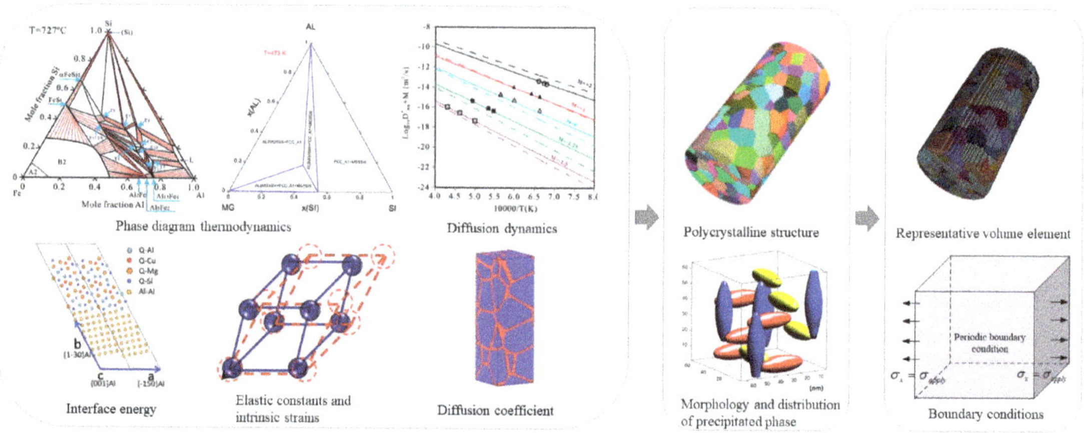

Figure 1. Schematic diagram of multi-scale calculations.

2. Knowledge Graph

The knowledge graph is a semantic network composed of nodes and edges that map the real world to the data world. Nodes represent entities or concepts in the physical world, and edges represent entities' attributes or relationships [7]. A knowledge graph was first proposed by Google in 2012 [8] for better serving searches. Nowadays, the application fields of knowledge graphs are becoming wider and wider. In the face of finance, medical and other industries, it is also possible to construct knowledge graphs belonging to specific fields. Through information acquisition, knowledge fusion, and knowledge processing, the facts in the original data are refined, analyzed, and formed into a graph. The machine can find the potential associations in the complex relationship and complete the work of case analysis and anti-fraud. In traditional material calculation, experimental data is scattered. The calculation data of a single process is stored and analyzed separately. We calculate the correlation between data by analyzing materials, obtaining material-oriented related laws and knowledge, and establishing a material knowledge graph. The specific implementation process is as follows: mining the input factors and result in performance/property sets and their corresponding relationships of each link in the material simulation calculation process, constructing the corresponding knowledge graph structure according to the association relationship, and filling the experimental calculation data into the knowledge graph structure through mining analysis and processing to form the knowledge graph. Based on the material knowledge graph, reasonable input parameter value recommendations can be provided for actual simulation calculations. Based on the analysis process shown in Figure 1, we divide the aluminum alloy simulation calculation process data into two categories: one is static data such as key performance, computational simulation methods, and basic simulation elements and simulation steps (Figure 2). The other is the input and output data of the software that can be changed in the processing flow, which we call dynamic data (Figure 3). The static data part mainly reflects the relationship between materials and software and calculation types. The active data part mainly records the calculation process data.

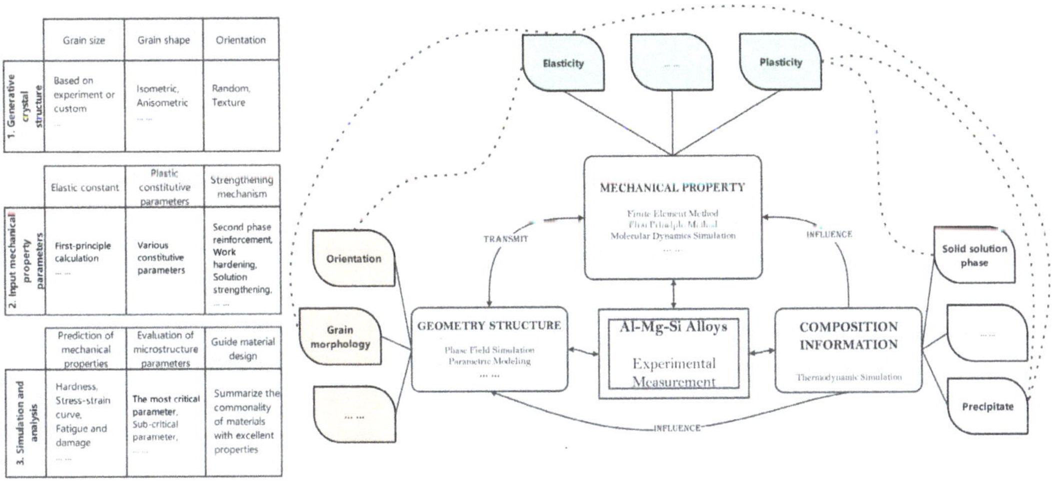

Figure 2. Static data, which calculations can be performed for a specific material, what method is used for each analysis, and the selection of material parameters used for the calculation.

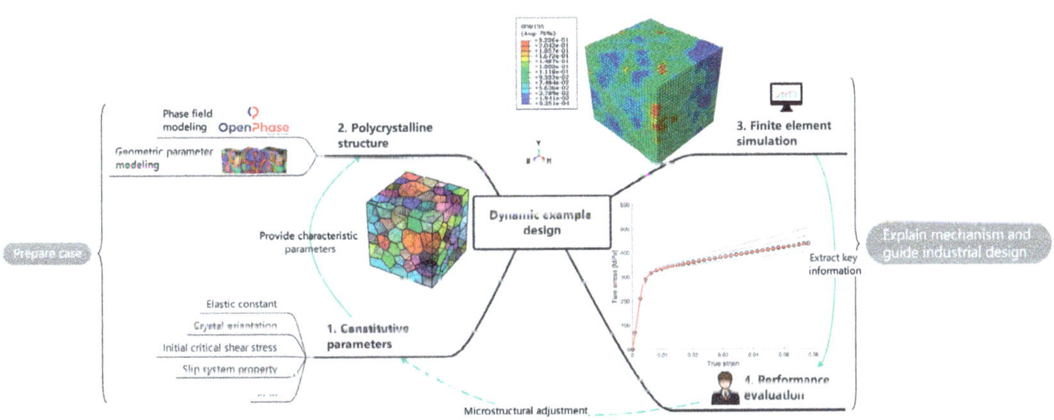

Figure 3. Schematic diagram of dynamic data, indicating specific simulation experiments based on static data of specific materials, what types of calculations are performed in the simulation experiments, and the relevant software and steps required for specific calculations.

For 6XXX series alloys, the components in the static data are extracted based on the concerned work hardening to form a complete calculation example process. Starting from the selection of material parameters, the geometric model is constructed, the calculation example is prepared, the simulation results are obtained, and the results and parameters are analyzed. Finally, the parameters are evaluated and a solution for material performance improvement is obtained. The calculation results of dynamic data will be used to increase or decrease static data, such as by introducing new methods and strategies, considering more microstructure information, or discarding unimportant microstructure information.

3. Material Modeling

In this section, the basic framework of micromechanical modeling is introduced in detail. The described model consists of a geometric description of the grain structure of polycrystals. The present construction model of plastic deformation in a single grain is realized by the crystal plasticity method through the user-defined material model (UMAT) of ABAQUS.

3.1. Representative Volume Elements

When simulating the properties of a material, the results are undoubtedly most accurate if the model constructed covers all the information about the material but requires a large number of calculations beyond the current level of computer development. The representative volume element (RVE) is a unit that is much smaller than the macroscopic system of the material but is large enough to capture the basic characteristics of the microstructure. The construction of representative volume cells with various structural features enables rapid analysis of the effects of material microstructural changes on performance.

There are two main strategies to construct representative volume elements, one is to use experimental characterization techniques such as electron backscatter diffraction (EBSD) to obtain real microstructures and thus build geometric models, and the other is to get microstructures using phase-field simulations, Monte Carlo methods, Voronoi, etc.

The first strategy is mainly used to reveal the relationship between microstructure and properties of specific materials. Depending on the research problem, it can be either by modeling the high matching of the observed region or by extracting statistical information from the EBSD observations to construct RVE. Based on the EBSD observations, we can obtain information about the grain shape, size, orientation, etc. Luo et al. [9,10] targeted the initial stages of fatigue cracking by constructing RVE directly based on microstructure scans of material samples in regions of high-stress concentration. This approach, which corresponds the RVE exactly to the modeled area, is inherently deterministic but requires a large amount of experimental data and complex preparation, as well as a significant computational effort, and is therefore only used when exploring specific micromechanical mechanisms. When studying physical quantities of macroscopic statistical significance, such as yield stress, tensile strength, and elongation at break at the visible level, information such as grain features, including orientation, disorientation, and grain size, can be extracted from microstructural observations to construct RVE. This statistical information-based modeling approach can be easily implemented by many software programs, such as Neper [11–13], DREAM.3D [14], and Kanapy [15]. Figure 4 shows two RVE with equiaxed crystal organization constructed by Neper based on the same seed point. Figure 4a shows the standard Voronoi polyhedral structure, and Figure 4b shows the stable system with higher grain sphericity. Figure 4c,d reflects the difference in the size distribution of grain morphology, with a more concentrated grain size distribution in the Voronoi polyhedral structure.

The second strategy is mainly used to find the correspondence between process parameters and properties of the material. The microstructures obtained by simulation-based means are more energy-efficient and faster than preparing alloy samples, and the uncontrollable factors are significantly reduced. Borukhovich et al. [16] combined the phase field approach with crystal plasticity theory to simulate the entire machining cycle from quenching, and over-tempering, to mechanical testing.

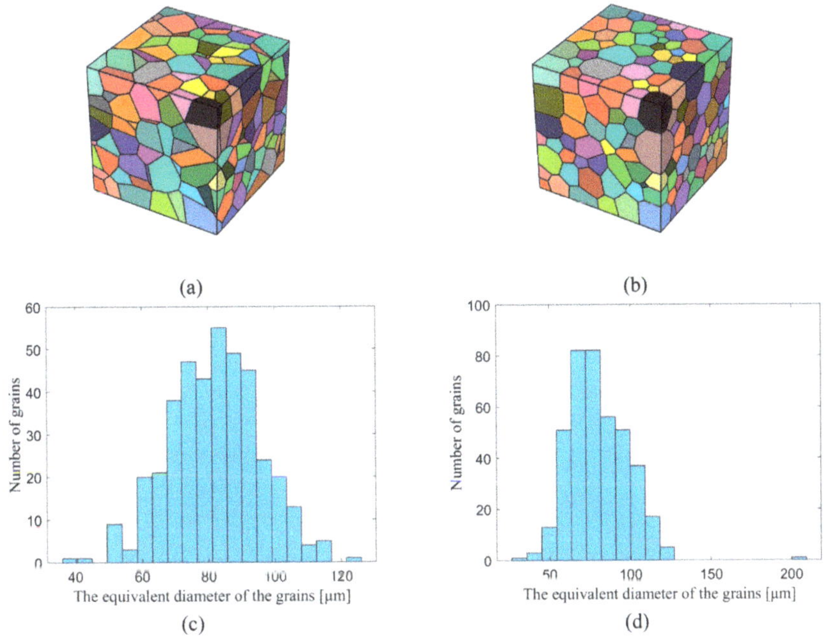

Figure 4. RVE of two different grain shapes, both cubes with 500 μm side length and containing 399 grains. (**a**) The shape of the grains is Voronoi polyhedral. (**b**) The shape of the grains is polyhedral with high sphericity. (**c**,**d**) are the equivalent diameter distributions of grains in RVE in (**a**,**b**), respectively.

In addition to constructing representative volume cells that match the real microstructure, it is also possible to explore the influence of material microstructure on properties by constructing representative volume cells with different characteristics, such as grains of specific morphology, the spatial distribution of grain size, and chemical composition, etc. Figure 5a shows the mechanical properties of non-isometric crystals by constructing grains with different aspect ratios, which correspond to aluminum alloys with elongated grains after rolling or unidirectional stretching. Figure 5b,c shows the non-uniform distribution of grains and precipitates composition, respectively.

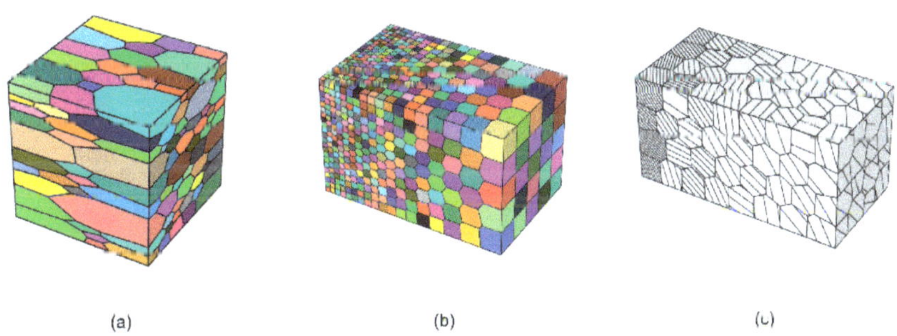

Figure 5. (**a**) Polycrystalline geometry model of nonequiaxed crystal. (**b**) Polycrystalline geometric model with gradient distribution of crystal size. (**c**) Polycrystalline geometric model with a gradient distribution of the number of precipitates.

In reality, aluminum alloys may not always exhibit such distinct gradient structures, but setting these structural variations to be obvious in the simulation allows for a more intuitive exploration of the effects of such structures.

By setting the corresponding boundary conditions for representative volume cells, the deformation processes of different materials, such as tension, shear, etc., can be simulated. The purpose of this work is to study the yielding and work-hardening behavior of the material by simulating the uniaxial stretching of the material. In total, the boundary conditions are set on four faces of the RVE, as shown in Figure 6.

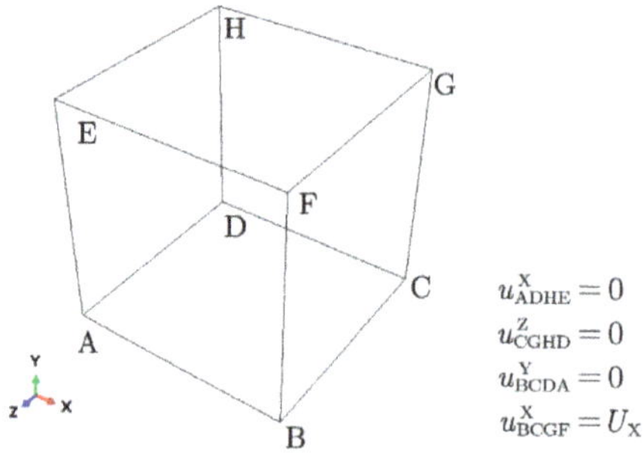

Figure 6. Boundary conditions of uniaxial tension.

3.2. *Crystal Plasticity Model*

The material behavior simulated by the finite element method is described by a phenomenological crystal plasticity model. In order to address the non-uniform deformation caused by abrupt changes in the mechanical behavior of polycrystalline grain boundaries and to consider the effect of crystallographic textures, the single crystal constitutive model proposed by Asaro [17] is used. In this paper, vectors (lowercase letters) and second-order tensor matrices (uppercase letters) are indicated in bold.

The deformation kinematics theory points out that the total deformation gradient $\boldsymbol{F}$ can be decomposed by multiplication and expressed as a combination of $\boldsymbol{F}^{e}$ and $\boldsymbol{F}^{p}$: $\boldsymbol{F} = \boldsymbol{F}^{e}\boldsymbol{F}^{p}$. The elastic deformation of the material follows Hooke's law. Plastic deformation is mainly calculated by the plastic strain gradient $\boldsymbol{l}^{p}$, which is a function of the plastic deformation gradient $\boldsymbol{F}^{p}$:

$$\boldsymbol{l}^{p} = \dot{\boldsymbol{F}}^{p}\boldsymbol{F}^{p-1} \tag{1}$$

Assuming that slip is the only displacement mechanism of plastic deformation, $\boldsymbol{l}^{p}$ can be expressed as the sum of the shear rates of all slip systems:

$$\boldsymbol{l}^{p} = \sum\nolimits_{\alpha} \dot{\gamma}^{\alpha} \boldsymbol{s}^{*\alpha} \otimes \boldsymbol{m}^{*\alpha}, \tag{2}$$

where $\dot{\gamma}^{\alpha}$ is the plastic shear rate and $\boldsymbol{s}^{*\alpha} \otimes \boldsymbol{m}^{*\alpha}$ is the Schmid tensor of the slip system α, which is obtained by dyadic operation between the slip direction $\boldsymbol{s}^{*\alpha}$ and the normal direction of the slip surface $\boldsymbol{m}^{*\alpha}$. Aluminum is Face-Centered Cubic (FCC) crystal, the value of α is 1 to 12. According to the power law model proposed by Asaro et al. [17], the plastic shear rate of α slip system can be expressed as:

$$\dot{\gamma}^{\alpha} = \dot{\gamma}_0 \left(\frac{|\tau^{\alpha}|}{g^{\alpha}} \right)^{1/m} \mathrm{sgn}(\tau^{\alpha}), \tag{3}$$

where $\dot{\gamma}_0$ is the reference shear rate, g^{α} is the plastic deformation resistance, and the resolved shear stress τ^{α} is the projection of the Kirchhoff stress tensor $\mathrm{def}(\boldsymbol{F})\sigma$ onto the slip surface. The parameter m controls the sensitivity of the strain rate. Assuming that the density of the material remains constant during the deformation process, τ^α can be expressed by the stress tensor $\boldsymbol{\sigma}$:

$$\tau^\alpha = \boldsymbol{m}^{*\alpha} \cdot \boldsymbol{\sigma} \cdot \boldsymbol{s}^{*\alpha} \quad (4)$$

The initial value of g^α is initial slip resistance τ_i, which is assumed to be the same for all slip systems. Work hardening is introduced by making the resistance to plastic deformation a function of plastic strain:

$$\dot{g}^\alpha = \sum_{\beta=1}^{12} h_{\alpha\beta} \left| \dot{\gamma}^\beta \right|, \quad (5)$$

where $h_{\alpha\beta}$ is the hardening modulus matrix, $h_{\alpha\alpha}$ denotes the hardening due to the slip of its own slip system, usually called the self-hardening coefficient, and $h_{\alpha\beta}(\alpha \neq \beta)$ denotes the hardening due to the slip of other slip systems, usually called the latent hardening coefficient. q is the ratio of the latent hardening coefficient to the self-hardening coefficient, 1 and 1.4 are the more common values. The $h_{\alpha\beta}$ can be expressed as a unified equation:

$$h_{\alpha\beta} = h(\alpha, \alpha)\left[q + (1-q)\delta_{\alpha\beta}\right] = \begin{cases} h(\alpha,\alpha) & \beta = \alpha \\ qh(\alpha,\alpha) & \beta \neq \alpha \end{cases} \quad (6)$$

Self-hardening coefficients using the model adopted by Peirce et al. [18]:

$$h(\alpha, \alpha) = h_0 \mathrm{sech}^2\left(\frac{h_0 \gamma}{\tau_\mathrm{s} - \tau_\mathrm{i}}\right) = h_0 \mathrm{sech}^2(k\gamma), \quad (7)$$

where h_0 is the hardening modulus at the beginning of yield, τ_s is the plastic flow breakthrough stress in the first stage of the material, and $\gamma = \sum_\alpha \int_0^t \left|\dot{\gamma}^\alpha\right| dt$ is the cumulative shear strain of each slip system. Noting that $h_0/(\tau_\mathrm{s} - \tau_\mathrm{i})$ is a constant value, the softening factor k can be introduced to visualize the effect of γ on the degree of hardening of the material.

The region of large plastic deformation during material deformation may become the location where cracks sprout, and the location of material failure can be predicted by introducing accumulated plastic deformation p [19]:

$$p = \int_0^T \left(\frac{2}{3} l^\mathrm{p} : l^\mathrm{p}\right)^{\frac{1}{2}} \mathrm{d}t \quad (8)$$

In addition, the local plastic dissipation energy E_p [20,21] can also provide a prediction of material damage:

$$E_\mathrm{p} = \int_0^T \boldsymbol{\sigma} : l^\mathrm{p} \mathrm{d}t = \sum_\alpha \int_0^T \tau^\alpha \dot{\gamma}^\alpha \mathrm{d}t \quad (9)$$

From the FE simulation, ABAQUS gives the value of each physical quantity at the center-of-mass for each element, $(\cdot)$. The black dots in parentheses indicate the homogenized parameters, i.e., stress and strain, etc. To obtain a global representative value for each time step, the center-of-mass values of each element are averaged through the element volume. For example, volume averaging of stresses and strains ($\overline{\sigma}_{ij}^{\mathrm{RVE}}$ and $\overline{\epsilon}_{ij}^{\mathrm{RVE}}$) can be performed for comparing the stress-strain curves obtained from the tests.

$$\begin{aligned} \overline{\sigma}_{ij}^{\mathrm{RVE}} &= \frac{1}{V^{\mathrm{RVE}}} \sum_{n=1}^{N} \left(\sigma_{ij}\right)_n \cdot V_n \\ \overline{\epsilon}_{ij}^{\mathrm{RVE}} &= \frac{1}{V^{\mathrm{RVE}}} \sum_{n=1}^{N} \left(\epsilon_{ij}\right)_n \cdot V_n \end{aligned}, \quad (10)$$

where the subscript n represents the value of each unit, and V^{RVE} represents the total volume of the entire RVE. This averaging strategy can be applied equally to p and E_p to measure the deformation properties of the material.

In addition, the Lode stress parameter μ_σ can be introduced to analyze the stress state of the material during deformation. Notice that μ_σ is equal to -1 in uniaxial tension and 0 in pure shear. The Lode stress parameter for an RVE $\overline{\mu}_\sigma^{RVE}$ is the volume average of each element μ_σ:

$$\mu_\sigma = \frac{\sigma_2 - (\sigma_1 + \sigma_3)/2}{(\sigma_1 - \sigma_3)/2}, \tag{11}$$

where $\sigma_1, \sigma_2, \sigma_3$ are the three principal stresses ($\sigma_1 \geqslant \sigma_2 \geqslant \sigma_3$), which can be obtained from the stress tensor $\boldsymbol{\sigma}$.

3.3. Strength Model

In the crystal plasticity model proposed in the previous section, the most important parameter is τ_i. There are many strengthening models [5,22,23] for 6XXX series aluminum alloys, which aim to relate microstructural parameters, such as grain size, texture, size distribution of precipitates, type, and content of solid solution phases, to macroscopic yield strength σ_y. Contributions are usually linearly additive [5,6]:

$$\sigma_y = \sigma_{Al} + \sigma_{ss} + \sigma_{ppt} \tag{12}$$

The intrinsic strength, solid solution contributions, and precipitates to the yield stress of aluminum are denoted as σ_{Al}, σ_{ss} and σ_{ppt}, respectively. Similarly, at the single crystal level, the initial slip resistance τ_i is determined by various microstructural parameters:

$$\tau_i = \tau_{Al} + \tau_{ppt} + \tau_{ss} \tag{13}$$

τ_{Al} stands for intrinsic strength of aluminum, which is numerically equal to 3/1 of the yield strength of pure aluminum with the same average particle size. Therefore, in this model, the grain size effect is also considered.

The solid solution strengthening term τ_{ss} is due to the strain field generated around the substitutional atoms dissolved in the matrix that can interact with dislocations and impede their movement, resulting in strengthening. Based on the principle that the contributions of different solute atoms to the yield strength can be linearly superimposed, the solid solution strengthening effect of the alloy can generally be expressed as:

$$\tau_{ss} = \sum_i k_i C_i^{2/3}, \tag{14}$$

where k_i is the scaling factor and C_i is the mass fraction (wt.%) of the specific element (Mg, Si) in the solid solution. The value of C_i is easily known based on the thermodynamics of the phase diagram or on the quantitative chemical analysis of the alloy structure. According to the work of Myhr et al. [24], k_{Mg} and k_{Si} take the values of 15.0MPa/wt.%$^{2/3}$ and 33.0MPa/wt.%$^{2/3}$, respectively.

Unlike τ_{Al} and τ_{ss}, which have clear and unambiguous expressions, modeling τ_{ppt} is always extremely difficult. Esmaeili et al. [22] proposed that the reinforcement of the precipitated phase is related to several microscopic variables, which can be expressed as:

$$\tau_{ppt} = \mathcal{F}(r, f, F, l, S) \tag{15}$$

where r and f are the average size and volume fraction of the precipitates, respectively, F is the maximum interaction force between the particle and dislocation with the average radius, l is the average distance between the particles of the precipitates with obstruction, and S is a series of microscopic parameters indicating the particle shape of the precipitates and the dislocation relationship between the particle and the matrix. This idea was widely adopted and promoted in the following decades [6,23,25,26]. Because the types, morphology and distribution of the precipitates depend on the processing technology and the composition of the raw materials, it is not easy to construct a clear functional relationship $\mathcal{F}$. Therefore, the τ_{ppt} can be considered to be determined preferentially by Equation (13):

$$\tau_{ppt} = \tau_i - \tau_{Al} - \tau_{ss} \tag{16}$$

In most studies, the yield stress and the initial slip resistance are considered to satisfy a linear relationship: $\sigma_y = M\tau_i$, where M is the Taylor factor, which is the most important parameter connecting the continuum plasticity theory and the crystal plasticity theory. For specific materials, the measurement of yield stress σ_y is very convenient, so τ_i can be estimated by Taylor factor M:

$$\tau_i = \frac{\sigma_y}{M} \tag{17}$$

For randomly oriented FCC structured metals, the value of M is 2.2 for the Sachs model [27] and 3.1 for the Taylor model [28]. The work of Zhang et al. [29] states that the value of M when using the crystal plasticity finite element model is about 2.7. In fact, M is closely related to the textures of the material and is an important physical quantity that characterizes the statistical significance of the crystallographic orientation.

3.4. Parameter Calibration

Although the above model has, as far as possible, covered all the factors most commonly considered in the study of the strength of 6XXX series aluminum alloys. However, due to the complex process conditions, composition ratios, and the complexity of the real microstructure of the material during production, some of the phenomenological or structure-sensitive parameters in the model need to be calibrated to the results of material-specific mechanical properties tests in order to subsequently predict the mechanical response of the material when its microstructural characteristics are altered or under more complex loading conditions. For parameters that are universal and structurally insensitive, such as elastic constants and power-law hardening parameters, the values in the literature are directly selected.

The 6XXX series aluminum alloy under under-aged and peak-aged states reported by Yang et al. [25] was selected for finite element simulation and some key parameters were compared accurately. The purpose of this work was chosen because the two materials reported in the paper only have different aging times, the remaining microstructure parameters are almost identical to the initial alloy composition, they have similar compositions in the knowledge graph, and differences in their mechanical properties are also easy to test whether they can be reflected by specific parameters. Each material is tested for uniaxial tensile, and multiple sets of test values are selected to average to ensure the effectiveness of the test. The specimen used for mechanical testing is cut from a randomly selected location in the casting sample, and the entire specimen is not significantly mechanically processed, so it can be considered that the orientation of the grains is randomly distributed, without texture. Considering that the average grain sizes of the two samples are the same, the same geometric model is constructed as well as the random grain orientation. The RVE is divided into a total of 39304 elements, 34 elements in each direction. The amount of stretch along the X direction is 6% of the side length.

In this study, the following parameters are based on the literature: $(C_{11}, C_{12}, C_{44}(\text{MPa})) = (106430, 60350, 28210)$, $[\dot{\gamma}_0(\text{s}^{-1}), m] = [0.001, 0.02]$. This information exists in the knowledge graph and is used during finite element simulations. Parameters such as $[M, h_0, k, q]$ are the key variables that reflect the differences in the properties of each alloy, reflecting the characteristics of the material organization and composition, and fundamentally correspond to different production and processing processes. In addition, τ_i is determined by Equation (17), where the value is taken from the work of Yang et al. [25], as 273.3 MPa for peak-aged alloy and 258.6 MPa for under-aged alloy. M and q are highly correlated with the crystal structure properties and have well-defined intervals, while h_0 and k are closely related to the precipitation phase properties in the alloy and tend to vary over a wide range, but not by more than one order of magnitude. Based on the trial-and-error method, the approximate range of the values of these parameters is obtained, and then the accurate values are determined by combining the neural network and the genetic algorithm. Figure 7 shows that the stress-strain curve obtained by the finite element simulation is basically consistent with the experiment based on the parameters calibrated by the experiment. Finally, the crystal plastic finite element simulation parameters calibrated based on experiments are summarized in Table 1.

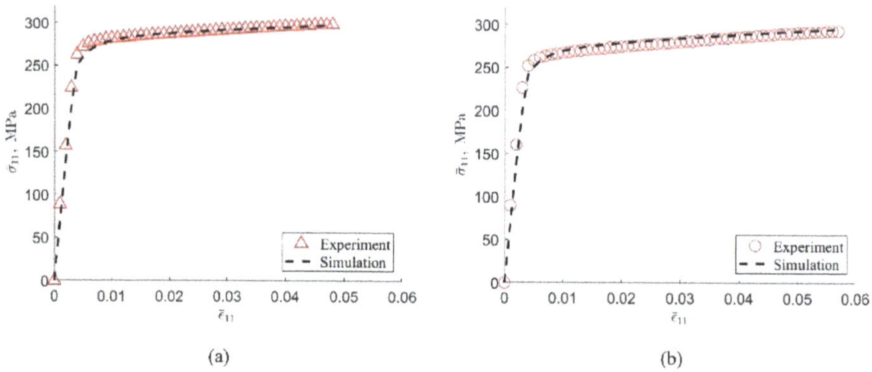

Figure 7. Experimental and simulation results of peak-aged (**a**) and under-aged (**b**) 6XXX series aluminum alloy stress-strain curves.

Table 1. Calibration of crystal plastic finite element simulation parameters based on experimental data.

Materials	C_{11},C_{12},C_{44}(MPa)	h_0(MPa)	τ_i(MPa)	τ_s(MPa)	$\dot{\gamma}_0$(s^{-1})	m	M	k	q
Peak-aged	106430, 60350, 28210	27.54	101.81	110.99	0.001	0.02	2.7	3.0	1.4
Under-aged	106430, 60350, 28210	44.25	97.75	110.39	0.001	0.02	2.7	3.5	1.4

4. High-Throughput Computing

The work in the previous chapter has shown that the mechanical behavior of materials can be accurately predicted based on microstructural modeling and parameter calibration. The materials under study can be quantified and recorded in the knowledge graph through physical quantities such as microstructural parameters and mechanical performance parameters. When more experimental data are considered in the future, the knowledge graph will be updated. The background will also continue to become clearer. Another important task of a knowledge graph is to expand new material information and cognition from existing material knowledge. For example, what effect will the combination of various microstructure information have on the performance of the material itself? Which microstructural information will play a more important role? Based on the goal to be explored, a variety of numerical examples can be constructed to study the law of the influence of various parameters on the material properties.

Based on the idea of high-throughput computing, we designed a series of examples to obtain many computational results by varying the values of some parameters of materials properties and considering different initial polycrystal structures for simulation, and summarize the influence weights of each parameter to build a comprehensive mechanical properties-microstructure knowledge graph. The same RVE, Figure 4a, is used for all the calculations. With the material properties either calculated by multi-scale calculations, or using existing material data, high-throughput computing is further performed to achieve efficient screening of composition/organization/performance, etc., to guide the process optimization issues and to accelerate the development of new materials and significantly reduce the cost of material development.

Uniaxial tensile simulations were performed for the model shown in Figure 4a by assigning the same random grain orientation and different material parameters. Table 2 shows the range of values for each parameter, and when a parameter is varied, the default values are set for the remaining parameters. In addition, the effect of crystal orientation on mechanical properties is analyzed by setting different initial textures with constant material parameters taking default values. Table 3 shows the typical common textures of FCC structured metals.

Table 2. Range of variation of material parameters.

Parameter	Default Value	Minimum	Maximum
M	2.7	2.2	3.1
h_0(MPa)	44.25	20	200
q	1.4	1.0	1.8
k	3.5	1	10

Table 3. Euler angles of typical textures in FCC metals.

Type	{*hkl*}	*uvw*	φ_1	ϕ	φ_2
Cube	001	100	0°	0°	0°
Goss	011	100	0°	45°	0°
Copper	112	111	90°	35°	45°
Brass	011	211	35°	45°	0°
S	123	634	59°	37°	63°

Figure 8 shows the tensile simulation results for polycrystalline aggregates with material parameters $[M, h_0, k, q]$ set to the default values in Table 2. From the Mises stress distribution shown in Figure 8b and the cumulative plastic strain distribution shown in Figure 8c, it can be seen that there are significant differences in stresses and strains among grains, and the stress concentrations are mainly found at grain boundaries. The differences in plastic deformation and stress response between grains are mainly caused by differences in grain orientation, while the differences within grains are mainly caused by grain arrangement and their interactions. The highly anisotropic elastic and plastic behavior of single grains allows the deformation or stress concentration at grain boundaries to satisfy both stress equilibrium and strain compatibility. Therefore, both stress and strain tend to occur at grain boundaries.

Figure 9 illustrates the calculated results of the stress-strain curve for a single parameter varying according to the values taken in Table 2. The calculated results illustrate that the boundary values of the parameters correspond to the extreme cases of the stress-strain curve, and the gray area in the figure indicates the position of the stress-strain curve of the material when the parameters take intermediate values. M affects the yield stress of the simulation results, h_0 reflects the strain hardening capacity of the material, k corresponds to the degree of strain hardening of the material, and the variation of q does not bring much difference in the results, thus the correlation between this parameter and the performance is less sensitive. By comparing the experimental values with the calculated results after parameter adjustment, the influence characteristics of these independent parameters on the mechanical properties can be more deeply understood, and the corresponding relationship between them and the knowledge of materials science can be interpreted. M reflects the anisotropy of the material and reflects the orientation correlation between the strength of single crystal and polycrystalline strength. It is more effective to calibrate the initial slip system yield strength based on M than other parameters. When the initial hardening modulus varies between orders of magnitude, the stress-strain response has limited changes in the initial plastic deformation, and then shows a significant difference in work hardening properties. Therefore, this parameter is effective for the characterization of material work hardening, and the difficulty of deformation at the initial stage of material work hardening will significantly correspond to the change in the value of this parameter. After the other parameters are calibrated, the two parameters q and k do not significantly affect the stress-strain response of the material within the common value range, so they may not be suitable for establishing the relationship between microstructure and performance.

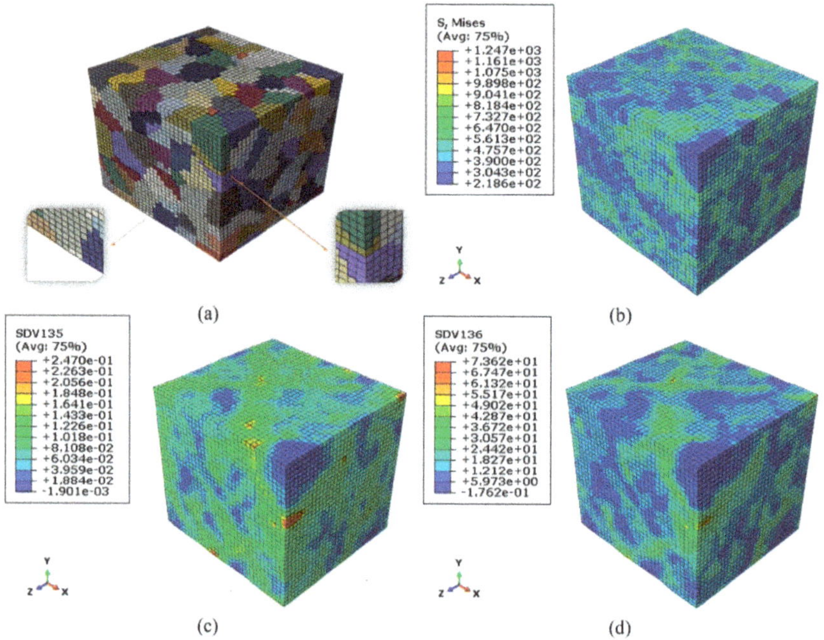

Figure 8. Tensile simulation results for polycrystalline aggregates. (**a**) Grains deformation diagram. (**b**) Distributions of Von Mises equivalent stress. (**c**) Distributions of accumulated plastic deformation. (**d**) Distributions of local plastic dissipation energy.

Figure 10a shows the results of stress-strain simulations for polycrystalline aggregates with different initial crystal orientations, where uniaxial stretching corresponds to the $\langle 100 \rangle$ orientation of the material and thus the ease of grain slip initiation differs for different orientations. This result is consistent with the study of Zhao et al. [30]. Figure 10b shows the effect of the initial texture on the Lode stress parameter-strain curve. The results show that the different oriented materials exhibit significant differences in mechanical properties during deformation in the specified directions.

The rich calculation results obtained by high-throughput calculation provide rich materials for the construction of an aluminum alloy knowledge graph. Based on the objective fact that the structure determines the performance, we set different structures and different material parameters to conduct finite element simulations, and the mechanical responses of various structures can be obtained. The simulation results are stored in the knowledge graph. In the subsequent material development, the key influencing parameters can be located according to the expected mechanical properties, and then key process schemes or material ratios can be found to achieve the accurate material design.

Finally, the widely used commercial alloys 6061-T4 and 6061-T6 in the 6XXX series were selected as the research objects, and the materials were selected from the alloy samples with a mass ratio of Mg to Si of 1.19 in the work of Kim et al. [31]. 6061-T4 is an alloy that is naturally aged after solution heat treatment. Its yield stress is approximately 122.0 MPa. This alloy is of average strength but has excellent machinability. The material fails under 20% tensile strain. 6061-T6 is an artificially aged alloy after solution heat treatment. The yield stress is about 325.0 MPa, but the processing performance is poor, and the tensile strain is about 5%. Since the two are obtained from the same batch of alloy samples processed by different aging processes, and the polycrystalline morphology and orientation of the material are basically unchanged during the aging process, it is still assumed that M is the same value, and based on the previous analysis, k and q are also similar. It is considered that it is not sensitive to the structure, and the same value is approximately taken. The final simulation result is shown in Figure 11. The simulation parameters determined based on semi-analytical and semi-experience are as follows. In order to reflect the material analysis strategy based on high-throughput calculation and knowledge graph. Therefore, methods such as machine learning are not used, and the simulation results and the parameters used to reflect the applicability of the previously concluded laws. It also reflects that the initial yield strength h_0 is a key performance index for 6XXX. The crystal plastic constitutive parameters of the two alloys are listed in Table 4, from which it can be seen that the τ_s of the 6061-T4 alloy is about twice that of the τ_i, indicating that the alloy after natural aging has better work hardening ability. If the material has an excellent degree of work hardening at the initial stage of plastic strain, the subsequent deformation performance will also be guaranteed.

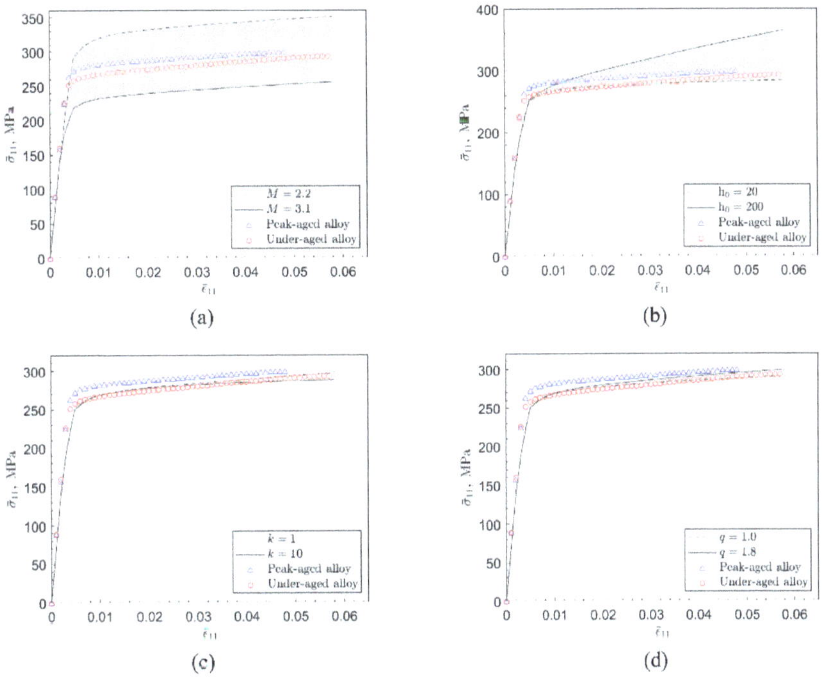

Figure 9. Stress-strain curves of polycrystalline aggregates with different intrinsic parameters ([M, h_0, k, q]). (**a**) The stress-strain response of the material when M varies from 2.2 to 3.1. (**b**) The stress-strain response of the material when h_0 varies from 20 MPa to 200 MPa. (**c**) The stress-strain response of the material when k varies from 1 to 10. (**d**) The stress-strain response of the material when q varies from 1.0 to 1.8.

Table 4. 6061 aluminum alloy crystal plasticity finite element simulation parameters.

Materials	C_{11},C_{12},C_{44}(MPa)	h_0(MPa)	τ_i(MPa)	τ_s(MPa)	$\dot{\gamma}_0$(s^{-1})	m	M	k	q
6061-T4	106430, 60350, 28210	200.0	47.0	87.0	0.001	0.02	2.6	5.0	1.4
6061-T6	106430, 60350, 28210	50.0	125.0	133.33	0.001	0.02	2.6	6.0	1.4

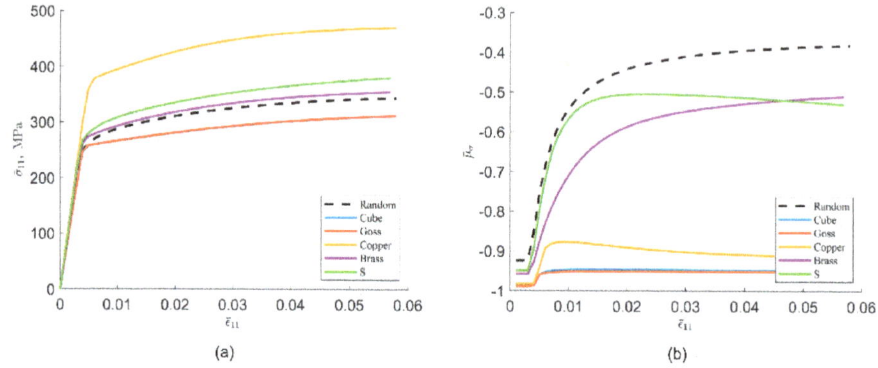

Figure 10. Stress-strain curves of polycrystalline aggregates with different textures ([M, h_0, k, q] uses the default values in Table 1). (**a**) Stress-strain curves. (**b**) Lode stress parameter-strain curves.

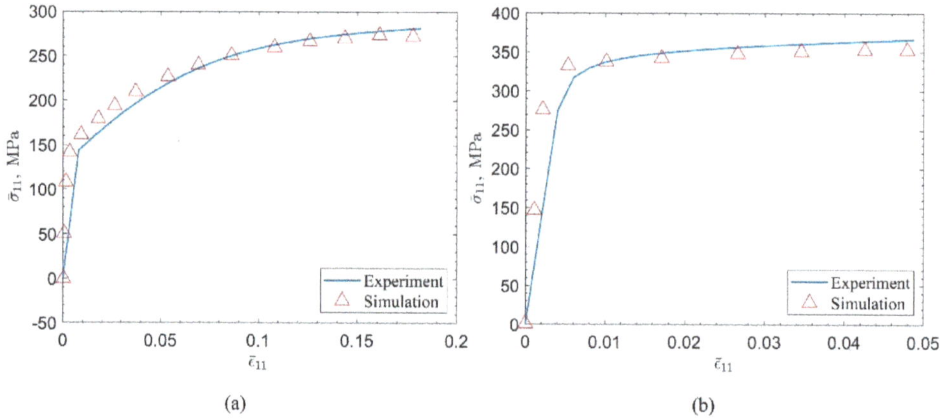

Figure 11. Experimental and simulation results of the stress-strain curve of 6061 aluminum alloy. (**a**) 6061-T4. (**b**) 6061-T6.

5. Conclusions

To comprehensively improve the performance of materials and deepen the knowledge of materials undoubtedly requires the ability to simultaneously combine multiple time scales and space scales in simulation calculations, but there is currently no universal method that can cover all time and space scales. Analyzing material properties by synthesizing various factors without screening will result in a huge amount of computation. Therefore, building a knowledge network of various microstructure information-performance of materials based on knowledge graphs will be a major mainstream analysis method in the future.

Integrated computational materials engineering and high-throughput computing will change the traditional empirical trial-and-error approach to alloy research and development (R&D) and become a fundamental R&D platform for collaborative knowledge innovation with the interconnection of multi-scale calculations, experiments, and databases. On this platform, as a three-dimensional data network capable of regularly linking information about the microstructure, properties, and computational methods of materials, the knowledge graph can be used to recommend reasonable input parameter values and store results, thus assisting high-throughput computation. To sum up specifically, for a certain research object, we first summarize and sort out important structural components or key performance control parameters based on previous materials science cognition and experiment-based summary, build a knowledge graph based on the performance of interest, and design a set of analysis and simulation process including these parameters. Multiple sets of examples were designed to study the influence weight of each variable on the performance. Comparing the calculated results with existing experimental observations can explore the relationship between structure and performance on a deeper level, thus reducing the consideration of secondary factors and optimizing the R&D strategy.

6XXX series aluminum alloys are chosen to demonstrate the proposed strategy of studying the correlation between microstructure and mechanical properties by high-throughput computing assisted by a knowledge graph. The simulation results show that the orientation distribution and initial hardening modulus of the material are the main factors affecting its performance, indicating that the previous assumption of isotropy is not suitable for 6XXX series, and how to improve the hardening ability at the initial stage of plastic strain is the focus of research. The crystal plasticity finite element method, as a bridge linking the micro-to-macro to quantitatively describe the relationship between the microstructure and properties of the alloy, is chosen to perform high-throughput computing with varying the parameters of materials properties and the characteristics of the structure models. The constructed knowledge graph is divided into two parts: static data and dynamic data, and can be used to guide material design for 6XXX Al-Mg-Si based alloys. Static data contains the basic characteristics and the most essential characteristic parameters of materials. The purpose of continuous generation and adjustment of dynamic data is to improve the cognition of static data. This research method has universality and popularization value.

Author Contributions: Data curation, Writing—original draft, X.Z.; Investigation, T.C. and X.L.; Conceptualization, Writing—review & editing, Supervision, Y.M.; Conceptualization, Writing—review & editing, Supervision, Funding acquisition, Project administration, Y.K.; Conceptualization, Results review, Writing—review, Y.D. All authors have read and agreed to the published version of the manuscript.

Funding: This research was funded by the National Key Research and Development Program of China (grant nos. 2018YFB0704003), and the National Natural Science Foundation of China (grant No. 51771234 and 51820105001).

Institutional Review Board Statement: Not applicable.

Informed Consent Statement: Not applicable.

Data Availability Statement: The data presented in this study are available on request from the corresponding author.

Conflicts of Interest: The authors declare no conflict of interest.

References

1. Miller, W.S.; Zhuang, L.; Bottema, J.; Wittebrood, A.J.; De Smet, P.; Haszler, A.; Vieregge, A. Recent development in aluminium alloys for the automotive industry. *Mater. Sci. Eng. A* **2000**, *280*, 37–49. [CrossRef]
2. Hirsch, J. Recent development in aluminium for automotive applications. *Trans. Nonferrous Met. Soc. China* **2014**, *24*, 1995–2002. [CrossRef]
3. Hirsch, J. Aluminium in innovative light-weight car design. *Mater. Trans.* **2011**, *52*, 818–824. [CrossRef]
4. Yi Wang, W.; Li, J.; Liu, W.; Liu, Z.-K. Integrated computational materials engineering for advanced materials: A brief review. *Comput. Mater. Sci.* **2019**, *158*, 42–48. [CrossRef]
5. Myhr, O.R.; Grong, O.; Andersen, S.J. Modelling of the age hardening behaviour of Al-Mg-Si alloys. *Acta Mater.* **2001**, *49*, 65–75. [CrossRef]
6. Myhr, O.R.; Grong, Ø.; Pedersen, K.O. A combined precipitation, yield strength, and work hardening model for Al-Mg-Si alloys. *Metall. Mater. Trans. A* **2010**, *41*, 2276–2289. [CrossRef]
7. Shadbolt, N.; Berners-Lee, T.; Hall, W. The semantic web revisited. *IEEE Intell. Syst.* **2006**, *21*, 96–101. [CrossRef]
8. Mrdjenovich, D.; Horton, M.K.; Montoya, J.H.; Legaspi, C.M.; Dwaraknath, S.; Tshitoyan, V.; Jain, A.; Persson, K.A. Propnet: A knowledge graph for materials science. *Matter* **2020**, *2*, 464–480. [CrossRef]

9. Luo, C.; Wei, J.; Parra-Garcia, M.; Chattopadhyay, A.; Peralta, P. Fatigue damage prediction in metallic materials based on multiscale modeling. *AIAA J.* **2009**, *47*, 2567–2576. [CrossRef]
10. Luo, C.; Chattopadhyay, A. Prediction of fatigue crack initial stage based on a multiscale damage criterion. *Int. J. Fatigue* **2011**, *33*, 403–413. [CrossRef]
11. Quey, R.; Dawson, P.R.; Barbe, F. Large-scale 3D random polycrystals for the finite element method: Generation, meshing and remeshing. *Comput. Methods Appl. Mech. Eng.* **2011**, *200*, 1729–1745. [CrossRef]
12. Quey, R.; Renversade, L. Optimal polyhedral description of 3D polycrystals: Method and application to statistical and synchrotron X-ray diffraction data. *Comput. Methods Appl. Mech. Eng.* **2018**, *330*, 308–333. [CrossRef]
13. Quey, R.; Villani, A.; Maurice, C. Nearly uniform sampling of crystal orientations. *J. Appl. Crystallogr.* **2018**, *51*, 1162–1173. [CrossRef]
14. Groeber, M.A.; Jackson, M.A. DREAM.3D: A digital representation environment for the analysis of microstructure in 3D. *Integr. Mater. Manuf. Innov.* **2014**, *3*, 56–72. [CrossRef]
15. Prasad, M.; Vajragupta, N.; Hartmaier, A. Kanapy: A Python package for generating complex synthetic polycrystalline microstructures. *J. Open Source Softw.* **2019**, *4*, 1732. [CrossRef]
16. Borukhovich, E.; Du, G.; Stratmann, M.; Boeff, M.; Shchyglo, O.; Hartmaier, A.; Steinbach, I. Microstructure design of tempered martensite by atomistically informed full-field simulation: From quenching to fracture. *Materials* **2016**, *9*, 673. [CrossRef] [PubMed]
17. Asaro, R.J. Micromechanics of crystals and polycrystals. *Adv. Appl. Mech.* **1983**, *23*, 1–115. [CrossRef]
18. Peirce, D.; Asaro, R.J.; Needleman, A. An analysis of nonuniform and localized deformation in ductile single crystals. *Acta Metall.* **1982**, *30*, 1087–1119. [CrossRef]
19. Manonukul, A.; Dunne, F.P.E. High– and low–cycle fatigue crack initiation using polycrystal plasticity. *Proc. R. Soc. London Ser. A Math. Phys. Eng. Sci.* **2004**, *460*, 1881–1903. [CrossRef]
20. Cheong, K.; Smillie, M.; Knowles, D. Predicting fatigue crack initiation through image-based micromechanical modeling. *Acta Mater.* **2007**, *55*, 1757–1768. [CrossRef]
21. Skelton, R.P. Cyclic hardening, softening, and crack growth during high temperature fatigue. *Mater. Sci. Technol.* **2013**, *9*, 1001–1008. [CrossRef]
22. Esmaeili, S.; Lloyd, D.J.; Poole, W.J. A yield strength model for the Al-Mg-Si-Cu alloy AA6111. *Acta Mater.* **2003**, *51*, 2243–2257. [CrossRef]
23. Holmedal, B. Strength contributions from precipitates. *Philos. Mag. Lett.* **2015**, *95*, 594–601. [CrossRef]
24. Myhr, O.R.; Grong, Ø.; Schäfer, C. An extended age-hardening model for Al-Mg-Si alloys incorporating the room-temperature storage and cold deformation process stages. *Metall. Mater. Trans. A* **2015**, *46*, 6018–6039. [CrossRef]
25. Yang, M.; Chen, H.; Orekhov, A.; Lu, Q.; Lan, X.; Li, K.; Zhang, S.; Song, M.; Kong, Y.; Schryvers, D.; et al. Quantified contribution of β'' and β' precipitates to the strengthening of an aged Al-Mg-Si alloy. *Mater. Sci. Eng. A* **2020**, *774*, 138776. [CrossRef]
26. Li, Y.L.; Kohar, C.P.; Mishra, R.K.; Inal, K. A new crystal plasticity constitutive model for simulating precipitation-hardenable aluminum alloys. *Int. J. Plast.* **2020**, *132*, 102759. [CrossRef]
27. Sachs, G. Plasticity problems in metals. *Trans. Faraday Soc.* **1928**, *24*, 84. [CrossRef]
28. Taylor, G.I. Plastic strain in metals. *J. Inst. Met.* **1938**, *62*, 307–324.
29. Zhang, K.; Holmedal, B.; Mánik, T.; Saai, A. Assessment of advanced Taylor models, the Taylor factor and yield-surface exponent for FCC metals. *Int. J. Plast.* **2019**, *114*, 144–160. [CrossRef]
30. Zhao, Q.; Abdel Wahab, M.; Ling, Y.; Liu, Z. Grain-orientation induced stress formation in AA2024 monocrystal and bicrystal using Crystal Plasticity Finite Element Method. *Mater. Des.* **2021**, *206*, 109794. [CrossRef]
31. Kim, Y.; Mishra, R.K.; Sachdev, A.K.; Kumar, K.S. A combined experimental-analytical modeling study of the artificial aging response of Al-Mg-Si alloys. *Mater. Sci. Eng. A* **2021**, *820*, 141566. [CrossRef]

materials MDPI

Article

Effect of Various Forms of Aluminum 6082 on the Mechanical Properties, Microstructure and Surface Modification of the Profile after Extrusion Process

Piotr Noga *, Andrzej Piotrowicz, Tomasz Skrzekut, Adam Zwoliński and Paweł Strzępek

Faculty of Non-Ferrous Metals, AGH University of Science and Technology, A. Mickiewicz Av. 30, 30-059 Cracow, Poland; andpio@agh.edu.pl (A.P.); skrzekut@agh.edu.pl (T.S.); zwolo@agh.edu.pl (A.Z.); strzepek@agh.edu.pl (P.S.)
* Correspondence: pionoga@agh.edu.pl

Abstract: This article presents a method of reusing aluminum scrap from alloy 6082 using the hot extrusion process. Aluminum chips from milling and turning processes, having different sizes and morphologies, were cold pressed into briquettes prior to hot pressing at 400 °C at a ram speed of 2 mm/s. The study of mechanical properties combined with observations of the microstructures, as well as tests of density, hardness and electrical conductivity were carried out. On the basis of the results, the possibility of using the plastic consolidation method and obtaining materials with similar to a solid ingot mechanical properties, density and electrical conductivity was proven. The possibility of modifying the surface of consolidated aluminum scrap was tested in processes examples: polishing, anodizing and coloring. For this purpose, a number of analyses and tests were carried out: comparison of colors on color histograms, roughness determination, SEM and chemical composition analysis. It has been proven there are differences in the surface treatment of the solid material and that of scrap consolidation, and as such, these differences may significantly affect the final quality.

Keywords: recycling; 6082 aluminum alloy; plastic consolidation; mechanical properties; microstructure; surface modification

Citation: Noga, P.; Piotrowicz, A.; Skrzekut, T.; Zwoliński, A.; Strzępek, P. Effect of Various Forms of Aluminum 6082 on the Mechanical Properties, Microstructure and Surface Modification of the Profile after Extrusion Process. *Materials* **2021**, *14*, 5066. https://doi.org/10.3390/ma14175066

Academic Editor: Lijun Zhang

Received: 8 August 2021
Accepted: 31 August 2021
Published: 4 September 2021

1. Introduction

Recycling of light metals is largely based on the remelting method. For this purpose, various types of melting furnaces are used, from low-cost gas furnaces to electric furnaces with a special system of loading the initial material into the liquid metal which significantly increases the recycling yield, but also increases the processing cost. Recycled metal is subjected to the treatment processes: refining, chemical composition correction and degassing, and then casting into ingots. The average recycling yield when melting such scrap metal ranges from 75 to 81%. The low recycling yield of the melting methods (especially small chips, tapes, thin wires) causes irreversible losses of metal (forms low-quality slags and drosses) and energy used for the aluminum production [1–6].

The concept of recycling based on the plastic consolidation method creates opportunities to reduce losses during recycling and significantly reduces the energy consumption. This process completely omits the melting phase, thus eliminating losses and troublesome waste (drosses). The entire recycling process takes place in the solid phase, which avoids gasification effects of the liquid metal and loss of alloy additives during the metallurgical synthesis. The upsides of the process are lowering of the process temperature to approx. 350–450 °C, shortening the operation time to a few minutes and reducing energy consumption for reheating of the charge material several times. It also eliminates the risk of material losses as a result of oxidation, as surface oxidation is limited in this temperature range by a tight passivation layer [7,8].

Extrusion of fine fractions, such as metal chip, dusts, powders, tapes, etc., allows an effective plastic consolidation and the obtaining of a coherent material which has good utility properties [9–15]. Consolidation of the 6082 alloy chips by pressing and extrusion makes it possible to obtain briquettes with better fatigue properties compared to the solid material [16–18]. One of the materials most often used in the industry is aluminum and its alloys, in particular, alloy 6082. It is characterized by good resistance to corrosion and cracking. An interesting approach to the use of recycling by the extrusion method was applied to the AlSi11 alloy chips. After the plastic consolidation process, the obtained flat bars are subsequently welded, which leads to the production of metallic foams [19]. It is susceptible to machining and welding; it has decent mechanical strength. The above-mentioned properties enable the use of this material, among others, in the shipbuilding and automotive industries, where one has to deal with a corrosive environment and where heavy loads are transferred [20,21].

Aluminum is an amphoteric element and can dissolve both in an acidic and an alkaline environment, to form Al^{3+} and $H_2AlO_3{}^-$, respectively. In the pH range from 4.45 to 8.38, aluminum is covered with durable oxide layer of $Al_2O_3{\cdot}H_2O$ (bemite). The formation of a thick protective coating is possible in a specially directed corrosion process in oxidizing environments, both acidic and alkaline, with pH values outside the above-mentioned area. Oxide layer formatted on aluminum or its alloys by electrolytic methods is characterized by some special properties such as protection against weakly aggressive environments [22,23].

One of the ways to improve the surface properties of metals is to produce the so-called conversion coatings. One of them is oxide coatings, which are produced by an electrolysis process called anodizing. In the case of aluminum, aluminum oxide films are obtained in the anodizing process carried out in an electrolysis cell, in which the anode is an oxidized aluminum, while the cathode is an electrolyte-resistant metal (e.g., lead). During anodic oxidation of aluminum, the anode reaction is described by the summary equation:

$$2Al_{solid} + 3H_2O_{solution} \rightarrow Al_2O_{3solid} + 6H^+ + 6e^- \tag{1}$$

The aluminum oxide layer is formed as a result of the reaction of Al^{3+} ions (formed by the direct current) with O^{2-} ions or OH^- formed by the decomposition of water. Evolution of hydrogen gas takes place at the cathode:

$$2H^+ + 2e^- \rightarrow H_{2gas} \uparrow \tag{2}$$

Aluminum oxide layer is composed of densely arranged hexagonal cells with a pore inside. Due to the characteristic structure of the oxide layer (with relatively regular and densely spaced pores), the aluminum anodized surface is highly developed and has decent adsorption capacity. It is used in the coloring process to protect and decorate aluminum products like aluminum components [23,24].

Surface treatment with anodizing process includes a series of technological steps: 1. Surface preparation, 2. Anodizing, 3. Coloring, 4. Sealing. Preparation of aluminum for anodizing is extremely important because it determines whether anodizing will be successful. Firstly, the surface of the aluminum should be degreased in organic solvents, such as ethanol, acetone or carbon tetrachloride, to remove organic impurities and improve electrolyte wetting. If a special surface appearance (e.g., high gloss) is required, then mechanical or chemical polishing are used [25].

For decorative purposes, anodized aluminum can be colorized by using organic or inorganic dyes. The dye infiltrate the pores of the oxide layer, giving the color of the surface. The colorizing of the anodized aluminum related to the precipitation of insoluble inorganic salts in the pores, hereinafter referred to as two-step colorizing, occurs according to an example reaction:

$$(CH_3COO)_2Pb_{solution} + (NH_4)_2CrO_{4solution} \rightarrow PbCrO_4 \downarrow_{solid} + 2CH_3COONH_{4solution} \tag{3}$$

Sealing is a process to eliminate open pores of the anodized aluminum. It is also the process by which the dyes or insoluble salts contained in the pores become stuck as a result of the closure of the pores. During sealing, the oxide layer changes its structure; thus, hydrated $Al_2O_3 \cdot H_2O$ (bemite) is formed. This reaction is accompanied by swelling (due to the addition of water of crystallization and changes in the structure of the crystal lattice), which leads to the closure of the pores and the creation of a smooth surface.

Various aluminum alloys are anodized differently [26]. The 6000 alloy series, with the high levels of magnesium and silicon, which influence the anodizing process, are suitable for anodizing and coloring, with a very homogeneous oxide layer. As described in [27], a satin finish causes the surface to become dull. The oxide layer was over 30 microns thick and some parts of the surface were dark gray. For the tests in [27], samples provided by COLOR METAL SRL though SC ANOROM SRL were used, and the 6000 aluminum alloy series aluminum was 6082 SF02 in the form of extruded bars. Tests were made of anodizing as well as black coloring.

In another study [28], the possibilities of black dyeing, inorganic coloring and electrolytic coloring by sulfuric acid of 6061 series were investigated. For black dyeing, jet black was used as a dye, inorganic coloring was performed in two steps with using cobalt acetate and ammonium sulphide solutions and an electrolytic coloring: sulfuric acid, stannous chloride and phenol sulphonic acid. The influence of surface treatment and sealing method on the structure of the oxide layer, corrosion, porosity, etc., was studied. Regardless of the color method, all the porosity values were very low, which confirms the smoothness of the final surface.

In study [25], a two-step anodizing process was performed, resulting in nanoporous anodized aluminum oxide layers using 6082 aluminum alloy. The aim of this study was to choose appropriate anodizing parameters, such as voltage and temperature, using 0.4 M oxalic acid. The effect of these parameters on the morphological characteristics were investigated. Based on [25], the anodizing in present article was performed under similar technological conditions. It is also known that anodizing with oxalic acid turns the anodized aluminum yellow, which should be taken into account in its further coloring. The yellow color of anodized aluminum means that such aluminum is dedicated to coloring yellow or its related colors: green, orange, etc. therefore, in this article, attempts were made to color the anodized material into orange and yellow.

Xylenol orange is used primarily in spectrophotometric determination of aluminum as a colored reagent [29,30]. In study [31], xylenol orange was tested as a corrosion inhibitor for aluminum in trichloroacetic acid. In study [32], with a variety of dyes, methyl orange was used as a corrosion inhibitor for mechanically pretreated aluminum. Yet another study [33] used methyl orange as a dye adsorbed onto films of anodized aluminum oxide for pH sensing purposes. Films were prepared by anodizing aluminum in oxalic acid. As it turned out, depending on the number of dips and the pH of the solution, the dye adsorbed differently, which resulted in a different color of the aluminum.

Lead chromate ($PbCrO_4$) is an inorganic salt (often informally written as a mixture of oxides). Lead chromate may exist as a lemon yellow rhombic form, a reddish yellow monoclinic form, and a scarlet tetragonal form, but only the monoclinic form is stable at room temperature. Lead chromes are noted for their excellent opacity, low oil absorption, very bright shades, and high chroma (i.e., they give deep or saturated shades), making them ideal for full shade yellow paints. They also possess excellent solvent resistance and moderately good heat stability, which can be further improved by chemical stabilization [34]. Lead chromate pigments, whose color index CI Pigment Yellow is 34, a popular yellow dye [34] traditionally used for pipe and cable applications, are linked with the phase out of lead-based stabilizers and automotive paints [35].

Although methods for modifying the aluminum surface by anodizing and further processing are known [23,24,27,28,32,33,36,37], such tests have not been performed on recycled alloy 6082. This article shows practical examples of surface treatment of 6082 alloy:

solid and its scrap. Coloring with xylenol orange, noticed for the first time, is shown. Innovative color analysis was performed digitally based on color histograms.

The goal of this research was also to study the differences in the quality of the Al6082 alloy: solid, with coarse and fine chips after chemical surface treatment. For this purpose, several exemplary types of chemical surface treatments were carried out: 1. etching, 2. chemical polishing (shining) by 2a. currentless and 2b. current process, 3. anodizing with using oxalic acid and 4. organic and inorganic colorizing, including 4a. one-step and 4b. two-step processes.

2. Materials and Methods

The research material was 6082 alloy in the form of a profile after extrusion with a diameter of 40 mm and a height of 1000 mm. The chemical composition of the initial material and its mechanical properties are presented in Tables 1 and 2.

Table 1. Chemical composition of 6082 alloy [38].

Element	Al	Si	Fe	Cu	Mn	Mg	Cr	Zn	Ti	Other
Required by EN 1706 for base metal	95.2–98.3	0.7–1.3	<0.50	<0.1	0.4–1.0	0.6–1.2	<0.25	<0.2	<0.2	<0.15

Table 2. Mechanical properties of base metal 6082 [38].

Element	UTS, MPa	YS, MPa	Elongation, %	Hardness, HB	Density, g/cm^3
Required by EN 1706 for base metal	≥150	≥85	≥14	≥40	2.7

The 6082 alloys with different morphology were produced using machining (milling and turning). The machining process was carried out in laboratory conditions without the use of coolant, with the cleanliness of the workstations in the process. The first material used for the tests were chips produced on the ROBGRAF 1 milling machine (Markus–Texi, Poland), and for the purposes of the tests, these chips were marked as fine chips. The chip fraction measured using the imageJ software (1.35j) [39] was 0.315–0.4 mm (Figure 1A). In the milling process, the spindle penetrated the material to a depth of 0.5 mm and moved along the sample surface with a spiral direction, starting from the center and ending with the outer edge. The milling process was carried out at a rotational speed of 28,000 rpm. Another type of material used for the research was the chip, which for the purposes of the research was called coarse chips, the average size of which was: 20 × 5 × 1.8 mm (Figure 1B), these chips were produced in the process of turning on a TUM 35 lathe (Famot, Pleszew, Poland) with a rotational speed of 300 rpm, and a feed rate of 0.2 mm/s.

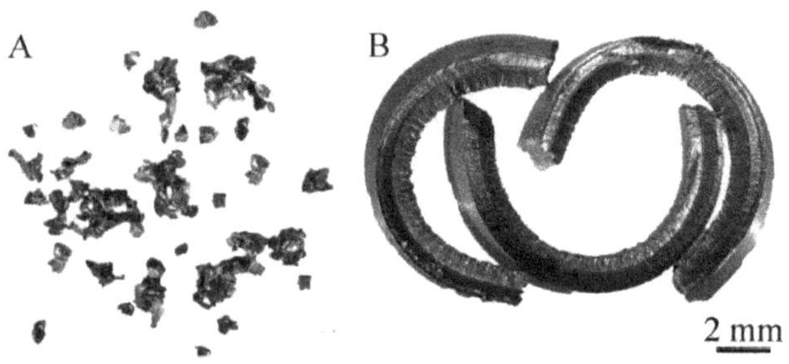

Figure 1. Chips after machining: (**A**) turning, (**B**) milling.

The chips obtained these ways were the input material for the cold compaction process without a protective atmosphere. Each of the batches weighing 25 g was individually placed in a special container and, using a stamp, was pre-pressed on a PS Logistics 100 presser under a pressure of 240 MPa, which resulted in moldings with a diameter of 38 mm and a height of 10 mm (Figure 2). To obtain solid bars intended for further tests, the materials obtained in the previous stage were subjected to the last operation, which was hot extrusion. The input for the process were briquettes/billets consisting of 6 compacts in the case of chips, and in the case of the reference material—a solid ingot with a diameter of 38 mm and a height of 60 mm. Each time, the batch was placed in a recipient heated to 400 °C and heated for 20 min in order to equalize the temperature in the entire volume of the material. It was followed by a co-extrusion process with a speed of 2 mm/s with the use of a rectangular die 3 mm high and 15 mm wide on a hydraulic press.

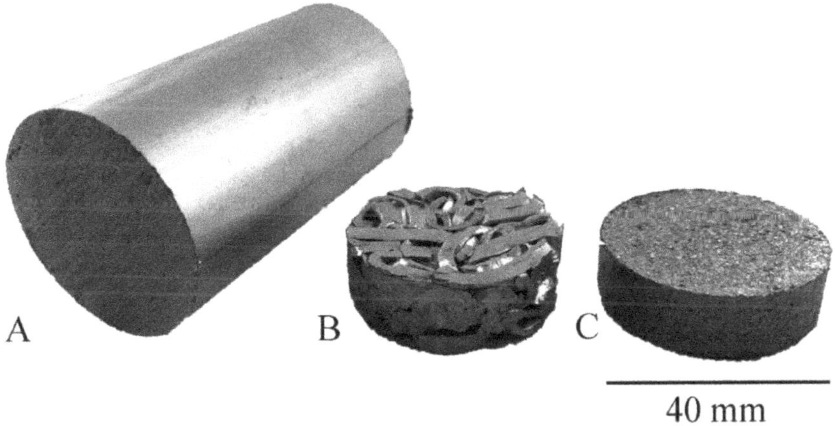

Figure 2. Test material: (**A**) after commercial extrusion, (**B**) after compacting the chip obtained from turning (**C**) after compacting the chip obtained from milling.

Samples for metallographic and anodizing tests obtained after the extrusion process were cut with a BP05d electro-erosion machine (Zakład Automatyki Przemysłowej B.P., Konskie, Poland). The dimensions of the samples for coating application were: height—12 mm, width—12 mm, thickness—3 mm. After cutting, they were subjected to grinding on sandpaper grades: 220, 500, 800, 1200, 2000 and 4000, and then polished with diamond paste with a particle size of 3 μm and 1 μm and finished with OPS polishing agent. The above operations were carried out on the Roto-Pol-11 device (Struers, Copenhagen, Denmark). Observations of the microstructure were performed using a Hitachi SU-70 scanning electron microscope (SEM) (Hitachi Ltd., Tokyo, Japan). The tests of mechanical properties by static tensile test were carried out at a speed of 8×10^{-3} s^{-1} using the Zwick Roell Z050 testing machine (Zwick/Roell Group, Ulm, Germany); sample dimensions: measuring base—50 mm, measuring base width—10 mm, thickness of the measurement base—3 mm. Density measurements by Archimedes method were made with a laboratory scales XA 120/250.4Y (Radwag, Radom, Poland). Hardness tests were carried out on a Shimadzu HMV-2 T device (Shimadzu Corporation, Kyoto, Japan) using the Vickers method. After surface treatment, the surface quality was investigated at a 4.8 mm gauge length and 80 μm maximum measuring value on a Hommel Tester T1000 profilographometer (Hommelwerke GmbH, Germany). The electrical conductivity was determined with a SigmaTest 2.069 device (Forester Instruments Inc., Pittsburgh, PA, USA). Electrical conductivity was tested at a constant temperature of 20 °C and reported as the mean of 10 measurements. The course of the experiment and the research scheme are shown in Figure 3. Chemical composition tests carried out on the Foundry-Master-Pro 2 device (Ueden, Germany).

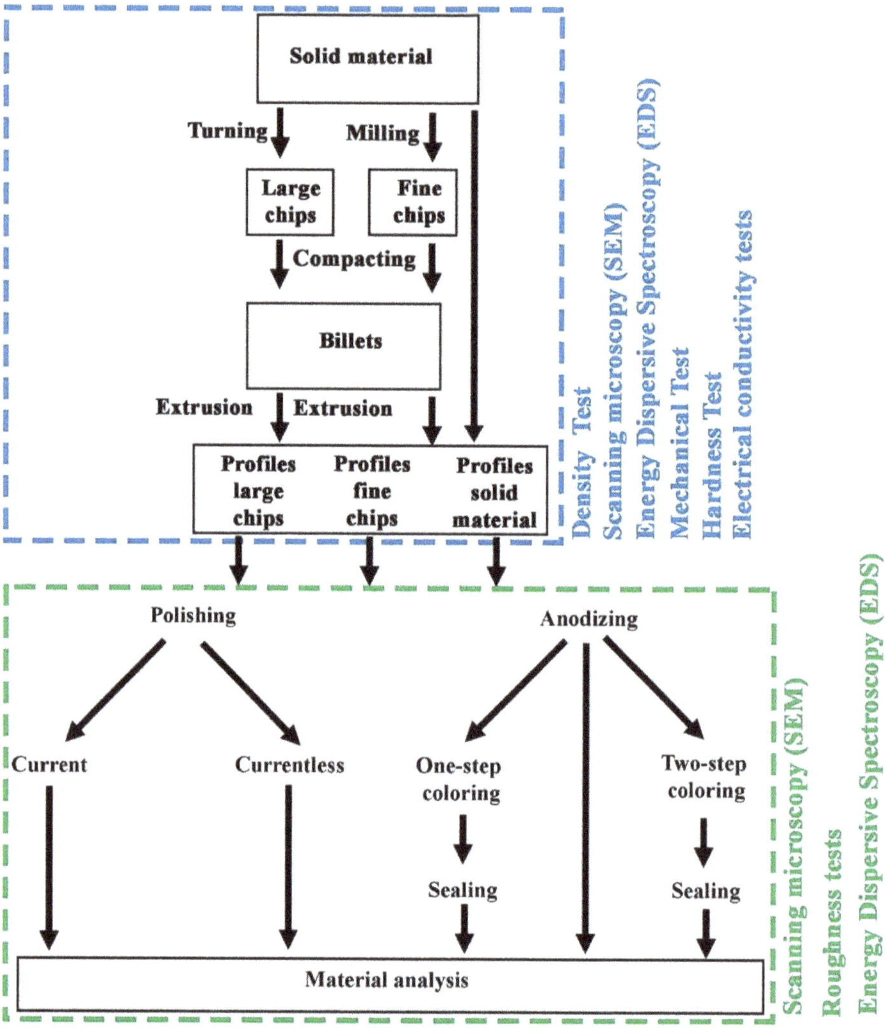

Figure 3. Scheme of experiment and research.

2.1. Chemical Surface Treatments of Al6082

For rinsing of samples and preparation of aqueous solutions, distilled water with a conductivity 6.10 µS was used. The following chemical reagents were used: sodium hydroxide—p.a., (POCH, Gliwice, Poland); acetone—99.5%, p.a., POCH; sodium chloride—p.a., POCH; sodium carbonate—p., POCH; sodium phosphate dodecahydrate—technical (98%); oxalic acid dihydrate—p. POCH, xylenol orange—p., POCH; lead acetate trihydrate—p., (WARCHEM, Warsaw, Poland); ammonium chromate(VII)—p. (CHEMPUR, Piekary Śląskie, Poland). The POLSONIC SONIC 0.5 ultrasonic bath (Warsaw, Poland) was used to clean the samples, and the POLSONIC SONIC 9 ultrasonic bath was used for the other treatments. RIGOL DP711 DC source (Warsaw, Poland) was used for current polishing and anodizing. During the anodizing, magnetic stirrer LGG LABWARE uniSTIRRER 7 (Meckenheim, Germany) was used. Water bath from WSL POLAND company (Świętochłowice,

Poland) was used for sealing. In all cases, drying in the desiccator (calcium chloride as moisture sorbent) took at least 24 h.

2.1.1. Etching and Cleaning

Samples of Al6082 alloy were etched in 1 M sodium hydroxide solution. The samples polished on sandpapers were immersed in the etching solution for 16 s, then washed with a strong stream of running tap water—the whole operation was repeated twice. The samples were finally washed with a stream of distilled water. The samples were then placed in acetone in an ultrasonic bath. Cleaning in acetone took 1 h. Cleaned samples were partially dried on a paper towel and then allowed to dry in a desiccator.

2.1.2. Chemical Currentless Polishing

Etched and cleaned samples of the Al6082 alloy were placed in a solution with the following composition: 100 g/L NaOH, 370 g/L NaCl (supersaturated solution). The currentless polishing was performed at 50 °C in an ultrasonic bath for 5.5 min. The polishing proceeded with the intense evolution of a large amount of gases (hydrogen). Upon completion of the currentless polishing, the samples were removed from the solution and washed in a stream of distilled water. The currentless polished samples were allowed to dry in a desiccator.

2.1.3. Chemical Current Polishing

Applied method of chemical current polishing is similar to the Brytal method [36]. Etched and cleaned samples of the Al6082 alloy were placed in a solution with the following composition: 15% Na_2CO_3 and 5% Na_3PO_4 (calculated as anhydrous). The samples were connected to a DC source as anodes. The cathodes were stainless steel plates. The current polishing was carried out at the temperature of 80 °C in an ultrasonic bath for 8 min. For each sample, the volume of the solution was 100 mL. The electrical parameters of the current polishing were 12 V and 3 A. After the current polishing was completed, the samples were removed from the solution and washed in a stream of distilled water. The current polished samples were allowed to dry in a desiccator.

2.1.4. Anodizing

Based on [25], the etched and cleaned samples of the Al6082 alloy were placed in a 0.4 M oxalic acid solution. The samples were connected to a DC source as anodes. The cathodes were lead plates with a 1% addition of silver. Anodizing was carried out for 1 h while stirring the solution with a magnetic dipole at a rotation speed of 500 rpm. Magnetic dipole was just below the anode. The electrical parameters of anodizing were 30 V and 3 A. After the anodizing was completed, the samples were removed from the solution and washed in a stream of distilled water. Anodized samples were allowed to dry in a desiccator.

2.1.5. One-Step Colorizing

Anodized samples of the Al6082 alloy were placed in an aqueous solution of xylenol orange at a concentration of 60 g/L. One-step colorizing was performed at the temperature of 40 °C in an ultrasonic bath for 20 min. Upon completion of the one-step colorizing, the samples were removed from the solution and washed in a stream of distilled water. The colorized samples were sealed (as presented at the end of the chapter).

2.1.6. Two-Step Colorizing

Anodized samples of the Al6082 alloy were first placed in a lead acetate solution at a concentration of 15 g/L. After 20 min, the samples were removed from the solution and washed under a stream of distilled water. Then, the samples were immersed in an ammonium chromate (VII) solution at a concentration of 15 g/L. After 20 min, the samples were removed from the solution and washed under a stream of distilled water. Two-step colorizing was performed at 60 °C in an ultrasonic bath for 40 min in total (2 × 20 min). The colorized samples were sealed (as discussed at the next paragraph).

2.1.7. Sealing

The colored samples of the Al6082 alloy were placed over a water bath in a stream of steam. Sealing was carried out at the temperature of 96 °C for 30 min. The sealed samples were washed in a stream of distilled water. The sealed samples were allowed to dry in a desiccator.

2.1.8. Color Analysis

It is not easy to register and compare colors: it depends on the individual predispositions of the observer (physiology and perception), on the conditions of color exposure (lighting, transparency of the medium surrounding the observed object, etc.) and the method of observation (unaided eye/aided eye). In order to accurately register and compare colors, one should choose one appropriate method, which, regardless of the observation conditions, will give unambiguous results, which will enable analysis and the possibility of making correct conclusions. For example, the photos in [27] can be compared inside this article, but cannot be compared with photos from other articles because the conditions under which these photos were taken are unknown (and probably not taken under the same conditions). Thus, to avoid this type of problem, the best solution may be to digitize the colors in the form of, e.g., color histograms. The validity of using color histograms in chemistry and related fields is described in [40]. Comparing color histograms is free from cognitive bias, but is based only on numerical analysis. This can be helpful when it is not possible to show a colorful photo (as for example in [25])—in this case, you can use the color histogram. Since aluminum coloring is primarily used to increase the artistic and functional value by changing the color, it is important to properly show the color of the obtained colored samples.

Color analysis of the Al6082 alloy: solid, with coarse and fine chips after surface treatment was performed based on the photographic images on a black background and hence color histograms were obtained in the Gimp 2.10.22 software [41].

3. Results and Discussions

Figure 4A–C shows the microstructures and the distribution maps of elements after the process of extruding flat bars obtained from solid material, coarse and fine chips. Images were taken in the backscatter electrons mode where the number of emitted electrons depends on the atomic number Z. Observation in the BSE mode allows one to visualize the differences in the composition of the sample with different levels of contrast; the map of the chemical elements was provided using EDX. The samples were observed in a transverse to the extrusion direction. In each of the analyzed materials, two types of phases can be distinguished: those containing magnesium and silicon (dark precipitates) and those containing aluminum, iron, manganese and silicon (light precipitates). The maps of the distribution of elements made it possible to identify the presence of given chemical elements in individual phases. Based on the chemical composition and the analysis of the literature [42–44], it might be concluded that the bright areas are the Al (FeMn) Si phase. Literature data show that in this type of alloy, depending on the chemical composition, there are phases with a different stoichiometric composition (e.g., Al_9Mn_3Si, Al_5FeSi, Al (FeMn) Si). These phases may also have a different morphology (they may have a columnar structure, polyhedral structure, as well as appearing in the form of the so-called "Chinese script"). Dark precipitation rich in Mg and Si is the Mg_2Si phase, the presence of which can also be confirmed in various scientific and research articles [44–46]. In the case of materials after the milling and turning process, fragmentation of the precipitates is observed in relation to the reference material; this fragmentation occurred as a result of the applied machining (turning/milling) and the hot extrusion process. For a solid material, the average diameter of the Al (FeMn) Si precipitates was 1.5 µm, and for samples obtained from coarse and fine chips, it was 1.2 µm and 0.9 µm, respectively. Microstructural observations of the materials after the extrusion process did not reveal the occurrence of defects inside the materials, which proves the well-chosen conditions of plastic consolidation. The results of the chemical composition presented in Table 3 confirm that as a result of machining (turning and milling) and hot extrusion, the materials were shredded without changing their chemical composition.

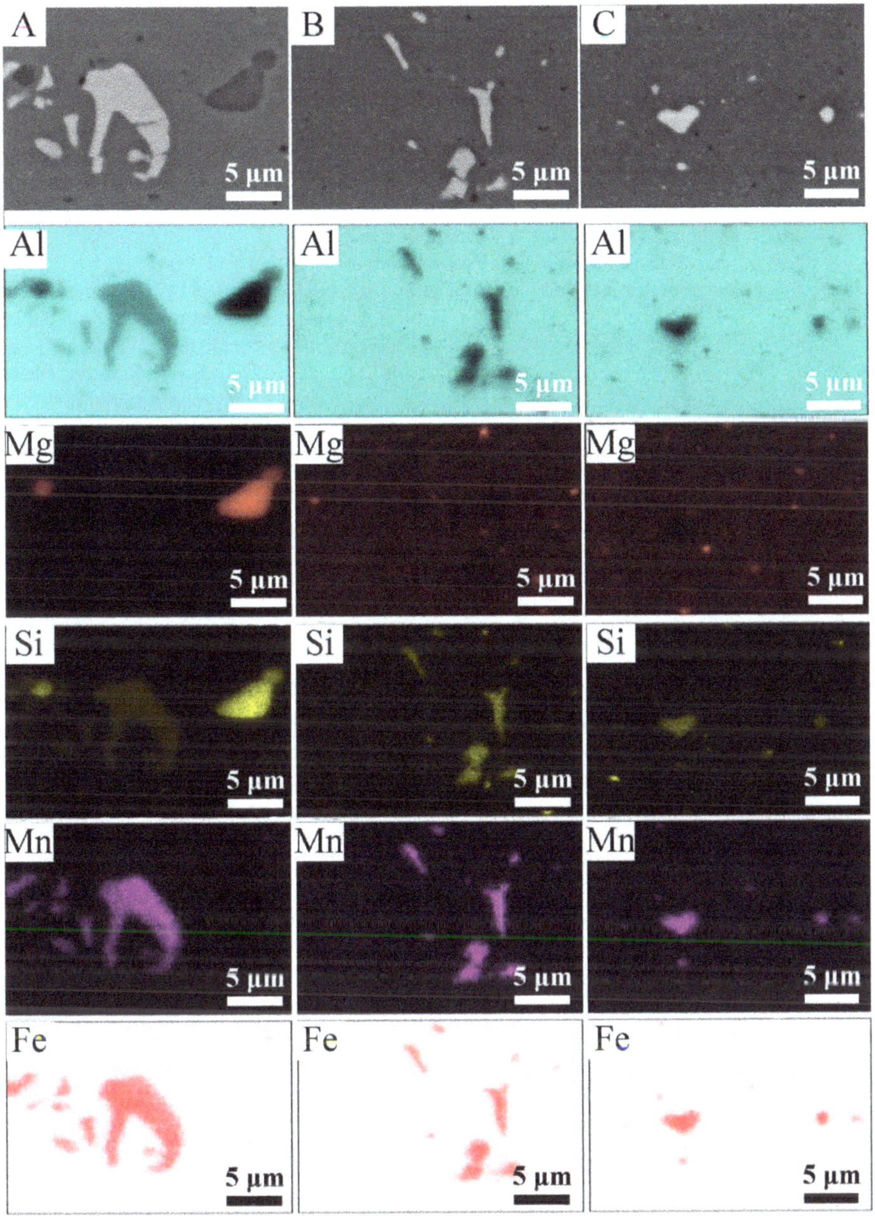

Figure 4. The microstructure in the cross-section of the materials after the extrusion process along with the EDS chemical mapping, (**A**) solid material, (**B**) coarse chips, (**C**) fine chips.

Table 3. Our own spectrometric tests of the chemical composition of solid material, coarse and fine chips.

Materials	Elements									
	Al	Si	Fe	Cu	Mn	Mg	Cr	Zn	Ti	Other
Solid Material	97.23	1.10	0.24	0.03	0.62	0.64	-	-	0.02	0.12
Coarse chips	97.14	1.15	0.22	0.02	0.70	0.69	-	-	0.02	0.06
Fine chips	97.41	1.02	0.21	0.02	0.59	0.64	-	-	0.02	0.09

The mechanical properties of the extruded profiles were tested in a uniaxial static tensile test. Figure 5 shows the tensile curves for the Al 6082 alloy samples obtained from solid material, coarse chips and fine chips. The solid material has the highest mechanical properties. The tensile strength in this case was 185 MPa and the yield point was 102 MPa. The other two materials reached the tensile strength of 162 MPa (large chips) and 173MPa (fine chips). The highest elongation values of 15% were obtained for the material made of fine chips; the material obtained from a solid ingot and coarse chips slightly deviated from the material made of fine chips and reached elongation of 13.8% and 14.2%, respectively. For comparison, Krolo et al., testing the same alloy, obtained UTS values of 150 MPa for consolidated shredded forms at the process temperature of 400 °C [47], while Tokarski, conducting similar tests, obtained UTS equal to 160 MPa [48]. Reduced strength properties of profiles after plastic consolidation as compared to solid material may be caused by the presence of oxygen layers on the chip surface, which was confirmed by the authors in [49]. They concluded that in order to ensure good strength properties of the chip after plastic consolidation, two conditions should be met: First, the oxide layers must be broken down to allow virgin metal-to-metal contact, and second, the cumulative value of the ratio of the mean stress to the flow stress must be greater than a constant value. In the stretching curves (Figure 5) for all materials, we can see the so-called "Serration"/Portevin Le Chaterlier (PLC) effect. This is the effect of a step, not a continuous change of stress in the material subjected to the static tensile test. The PLC effect is the result of dynamic strain aging (DSO) which is a dynamic interaction between sliding dislocations and free atoms [50]. The PLC effect is classified [51,52] into three types: A, B and C depending on the nature of the temporal–spatial organization of the deformation bands. Type A corresponds to deformations that propagate continuously along the stretching axis (representing isolated plastic waves). Type B means intermittent (time-oscillating) strain propagation (stop-and-go). Finally, type C means a deformation band that appears randomly/stochastically and does not propagate along the sample subjected to stretching [53]. In the case of the tested materials, a similar degree of intensity and frequency of occurrence of stress instability was observed. The conducted observations allow classifying of the deformation instability to the variant A due to relatively small stress drops. The mechanical and physical properties presented in Table 4, such as density, hardness and electrical conductivity, confirm the effective and efficient plastic consolidation. In the case of the obtained results of physical properties, the obtained results are in the range of the measurement error.

The color histogram is a representation by the number and brightness function of the pixels of the photos. By moving to the right or left of the graph, the function applies to the brighter or darker pixels, respectively. The peak heights in the diagrams show the number of pixels corresponding to a given color value. The presented histograms are linear with values in linear space, both for a single color and for a mixture (RGB).

Histograms were constructed on the basis of photographic images on a black background (Figure 6). In all of these images, the material impact on the appearance of the surface is visible by unaided eye. In the images of the material from coarse chips, regardless of the surface treatment, coarse pieces of chip reproduced on its surface may be seen. Moreover, in this case, the surface is slightly darker compared to the corresponding other materials. Both currentless and current polishing resulted in parallel streaks on the surface of the metal. It happened due to the intense gas generation during polishing, where gas bubbles (hydrogen) cavitated and caused mechanical corrosion of aluminum. During one-step coloring, samples of red-orange color were obtained; the resulting color is the effect of xylenol orange. The freshly colored samples were golden in color; however, they turned red-orange after sealing. Coarse and fine samples are more gray (less colored) compared to the solid sample; hence, the conclusion that coarse and fine are less susceptible to dyeing. As for the two-step coloring, it can be said that the color is uniform over the entire surface of the samples. The exception is material from coarse chips, which, depending on the chip position, is locally darker or lighter.

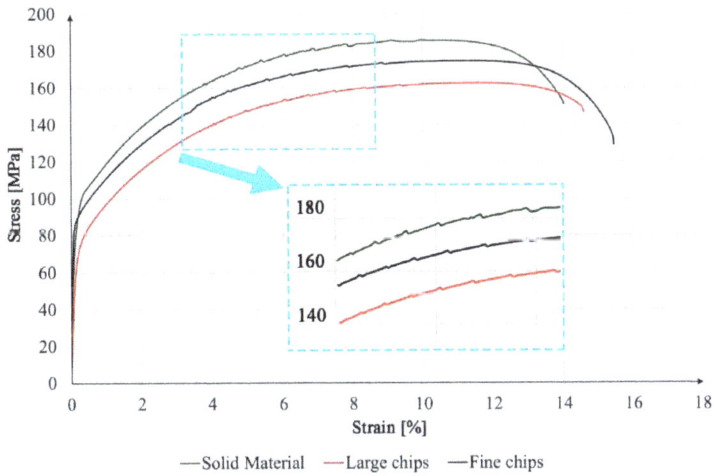

Figure 5. Collective graph of tension curves for solid material, coarse and fine chips.

Table 4. Mechanical and physical properties of materials after extrusion.

Method	Solid Material	St. Dev.	Coarse Chips	St. Dev.	Fine Chips	St. Dev.
UTS, MPa	185.00	2.03	162.00	1.50	175.00	0.70
YS, MPa	102.00	3.12	72.00	1.70	87.00	1.10
Elongation, %	14.00	2.03	15.00	1.60	15.00	1.20
Density, g/cm^3	2.70	0.03	2.69	0.03	2.70	0.02
Electrical conductivity, MS/m	29.93	0.05	30.00	0.02	28.93	0.05
Hardness, HV2	52.0	1.1	45.0	1.7	47.0	1.7

Observations on the macroscopic scale are well represented by the digital way. Through one-step coloring, solid and fine chips materials were the most homogenous in terms of color, as shown by the color histograms (Figure 7). Generally, in the case of the samples colored using xylenol orange, for the average values of 0.259–0.323, the standard deviation ranged from 0.61 to 0.103, which results from the fact that the subsequent color was the resultant of several primary colors. Coarse chip material gave the least red and more yellow surface. After two-step coloring, yellow color of surfaces was expected, according to the color of lead(II) chromate (reaction 3); however, as the color histograms show, a yellowish-green color was obtained. This happened probably because the anodized samples (and coloring was preceded by anodizing) were dark and eventually yellow on the dark grey which looks like green. Nevertheless, it shows that the coloring was successful in this case as well, albeit with a slightly different end result. The average and median values for the solid and fine chips materials are similar, 0.57, 0.510 and 0.516, 0.514, respectively, while for the coarse chips material they are significantly different, i.e., 0.438, 0.439. Visibly, samples of solid and fine chips materials have a more similar color to each other than to coarse chips material. The color of the coarse chips material is closer to yellow (mean and median values are closer to yellow).

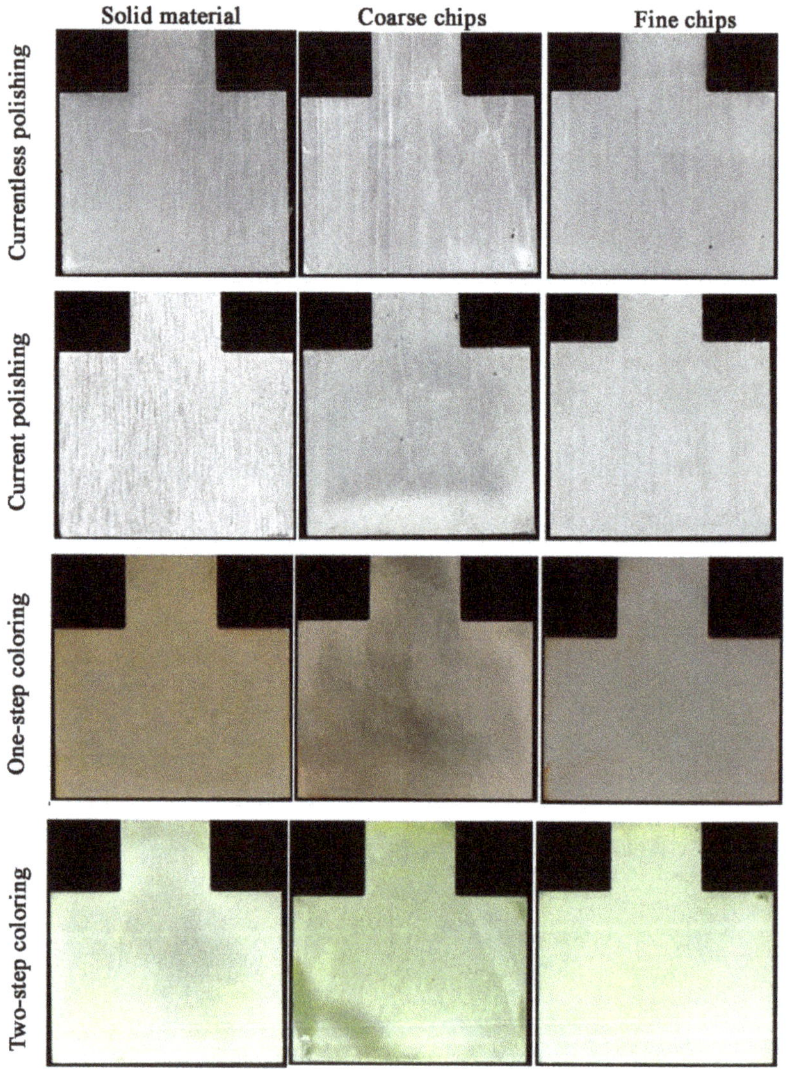

Figure 6. The appearance (unaided eye) of samples after surface treatment of Al6082 alloy: solid, with coarse and fine chips.

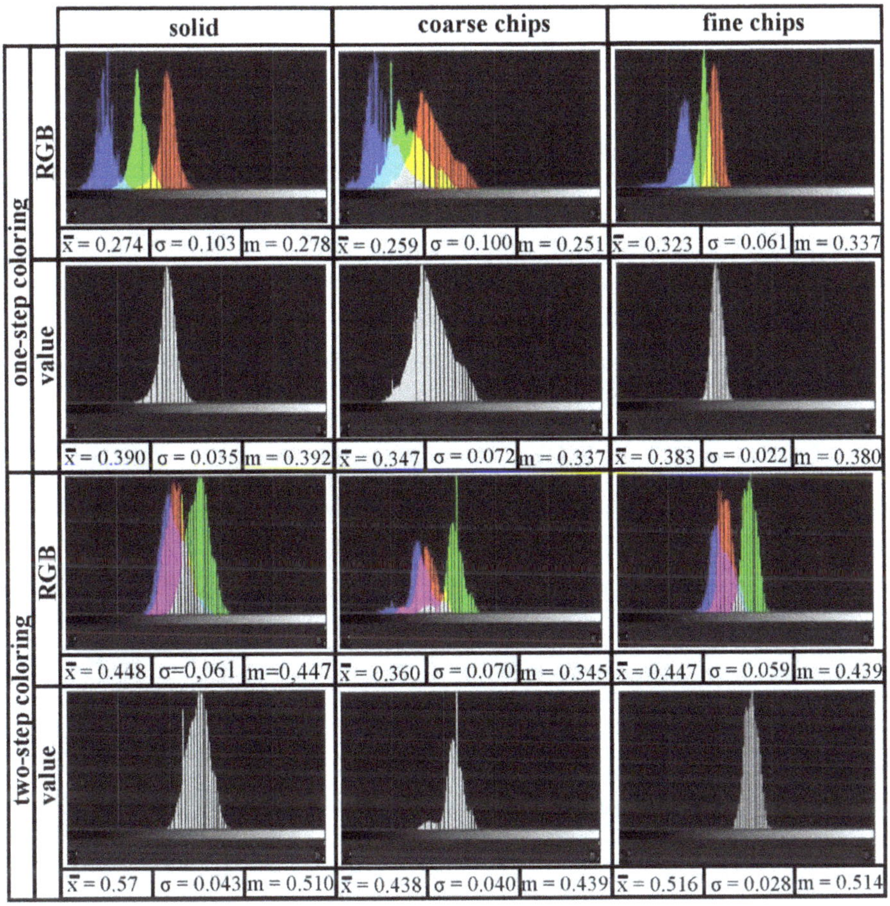

Figure 7. Color histograms of one- and two-step coloring Al6082 alloy: solid, with coarse and fine chips.

Figure 8 shows microscopic photos of the samples after surface treatment; the images were taken using the SE detector. The polishing samples show fine light particles which do not give off chemical polishing. Based on Figure 4, it is known that these phases are rich in iron, magnesium, manganese and silicon. Both during currentless and current polishing, craters can be observed on the surface, but in current polished samples, there are more of them. These craters were created as a result of cavitation—escaping gases caused mechanical corrosion. To avoid this phenomenon, the polishing process should be carried out at lower values of parameters. The roughness coefficient values were the largest of the remaining samples and ranged from 0.29–1.06; hence, it can be concluded that polishing in the given parameters was not successful, but taking into account the visual aspect, the polishing contributes to a significant brightening of the surface of the samples. Roughness for anodized and colored samples was low, about 0.1, and coloring and sealing did not affect the roughness of the samples. Probably this is due to the sealing, which contributes to the smoothing of the colored oxide film. Anodized sample is characterized by a developed surface; hence, it appears darker on a macroscopic scale compared to raw aluminum. One-step coloring of microscopic photos do not differ much from photos of anodized samples. The red dye, which is xylenol orange, does not change the surface microstructure, but only its macroscopic color. It is different in the case of two-step coloring.

Characteristic needles appear in the microstructure. These needles are lead-chromium oxide (similar to [54] or [37], in which lead-chromium oxide was crocoite in the form of longitudinal crystallites), which gives the surface a greenish-yellowish color.

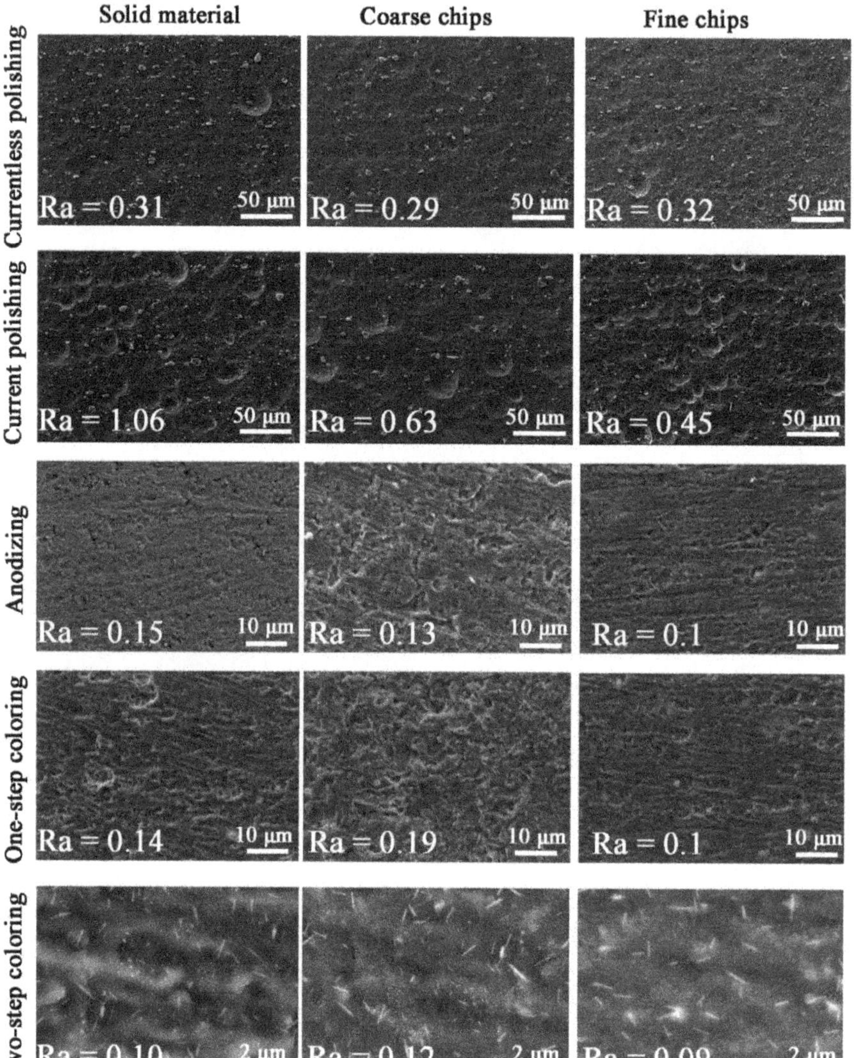

Figure 8. Surface of samples after surface treatment of Al6082 alloy: solid, with coarse and fine chips.

From the photos of the cross-sections of the samples near the surface (Figure 9, the images were taken using the BSE detector), the roughness of the currentless and current polished samples can be seen: for the current polishing, the roughness is greater. The anodized and colored samples show a characteristic barrier layer, which is aluminum oxide. This is a confirmation of the effective formation of the oxide layer during anodizing. This layer is compact, with a thickness of approximately 5–6 microns (Table 5). Its chemical composition, which is presented in Table 5, is approximate to the 1 Al: 1 O relationship (for anodized samples: 0.90–1.06), but the ratio slightly decreases for colored samples

(0.85–1.03). A few percent of the carbon value is most likely an impurity of aluminum. In the case of samples colored in a one-step process with an organic compound, an increased carbon content is noted compared to the anodized samples; 4.2–4.7 vs. 3.1–4.2, respectively. The colorant increases the carbon content, but not much, i.e., about 1%. However, this confirms that there is a carbon-containing component in the oxide layer. In the case of two-step coloring, $PbCrO_4$ needles are bright precipitates in the cross-section. This is also due to the chemical composition of the layer; there is a slight chromium content, 0.10–0.30 wt.%. and a lead content of several percent, 3.40–3.80 wt.%. The theoretical ratio of lead to chromium mass in $PbCrO_4$ is about 3.985, while according to the results from the table, for samples colored with precipitation of lead-chromium oxide, the ratio is 12.355–49.286. The reason for this is probably the cementation phenomenon, which causes metallic lead to precipitate out of the solution. However, this phenomenon may be so slight that it does not interfere with the coloring process.

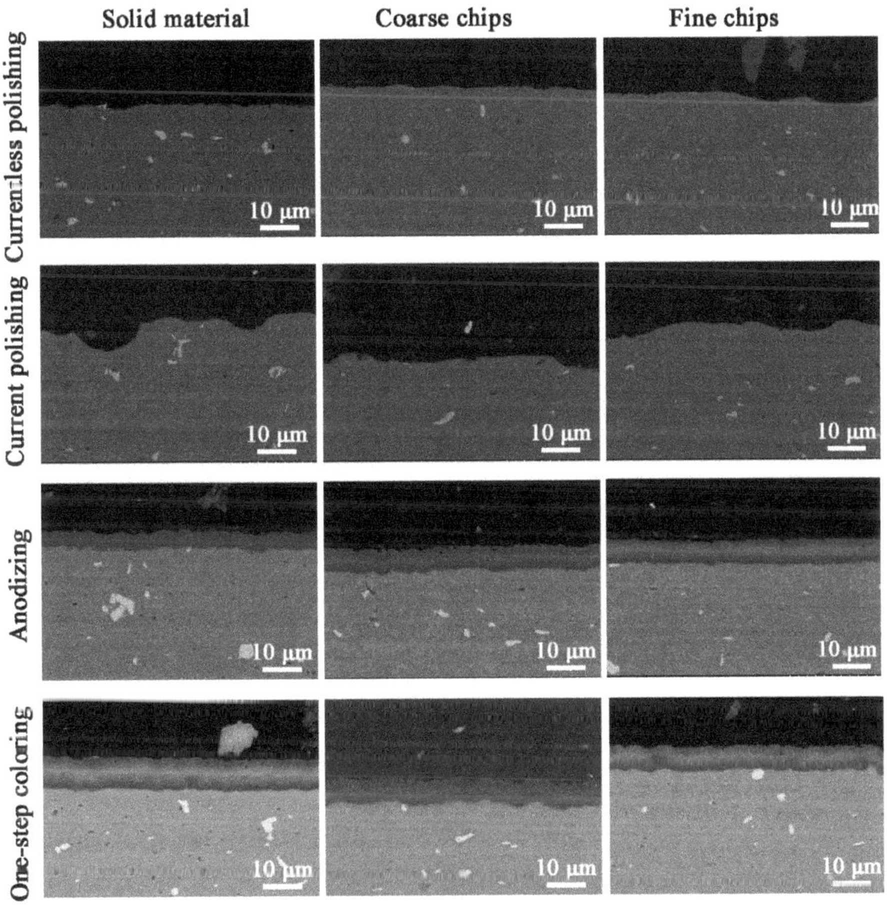

Figure 9. *Conts.*

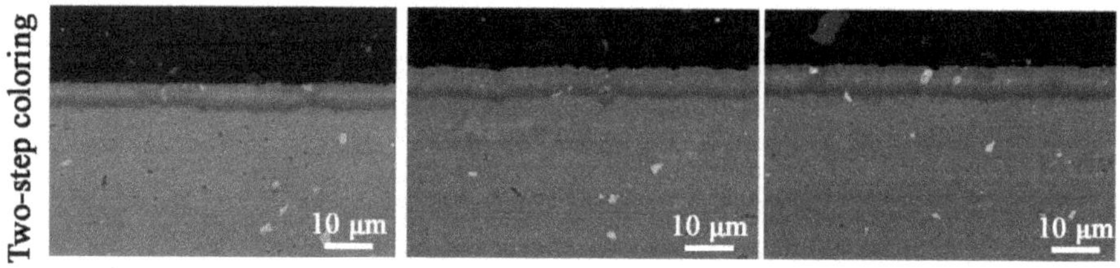

Figure 9. Microstructure of the cross-section of samples after surface treatment of Al6082 alloy: solid, with coarse and fine chips.

Table 5. The thickness of the oxide layer and the chemical composition.

	Material	Thickness, µm	Standard Deviation, µm	Weight % of Elements				
				Al	O	C	Cr	Pb
Anodized	Solid material	3.965	0.195	49.2	46.6	4.2	-	-
	Coarse chips	4.579	0.159	46.4	50.5	3.1	-	-
	Fine chips	5.788	0.343	45.5	50.7	3.8	-	-
One-step coloring	Solid material	8.125	0.275	45.2	50.1	4.7	-	-
	Coarse chips	5.867	0.152	48.7	47.1	4.2	-	-
	Fine chips	5.845	0.098	45.1	50.3	4.6	-	-
Two-step coloring	Solid material	5.179	0.268	43.6	49.4	3.5	0.10	3.4
	Coarse chips	5.737	0.292	42.9	50.3	2.7	0.30	3.8
	Fine chips	6.131	0.175	42.8	50.0	3.2	0.20	3.8

4. Conclusions

Based on the presented research results, the following conclusions can be drawn:

- it was found that using the plastic consolidation method, it is possible to produce a material with a similar density and electrical conductivity from the fragmented fractions to a press made of a solid ingot;
- using the plastic consolidation method, it is possible to obtain solid rods from waste materials that meet the requirements of mechanical properties set out in the standards [38]. The tensile strength for the material obtained from fine and coarse chips was lower compared to the solid material by 5.41% and 12.44%, respectively;
- materials after plastic consolidation show increased plasticity compared to the material obtained from a solid ingot with the same chemical composition;
- the microstructural tests carried out did not reveal any defects inside the analyzed materials, such as cracks or delamination, which is confirmed by appropriately selected parameters of the extrusion process, such as: temperature, speed or the degree of processing;
- machining treatment processes (milling/turning) and the extrusion process resulted in the fragmentation of brittle intermetallic phases of the Al (FeMn) Si type, which resulted in an increase in elongation. The smallest particle size of Al (FeMn) Si (0.9 µm) was obtained for fine chips;
- the surface treatment of the recycled 6082 alloy materials is similar to that of the raw material, but on the macroscopic scale, there are differences. The most important difference is the surface quality of coarse chip samples, the surface of which is darker and reflects the shape of the chips;
- one-step coloring of solid material gives a more intense color, in comparison to the recycled materials. Coloring with the precipitation of lead chromium oxide gives a color more green than yellow, which was originally assumed;

- polishing of 6082 alloy samples with given parameters causes excessive roughness. Despite this, the surfaces of the samples are bright and aesthetic;
- on the microscopic scale, the anodized and xylenol orange colored samples do not differ, as the colorant is not visible under the microscope. $PbCrO_4$ precipitation coloring causes precipitation of the characteristic needles of this compound;
- anodizing of 6082 aluminum alloy by using oxalic acid creates a dense oxide layer with a regular thickness of about 5–6 microns.

Author Contributions: Conceptualization, P.N., A.P., T.S., A.Z. and P.S.; methodology, P.N., A.P.; validation, P.N., T.S.; investigation, A.P.; writing—original draft preparation, P.N., T.S. and A.Z.; writing—review and editing, P.N., A.P.; visualization, P.N., A.P., T.S., A.Z. and P.S. All authors have read and agreed to the published version of the manuscript.

Funding: This research was funded by a subsidy from the Faculty of Non-Ferrous Metals of AGH University of science and Technology 16.16.180.006.

Institutional Review Board Statement: Not applicable.

Informed Consent Statement: Not applicable.

Data Availability Statement: All data are provided in full in the results section of this paper.

Conflicts of Interest: The authors declare no conflict interest.

References

1. Mathisena, A.; Sørensena, H.; Eldrupa, N.; Skagestada, R.; Melaaenb, M.; Müller, G.I. Cost optimised CO_2 capture from aluminium production. *Energy Procedia* **2014**, *51*, 184–190. [CrossRef]
2. Sverdrup, H.U.; Ragnarsdottir, K.V.; Koca, D. Aluminium for the future: Modelling the global production, market supply, demand, price and long term development of the global reserves. *Resour. Conserv. Recycl.* **2015**, *103*, 139–154. [CrossRef]
3. Wana, B.; Chena, W.; Lua, T.; Liua, F.; Jianga, Z.; Maoa, M. Review of solid state recycling of aluminum chips. *Resour. Conserv. Recycl.* **2017**, *125*, 37–47. [CrossRef]
4. Ab Rahim, S.N.; Lajis, M.A.; Ariffin, S. A Review on Recycling Aluminum Chips by Hot Extrusion Process. *Procedia CTRP* **2015**, *26*, 761–766. [CrossRef]
5. Capuzzi, S.; Timelli, G. Preparation and melting of scrap in aluminum recycling: A review. *Metals* **2018**, *8*, 249. [CrossRef]
6. Samuel, M. A new technique for recycling aluminum scrap. *J. Mater. Process. Technol.* **2003**, *135*, 117–124. [CrossRef]
7. Wędrychowicz, M.; Wzorek, Ł.; Tokarski, T.; Noga, P.; Wiewióra, J. Recycling without melting: An alternative approach to aluminum scrap recovery. *Key Eng. Mater.* **2016**, *682*, 284–289. [CrossRef]
8. Gronostajski, J.; Kaczmar, J.; Matuszak, A. Direct recycling of aluminium chips into extruded products. *J. Mater. Process. Technol.* **1997**, *64*, 149–156. [CrossRef]
9. Dybiec, H. Plastic consolidation of metallic powders. *Arch. Metall. Mater.* **2007**, *52*, 161–170.
10. Chmura, W.; Gronostajski, J. Mechanical and tribological properties of aluminum-base composites produced by the recycling of chips. *J. Mater. Procc. Technol.* **2000**, *106*, 23–27. [CrossRef]
11. Noga, P.; Zwolinski, A. Effect of work hardening and thermal stability of metal chips after machining on properties of plastically consolidated profiles. In Proceedings of the METAL 2019—28 International Conference on Metallurgy and Materials, Brno, Czech Republic, 22–24 May 2019; pp. 1241–1246
12. Włoch, G.; Skrzekut, T.; Sobota, J.; Woźnicki, A.; Cisoń, J. The structure and mechanical properties of plastically consolidated Al-Ni alloy. *Key Eng. Mater.* **2016**, *682*, 245–251. [CrossRef]
13. Skrzekut, T.; Blaz, L. Examination of aluminum matrix composites obtained by powder metallurgy and strenghtened by AgO and CeO2 particles. In Proceedings of the METAL 2018—27th International Conference on Metallurgy and Materials, Brno, Czech Republic, 23–25 May 2018, pp. 1722–1728.
14. Gronostajski, J.; Chmura, W.; Gronostaajski, Z. Bearing materials obtained by recycling of aluminium and aluminium bronze chips. *J. Mater. Process. Technol.* **2002**, *125–126*, 483–490. [CrossRef]
15. Skrzekut, T.; Kula, A.; Blaz, L.; Wloch, G.; Sugamata, M. Structural characterization of mechanically alloyed AlMg-CeO_2 composite. *Key Eng. Mater.* **2015**, *641*, 10–16. [CrossRef]
16. Koch, A.; Bonhage, M.; Teschke, M.; Lucker, L.; Behrens, B.A.; Walther, F. Electrical resistance-based fatigue assessment and capability prediction of extrudates from recycled field-assisted sintered EN AW-6082 aluminium chips. *Mater. Charact.* **2020**, *169*, 110644. [CrossRef]
17. Smyrak, B.; Jurkiewicz, B.; Zasadzińska, M.; Gniełczyk, M.; Jałowy, P. The Effect of Al-Mg-Si Wire Rod Heat Treatment on Its Electrical Conductivity and Strength. *Metals* **2020**, *10*, 1027. [CrossRef]
18. Jurkiewicz, B.; Smyrak, B.; Zasadzińska, M.; Franczak, K.; Strzępek, P. The researches of influence of strengthening on fatigue strength of aluminium wires for OHL conductors. *Arch. Civ. Mech. Eng.* **2019**, *19*, 862–870.

19. Noga, P.; Tuz, L.; Żaba, K.; Zwolinski, A. Analysis of Microstructure and Mechanical Properties of AlSi11 after Chip Recycling, Co-Extrusion, and Arc Welding. *Materials* **2021**, *14*, 3124. [CrossRef] [PubMed]
20. Aluminium Alloys – Aluminium 6082 Properties, Fabrication and Application. Available online: https://www.azom.com/article.aspx?ArticleID=2813 (accessed on 3 August 2021).
21. Aluminium 6082-T6. Available online: http://www.matweb.com/search/datasheet_print.aspx?maguid=fad29be6e64d4e95a241690f1f6e1eb7 (accessed on 3 August 2021).
22. L'Haridon-Quaireau, S.; Laot, M.; Colas, K.; Kapusta, B.; Delpech, S.; Gosset, D. Effects of temperature and pH on uniform and pitting corrosion of aluminium alloy 6061-T6 and characterisation of the hydroxide layers. *J. Alloys Compd.* **2020**, *833*, 155146. [CrossRef]
23. Martínez-Viademonte, P.; Abrahami, M.; Hack, S.; Burchardt, T.; Terryn, M. A Review on Anodizing of Aerospace Aluminum Alloys for Corrosion Protection. *Coatings* **2020**, *10*, 1106. [CrossRef]
24. Grubbs, A. Anodizing of aluminum. *Met. Finish.* **1999**, *97* (Suppl. S1), 476–493. [CrossRef]
25. Kozhukhova, A.E.; du Preez, S.P.; Bessarabov, D.G. Preparation of anodized aluminium oxide at high temperatures using low purity aluminium (Al6082). *Surf. Coat. Technol.* **2019**, *378*, 124970. [CrossRef]
26. Anorom—Surface Treatments for Aluminum. Available online: www.anorom.ro (accessed on 3 August 2021).
27. Ardelean, M. Surface treatments for aluminium alloys. *IOP Conf. Ser. Mater. Sci. Eng.* **2018**, *294*, 012042. [CrossRef]
28. Franco, M.; Anoop, S.; Uma Rani, R.; Sharma, A. Porous Layer Characterization of Anodized and Black-Anodized Aluminium by Electrochemical Studies. *Int. Sch. Res. Netw. ISRN Corros.* **2012**, *12*, 323676. [CrossRef]
29. Otomo, M. The Spectrophotometric Determination of Aluminum with Xylenol Orange. *Bull. Chem. Soc. Jpn.* **1963**, *36*, 809–813. [CrossRef]
30. Zolgharnein, J.; Shahrjerdi, A.; Azimi, G.; Ghasemi, J. Spectrophotometric determination of trace amounts of fluoride using an Al-xylenol orange complex as a colored reagent. *Anal. Sci.* **2009**, *25*, 1249–1253. [CrossRef] [PubMed]
31. Desai, P.; Vashi, R. Efficiency of xylenol orange as corrosion inhibitor for aluminium in trichloroacetic acid. *Indian J. Chem. Technol.* **2010**, *17*, 50–55.
32. Tsangaraki-Kaplanoglou, I.; Kanta, A.; Theohari, S.; Ninni, V. Acid-dyes as corrosion inhibitors for mechanically pretreated aluminum. *Anti-Corros. Methods Mater.* **2010**, *57*, 6–12. [CrossRef]
33. Silina, Y.; Kuchmenko, T.; Volmer, D. Sorption of hydrophilic dyes on anodic aluminium oxide films and application to pH sensing. *Analyst* **2015**, *140*, 771–778. [CrossRef]
34. Abel, A. 3-Pigments for paint. In *Woodhead Publishing Series in Metals and Surface Engineering, Paint and Surface Coatings, 2nd Edition*; Woodhead Publishing: Sawston, UK, 1999; pp. 91–165.
35. Ansdell, D. 10-Automotive paints. In *Woodhead Publishing Series in Metals and Surface Engineering, Paint and Surface Coatings, 2nd Edition*; Woodhead Publishing: Sawston, UK, 1999; pp. 411–491.
36. Wernick, S. *The Surface Treatment and Finishing of Aluminum and Its Alloys*; ASM International: Novelty, OH, USA, 1987; pp. 1, 73–81, 151–153.
37. Abou-Gamra, Z.; Ahmed, M.; Adel Hamza, M. Investigation of commercial $PbCrO_4/TiO_2$ for photodegradation of rhodamine B in aqueous solution by visible light. *Nanotechnol. Environ. Eng.* **2017**, *2*, 12. [CrossRef]
38. EN 1706:2020-10—Aluminum and Aluminum Alloys—Castings—Chemical Composition and Mechanical Properties.
39. *ImageJ, 1.35j*; U. S. National Institutes of Health: Bethesda, MD, USA,, 1997.
40. Gonçalves Dias Diniz, P. Chemometrics-assisted color histogram-based analytical systems. *J. Chemom.* **2020**, *34*, e3242. [CrossRef]
41. *GNU Image Manipulator Program, 2.10.22*; The GIMP Development Team: Berkeley, MD, USA, 2019.
42. Mrówka-Nowotnik, G.; Sieniawski, J.; Wierzbińska, M. Influence of long-term homogenization annealing on the microstructure and properties of 6066 aluminum alloy. *Inżynieria Mater.* **2015**, *6*, 381–385.
43. Lesniak, D.; Woznicki, A.; Wojtyna, A.; Włoch, G.; Jurczak, H. Influence of the forming temperature and cooling method during supersaturation on the hardness of alloy 6082. *Rudy i Met. Nieżelazne* **2016**, *61*, 159–163.
44. Nowotnik, G.M.; Sieniawski, J.; Wierzbińska, M.; Mrówka-Nowotnik, G. Intermetallic phase particles in 6082 aluminium alloy. *Arch. Mater. Sci. Eng.* **2007**, *28*, 69–76.
45. Wimmer, A.; Lee, J.; Schumacher, P. Phase Selection in 6082 Al-Mg-Si Alloys. *BHM Berg-und Hüttenmännische Monatshefte.* **2012**, *157*, 301–305. [CrossRef]
46. Mroówka-Nowotnik, G.; Sieniawski, J. Influence of precipitation hardening conditions on the microstructure and mechanical properties of an aluminum alloynium. *Inżyniera Mater.* **2006**, *3*, 217–220.
47. Krolo, J.; Lela, B.; Ljumovic, P.; Bagavac, P. Enhanced Mechanical Properties of Aluminium Alloy EN AW 6082 Recycled without Remelting. *Tech. Vjesn.* **2019**, *26*, 1253–1259.
48. Tokarski, T. Mechanical properties of solid-state recycled 4xxx aluminum alloy chips. *J. Mater. Eng. Perform.* **2016**, *25*, 3252–3259. [CrossRef]
49. Gu¨ley, V.; Gu¨zel, A.; Ja¨ger, A.; Khalifa, N.; Tekkaya, A.; Misiolek, W. Effect of Die Design on the Welding Quality During Solid State Recycling of AA6060 Chips by Hot Extrusion. *Mater. Sci. Eng. A* **2013**, *574*, 163–175. [CrossRef]
50. Dybiec, H. Versions, aversions and controversies, that is what a Portevin LeChatelier effect is? *Inżynieria Mater.* **2012**, *33*, 115–121.
51. Rizzi, E.; Hahner, P. On the Portvein-Le Chatelier effect: Theoretical modeling and numerical results. *Int. J. Plast.* **2004**, *20*, 121–165. [CrossRef]

52. Casarotto, L.; Tutsch, R.; Ritter, R.; Weidenmuller, J.; Ziegenbein, A.; Klose, F.; Neuhauser, H. Propagation of deformation bands investigated by laser scanning extensometry. *Comput. Mater. Sci.* **2003**, *26*, 210–218. [CrossRef]
53. Hahner, P.; Zaiser, M. From mesoscopic heterogeneity of slip to macroscopic fluctuations of stress and strain. *Acta Mater.* **1997**, *45*, 1067–1075. [CrossRef]
54. White, K.; Detherage, T.; Verellen, M.; Tully, J. An investigation of lead chromate (crocoite-$PbCrO_4$) and other inorganic pigments in aged traffic paint samples from Hamilton, Ohio: Implications for lead in the environment. *Environ. Earth Sci.* **2014**, *71*, 3517–3528. [CrossRef]

materials

MDPI

Article

Microstructures and Mechanical Properties of H13 Tool Steel Fabricated by Selective Laser Melting

Fei Lei [1,†], Tao Wen [1,†], Feipeng Yang [1], Jianying Wang [1], Junwei Fu [2], Hailin Yang [1,*], Jiong Wang [1,*], Jianming Ruan [1] and Shouxun Ji [3]

1 State Key Laboratory of Powder Metallurgy, Central South University, Changsha 410083, China; 1533170003@csu.edu.cn (F.L.); 213301068@csu.edu.cn (T.W.); 203301064@csu.edu.cn (F.Y.); jianying.wang@csu.edu.cn (J.W.); jianming@csu.edu.cn (J.R.)
2 Key Laboratory of Marine Environmental Corrosion and Bio-Fouling, Institute of Oceanology, Chinese Academy of Sciences, Qingdao 266071, China; hitfujw@163.com
3 Brunel Centre for Advanced Solidification Technology (BCAST), Brunel University London, Uxbridge UB8 3PH, Middlesex, UK; shouxun.ji@brunel.ac.uk
* Correspondence: y-hailin@csu.edu.cn (H.Y.); wangjionga@csu.edu.cn (J.W.)
† These authors contributed equally to this work.

Abstract: H13 stool steel processed by selective laser melting (SLM) suffered from severe brittleness and scatter distribution of mechanical properties. We optimized the mechanical response of as-SLMed H13 by tailoring the optimisation of process parameters and established the correlation between microstructure and mechanical properties in this work. Microstructures were examined using XRD, SEM, EBSD and TEM. The results showed that the microstructures were predominantly featured by cellular structures and columnar grains, which consisted of lath martensite and retained austenite with numerous nanoscale carbides being distributed at and within sub-grain boundaries. The average size of cellular structure was ~500 nm and Cr and Mo element were enriched toward the cell wall of each cellular structure. The as-SLMed H13 offered the yield strength (YS) of 1468 MPa, the ultimate tensile strength (UTS) of 1837 MPa and the fracture strain of 8.48%. The excellent strength-ductility synergy can be attributed to the refined hierarchical microstructures with fine grains, the unique cellular structures and the presence of dislocations. In addition, the enrichment of solute elements along cellular walls and carbides at sub-grain boundaries improve the grain boundary strengthening.

Keywords: H13 tool steel; selective laser melting; microstructure; mechanical properties

Citation: Lei, F.; Wen, T.; Yang, F.; Wang, J.; Fu, J.; Yang, H.; Wang, J.; Ruan, J.; Ji, S. Microstructures and Mechanical Properties of H13 Tool Steel Fabricated by Selective Laser Melting. *Materials* **2022**, *15*, 2686. https://doi.org/10.3390/ma15072686

Academic Editor: Elena Pereloma

Received: 10 March 2022
Accepted: 1 April 2022
Published: 6 April 2022

1. Introduction

Selective laser melting (SLM) is an additive manufacturing (AM) process to make components from melting alloy powder at specific locations layer by layer, which has received significant attention in recent years because of enabling the customized metallic components with complex geometries [1–3]. Compared with traditional casting methods, the highly localized melting and ultrafast cooling rate can generate unique non-equilibrium microstructures and superior mechanical properties [4,5].

The H13 tool steel has been widely used in industry for making moulds/dies for elevated applications because of the advantages in high strength, excellent ductility, good wear resistance and machinability [6]. Generally, the moulds/dies feature uniqueness and a low number of production. Therefore, it is very suitable for additive manufacturing. The microstructure and mechanical properties have been studied systematically for the SLMed H13. But the mechanical properties of as-SLMed H13 are often very scattered and unsatisfactory in applications. The variation of YS are normally from 830 to 1342 MPa and that of UTS are from 900 to 1712 MPa in the as-SLMed H13 [6–11]. More importantly, the elongation of as-SLMed H13 is normally lower than that obtained by conventional methods [12–14]. It was addressed that the high fluctuation of mechanical properties are mainly a response for the different contents of residual austenite [15,16], high residual stresses [17–19] and defects [20,21].

To address the key issues of severe brittleness and scatter distribution of mechanical properties of the H13 produced by SLM, various methods have been utilized, including the optimization of process parameters, substrate preheating and heat treatment of as-SLMed H13. The optimization of process parameters is always the key to obtain high density in as-SLMed alloys. Recent works have obtained the densities of ≥99% by modifying scan speed, laser power, powder federate and hatch spacing [22–25]. The substrate preheating has also been found as an effective technique in eliminating cracks, residual thermal stresses and can prevent the delamination in the as-SLMed H13 steel resulting from the reduction of cooling rates [26,27]. Mertens et al. [26] reported that the as-SLMed H13 showed a homogeneous morphology and superior mechanical properties when preheating the substrate at 400 °C. The post-heat treatment of as-SLMed H13 was also found as an effective method to remove the residual stresses, tune the microstructure and improve the mechanical properties. It has been reported that the as-SLMed tool steel can achieve a significant secondary hardening when tempered at 300–600 °C [28–30]. It was found that the as-SLMed H13 could achieve the yield strength of 1483 ± 48 MPa, the high ultimate strength of 1938 ± 62 MPa and the fracture strain of 5.8% after being tempered at 600 °C [5]. However, heat treatment further increases the cost and potentially destroys the fine microstructure produced by the as-SLMed process, especially the cellular structures. Zhong et al. [31] reported that the microstructures with intergranular cellular segregation network could increase the yield strength without scarifying the ductility. Zhu et al. [32] and Wang et al. [33] demonstrated that the cellular structure significantly increased the strength via dislocation hardening. The strategies to achieve high strength and ductility simultaneously in as-SLMed H13 steel are becoming attractive and critical for industrial applications. However, the strengthening mechanisms have not been well understood for cellular structures, dislocations, microstructural refinement and precipitates in the as-SLMed H13 [34–38]. More technical evidences are needed to confirm the achievement in high strength and ductility for the as-SLMed H13, as well as the detailed strengthening mechanisms.

In this work, we aimed to study the microstructure and mechanical properties of as-SLMed H13. The process optimization, the characterization of powders, the microstructure and mechanical properties in the as-SLMed H13 were systemically studied. The discussion focuses on the correlation between the cellular structures, the mechanical response and the dominant mechanisms responsible for strengthening the as-SLMed H13.

2. Experimental

2.1. Powder Preparation

H13 powders **were** provided by Hunan Hualiu New Materials Co., Ltd. (Hunan, China). Inductively coupled plasma atomic emission spectrometry (ICAP 7000 Series, Waltham, MA, USA) was used to determine the chemical composition of the H13 powders in Table 1. The sizes of H13 powders were tested via laser particle size analyser (Matersizer, Malvern, UK). The size distribution ranged from 15 to 53 μm, with D_{10} = 16.8 μm, D_{90} = 47.3 μm and D_{50} = 28.4 μm. The powder micrograph and particle size distribution are displayed in Figure 1.

Table 1. Chemical Compositions of the H13 powders (wt.%).

Element	Cr	Mo	V	Mn	Si	C	Fe
wt.%	5.12	1.26	1.03	0.39	0.98	0.42	Bal.

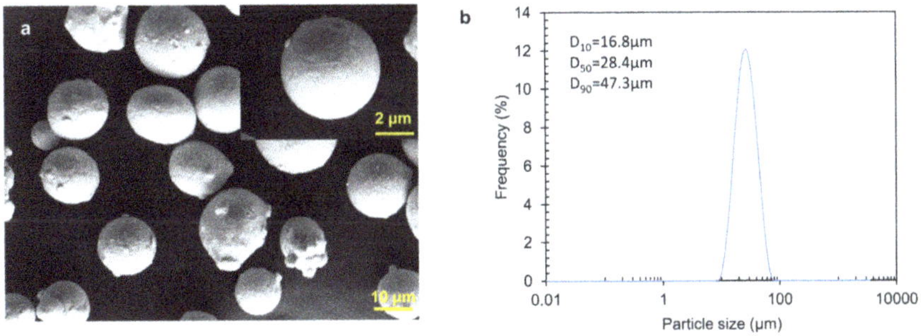

Figure 1. (**a**) SEM micrograph showing the micrograph of H13 powders; (**b**) distribution of particle sizes.

2.2. SLM Manufacturing Process

The SLM process was conducted using an FS271M selective laser melting system (Farsoon, Inc, Hunan, China). The samples were processed layer-by-layer on an H13 steel plate with 67° rotating scanning (Refer to [39,40]). The orientation changes for each layer and the angle difference between adjacent layers is 67°, as depicted in Figure 2a. The cubic samples (10 × 10 × 10 mm^3) and cuboid samples (80 × 10 × 10 mm^3) were fabricated based on the following processing parameters: hatch spacing of 0.1 mm, layer thickness of 0.03 mm, laser power of 170, 200, 230 and 260 W, scan speed of 600, 800 and 1400 mm/s. Figure 2b shows the samples of cuboid and cubes for microstructure and mechanical property testing. Additionally, the dog-bone-shaped sample for the test was with a spacing length of 30 mm and a cross-section of 4 × 2 mm^2 is shown in Figure 2c. Correspondingly, a representative schematic diagram showing the experimental procedures of H13 steel is illustrated, as shown in Figure 3.

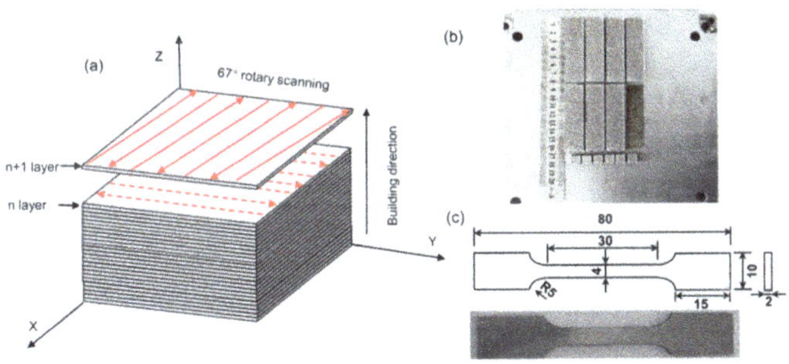

Figure 2. (**a**) Schematic diagram showing laser scanning approach during additive manufacturing; (**b**) Produced samples of cuboid and cubes; (**c**) Sample size for tensile test.

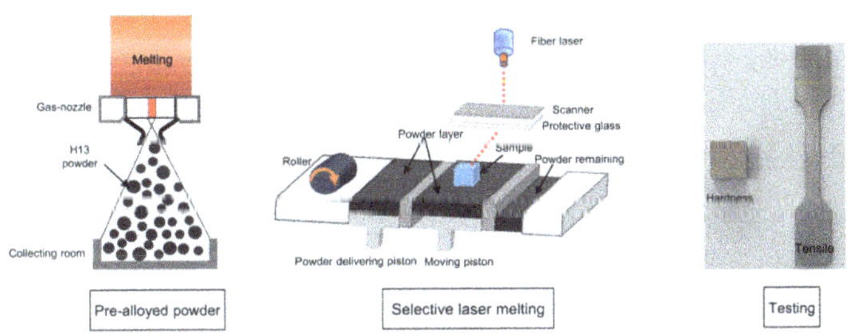

Figure 3. Schematic diagram showing the experimental procedures of H13 steel, including powder preparation, selective laser melting and the geometry of testing samples.

2.3. Microstructural Characterization

The densities of as-SLMed H13 samples were tested via Archimedes method [5]. All the featured data are based on the mean value of at least 5 measurements. The specimen density was calculated by the following equation:

$$\rho = \frac{\rho_0 m}{m - m_0} \quad (1)$$

where ρ is specimen density (g/cm^3), ρ_0 is distilled water density (g/cm^3), m is specimen weight in air and m_0 is specimen weight in distilled water. The relative density was obtained by comparing the specimen density with the theoretical density.

The volumetric energy density (*VED*) was computed via the Equation (2):

$$VED = \frac{P}{v \times h \times l} \quad (2)$$

where P is laser power (W), v is scanning speed (mm/s), h is hatch spacing (mm), and l is layer thickness (mm). Figure 4 shows the relative density of as-SLMed H13 specimens with *VED*.

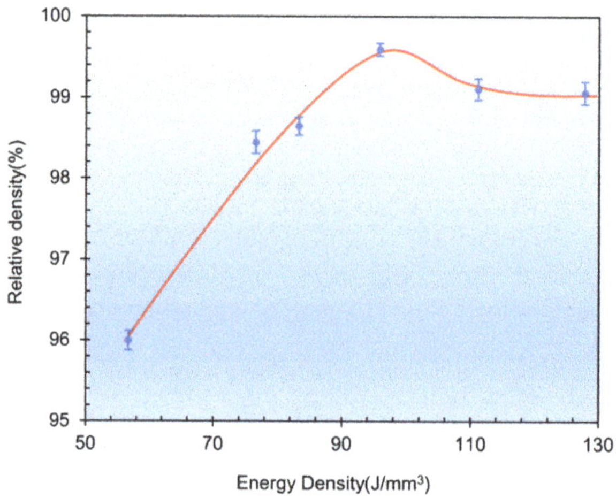

Figure 4. The relative density of as-SLMed H13 samples with different VEDs.

The phase constituent crystal structure of the as-SLMed H13 was examined in terms of X-ray diffraction with Cu Kα radiation (XRD, SmartLab 3Kw, Rigaku, Tokyo, Japan). Specimens for microstructural observations were etched with 4 vol% nitric acid solution. Microstructural features were characterized using a scanning electron microscopy (SEM, Quanta 250 FEG, FEI, Brno, Czech Republic). The grain size and grain orientation were detected by using electron backscattered diffraction (EBSD, Helios NanoLab G3 UC, FEI, Hillsboro, OR, USA) equipped with the TSL OIM data analysis. Furthermore, the transmission electron microscope (TEM; Tecnai G2F20, FEI, Hillsboro, OR, USA) was used for detailed microstructure examination. TEM specimens were fabricated using the precision ion polishing system (PIPS, Gatan691, Gatan, Pleasanton, CA, USA) at a voltage of 5 kV and an incident angle of 3–8°.

2.4. Evaluation of Mechanical Properties

Micro-hardness was measured using a micro-Vickers hardness instrument (HMV-2T, Shimadzu, Kyoto, Japan) with 300 g load for 15 s, and the average value was taken from at least 8 points of each sample. The dog-bone-shaped tensile specimens were cut by electrical discharge machining (EDM) from the as-SLMed specimens. Uniaxial tensile tests were evaluated via material testing system (Alliance RT30, MTS, Shanghai, China) with an engineering strain rate of 1×10^{-3} s^{-1} at room temperature. The tensile data were the mean of at least 5 measurements.

3. Results

3.1. Mechanical Properties

Figure 5 shows micro-hardness of the as-SLMed H13 samples. Obviously, the micro-hardness increases and then decreases with increasing VED. In addition, the micro-harnesses were lower in the building direction than those in the horizontal direction. The peak micro-hardness was obtained at the VED of 95.8 J/mm^3 and the corresponding micro-hardness was 537.5 Hv in the building and 560.9 Hv in the horizontal direction. According to the relative density and micro-hardness, the optimal process parameters were: P = 230 W, Vs = 800 mm/s and VED = 95.8 J/mm^3.

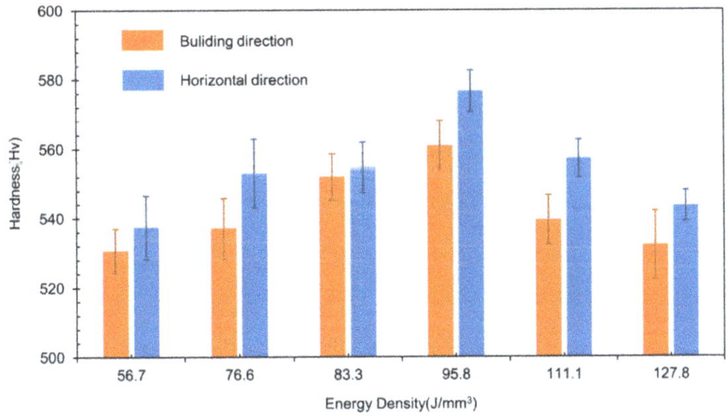

Figure 5. Micro-hardness of the H13 specimens with different VED.

Figure 6 shows the tensile stress-strain curves of the as-SLMed H13 samples. The mechanical properties are listed in Table 2. It is seen that an excellent combination with the YS of 1468 MPa, the UTS of 1837 MPa and the elongation of 8.5% were achieved at the VED of 95.8 J/mm^3. Figure 6b shows a comparison of tensile properties of H13 steel specimens fabricated by different techniques, including conventional casting methods [8,9,41], SLM [6,7,10,13], SLM + heat treatment [5,8]. Obviously, the method used in this work with optimised parameters could produce the optimal combination of strength and ductility for H13 steel.

Figure 7 shows the fractured morphologies of the as-SLMed H13 specimens processed at the VED of 95.8 J/mm^3. Little defects, such as unmelted powder and microvoids in the matrix at low magnification, were observed. A large number of dimples were observed in Figure 7b, which confirmed the existence of brittle and ductile fracture mixture.

Table 2. Mechanical properties of the SLMed H13 samples under different conditions.

VED	YS (MPa)	UTS (MPa)	Elongation (%)
83.3 J/mm^3	1456 ± 32	1828 ± 48	6.6 ± 0.3
95.8 J/mm^3	1468 ± 27	1837 ± 23	8.5 ± 0.6
111.1 J/mm^3	1408 ± 29	1817 ± 31	7.0 ± 0.4

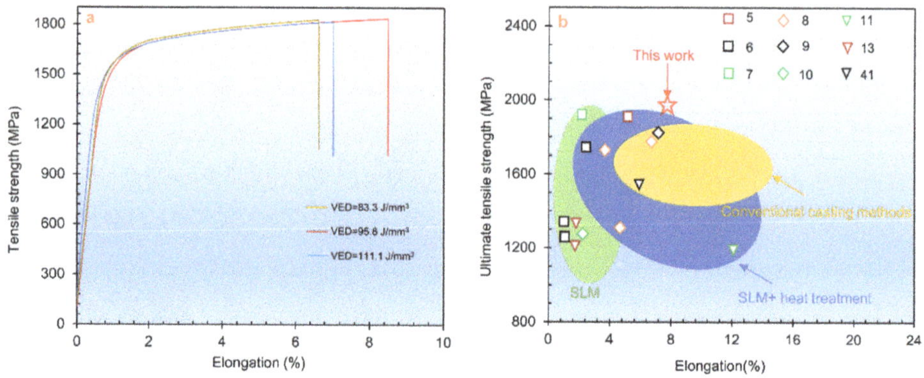

Figure 6. (**a**) Tensile stress–strain curves of SLMed H13; (**b**) Comparison of tensile strength of the H13 specimens manufactured by different techniques, including conventional casting methods, SLM, SLM + heat treatment.

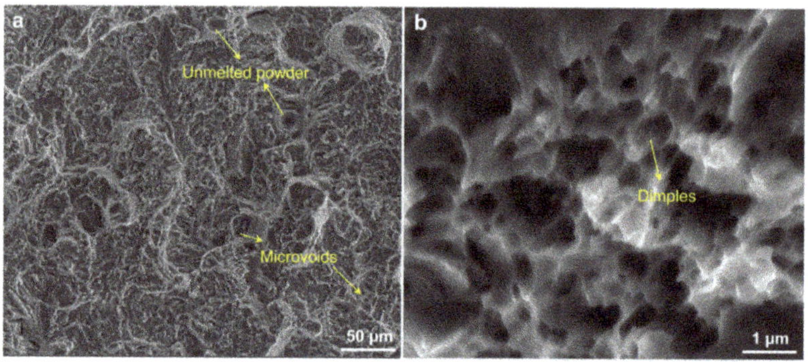

Figure 7. (**a**) SEM micrographs showing the unmelted and microvoids on fracture of tensile specimens possessed at the VED of 95.8 J/mm^3; (**b**) Detailed fracture morphology showing fine dimples.

3.2. Microstructural Characterization

Figure 8 shows the XRD spectra obtained from H13 powder and as-SLMed H13 specimens at the VED of 95.8 J/mm^3. The XRD patterns only presented the peaks of bcc-structured martensite phase (α-Fe) in the H13 steel powder. However, the peaks of bcc-structured martensite (α-Fe) and fcc-structured retained austenite phase (γ-Fe) were detected in the as-SLMed H13 specimens. The presence of residual austenite can be attributed to the fact that SLM process offers a rapid solidification, resulting in the incomplete transformation from austenite to martensite [42,43].

Figure 9 shows the optical micrographs (OM) along horizontal direction and building direction with isometric view. In the horizontal direction, continuous laser tracks were formed with 67° rotating scanning (Figure 9a,b). The melt poor boundaries were clearly observed along the build direction (Figure 9c,d). The defects were likely formed by the entrapment of inert gas or the evaporation of alloy elements. These micropores decreased the strength and elongation because of the easy crack initiation in the pore edge under loading [21]. To further characterize the features of as-SLMed H13, Figure 10 shows the detailed microstructure. The typical microstructure consisted of melt pool (MP) coarse, MP fine and heat affected zone (HAZ). The HAZ was considered as a transition zone between MP coarse and MP fine zones. The columnar grain and equiaxed grain are shown in Figure 10b,c. The equiaxed grain area was observed in the cellular structure. Lath martensite are shown in Figure 10d. The size of cellular structure was about 500 nm.

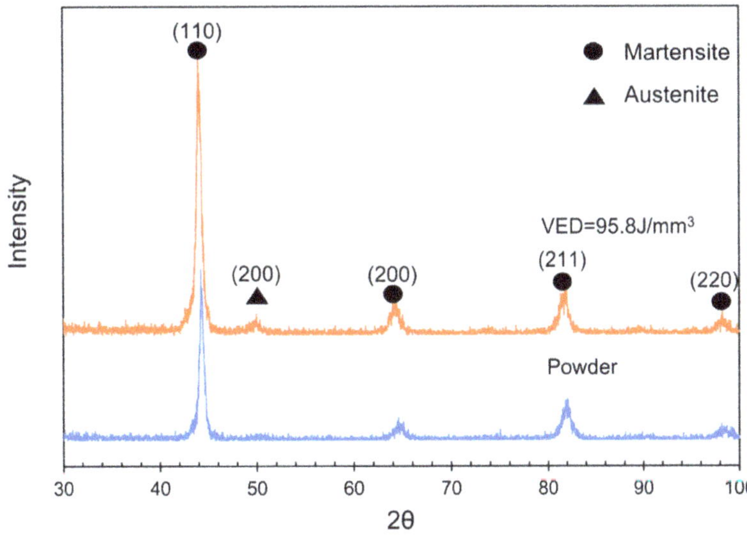

Figure 8. XRD spectra of H13 powder and as-SLMed H13 specimens at the VED of 95.8 J/mm^3.

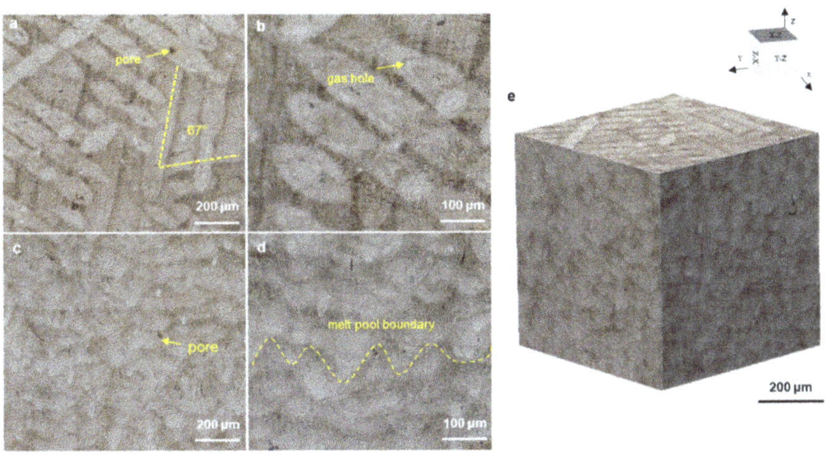

Figure 9. Optical micrographs showing the microstructure along (**a**,**b**) horizontal direction and (**c**,**d**) building direction; and (**e**) isometric view.

Figure 11 shows the inverse pole figure (IPF) maps of the as-SLMed H13 specimens. In the horizontal direction (Figure 11a) and building direction (Figure 11b), the IPF maps showed that the microstructure of the as-SLMed H13 specimens was dominated by fine grains with random distribution. In Figure 11c, the martensite content and residual austenite content were 94% and 6% along horizontal direction, respectively. However, it is noted that the residual austenite content of specimens along the building direction (8.7%) was higher than that specimens the along the horizontal direction (6%). The difference in content of residual austenite could be attributed to the different cooling rates during the SLM process [44]. In Figure 11e,g, it is also seen that the mean grain size of martensite and retained austenite were 1.67 μm and 1.15 μm along the horizontal direction, respectively. Meanwhile, the mean grain size of martensite and retained austenite in Figure 10f,h were 1.97 μm and 0.86 μm along the building direction, respectively. The variation is induced by the different cooling rates during processing and different phase contents formed in the alloy. Figure 12 presents the Kernel Average Misorientation (KAM) maps of the as-SLMed H13 specimens and misorientation angle with low-angle grain boundaries (LAGBs) and high-angle grain boundaries (HAGBs). The average KAM value of the building direction and the horizontal direction of the as-SLMed H13 specimens were 0.863 and 0.971, respectively. The variations in average KAM qualitatively reflected the degree of plastic deformation or defect density. According to Equation (1), the KAM was applied to calculate geometrically necessary dislocation (GND):

$$\rho_{GND} = 2\vartheta/(\mu b) \tag{3}$$

where ϑ is the average value of KAM, μ is the step size of EBSD measurement (0.2 μm), and b is the Burger vector (0.25 nm [5]). Correspondingly, ρ_{GND} of as-SLMed H13 steel along the horizontal direction and building direction were estimated to be ~3.45 × 10^{16} m^{-2} and ~3.88 × 10^{16} m^{-2}, respectively. The high density of GNDs would build a solid foundation for high strength. The distributions of misorientation angles map in Figure 12c indicates that the factions of HAGB (≥15°) and LAGB (<15°) were 74.5% and 25.5% along the building direction, respectively.

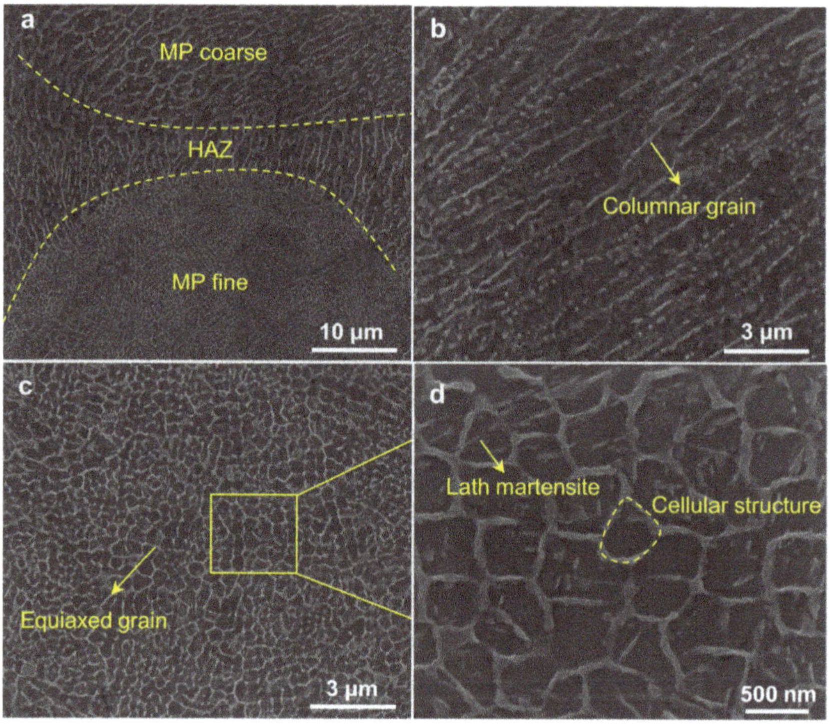

Figure 10. SEM images displaying the microstructure features of as-SLMed H13 specimens, (**a**) molten pool structure; (**b**) columnar grains; (**c**) equiaxed grains; (**d**) cellular structures and lath martensite.

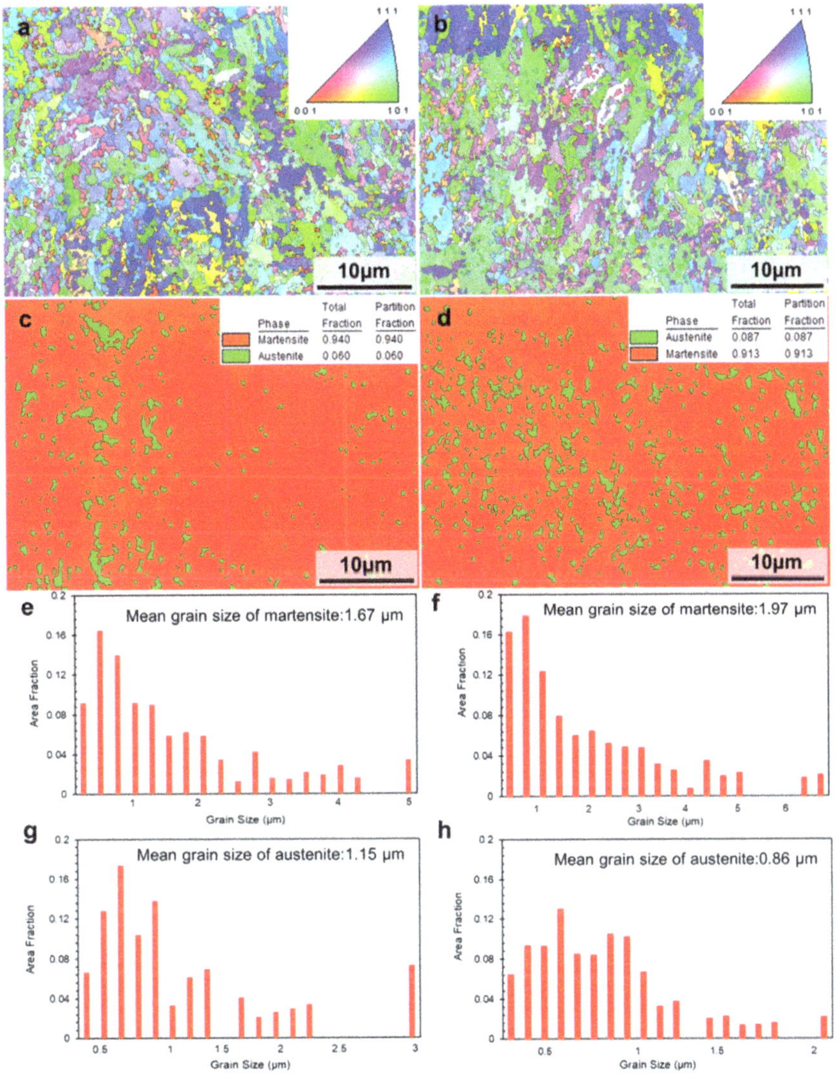

Figure 11. Inverse-pole figure (IPF) maps of the as-SLMed H13 specimens showing the grain orientations, phase maps and grain size distribution maps: (**a**,**c**,**e**,**g**) horizontal direction; (**b**,**d**,**f**,**h**) building direction.

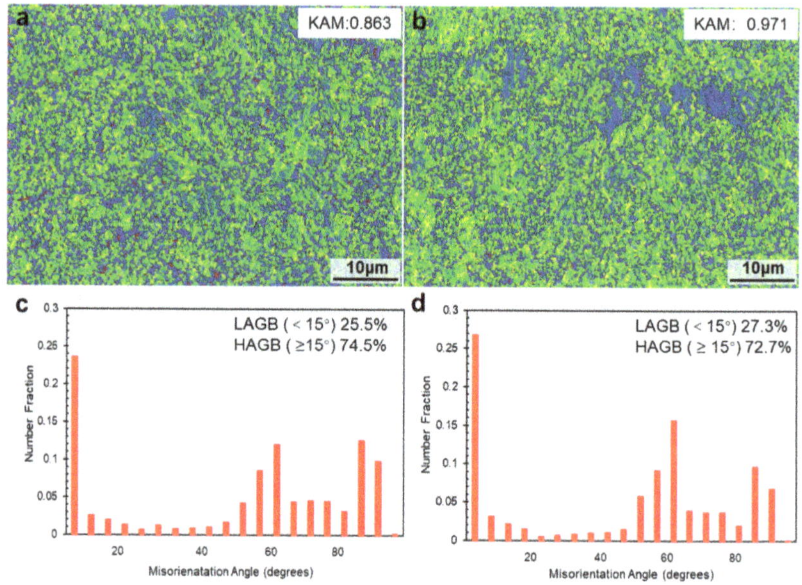

Figure 12. Kernel average misorientation (KAM) maps of the as-SLMed H13 specimens and misorientation angle with low-angle grain boundaries (LAGBs) and high-angle grain boundaries (HAGBs): (**a**,**c**) horizontal direction; (**b**,**d**) building direction.

4. Discussions

4.1. Relationship in between Microstructure and Mechanical Properties

In general, the SLMed H13 is typically featured by martensite as matrix with a small amount of retained austenite, similar to the conventional casting. Correspondingly, the bcc-structured martensite and the fcc-structured residual austenite in the $[\bar{1}11]$ direction are presented in Figure 13a,b. Due to the different cooling rates in different regions of the specimens and the cyclic thermal effect of the continuous layers, high volume fraction of subcooled austenite distributed in the melt pools can be rapidly cooled below the martensite transformation temperature (M_f) before martensite transformation, and therefore the transformation of some martensite to retained austenite is understandable. Although the generation of retained austenite reduce the strength, the retained austenite induces martensitic transformation during the stretching process, which absorbs stress and reduces the sprouting of cracks. As a result, the tensile strength and toughness are improved [45,46].

Compared to H13 alloys fabricated by other methods such as casting, the corresponding strengthening contributions of H13 alloys mainly depend on microstructure refinement, cellular structure, etc. In Figure 11e, the as-SLMed H13 specimens mainly contain fine grains with a mean grain size of 1.67 μm. Previous studies have shown that the average grain sizes of as-cast H13 steel, as-spray formed H13 steel and as-SLMed H13 steel are in the range of 100~150 μm, 10~20 μm and 2~3 μm [5], respectively. Thus, the contribution of grain boundary strengthening is higher than that of the H13 alloys fabricated by casting and spray forming. Most importantly, the formation of the unique cellular structures in the SLMed H13 alloys possesses outstanding strength. In Figures 10d and 13c, the size of these cellular structures in H13 steel is less than 500 nm. The size and volume fraction of cellular structures varied from sample to sample, in consistent with previous studies [47]. Meanwhile, the solidification conditions are closely related to the formation of cellular structures [48]. Based on the constitutional supercooling theory, the morphology features depends on the thermal gradients (G) and the grain growth rates (R). Specifically, with a decreasing of the ratio of G/R, the morphology transforms from isometric to cellular, columnar dendritic or equiaxed dendritic. Therefore, the solidification condition prefers to promote the formation of cellular structures during SLM processing. The ability of cellular structures to improve the strength of as-SLMed alloys has been demonstrated. Wang et al. [49] showed that the cellular structure was the main contributor to the increment of YS of as-SLMed 316L steel. Particularly, the average diameter of cellular structure has been used to measure the strength [34]. To further evaluate the element distribution of the cellular structure, the detail high-angle annular dark field scanning TEM (HAADF-STEM) images are shown in Figure 13d,g. However, due to the strong magnetic properties of the steel sample, the mapping results are not sufficiently clear. It is easy to distinguish that the existence of compositional enrichment of Cr and Mo along the cellular structure, which is consistent with previous investigations [50,51]. Furthermore, in combination with Figure 13e, it can be seen that the part of carbides pinned on the sub-grain boundary, hindering the grain boundary and dislocation movement, and the other part of carbides are uniformly distributed in the sub-grain boundaries as second phase particles. Finally, dislocations can be retained at grain boundaries, as shown in Figure 13f.

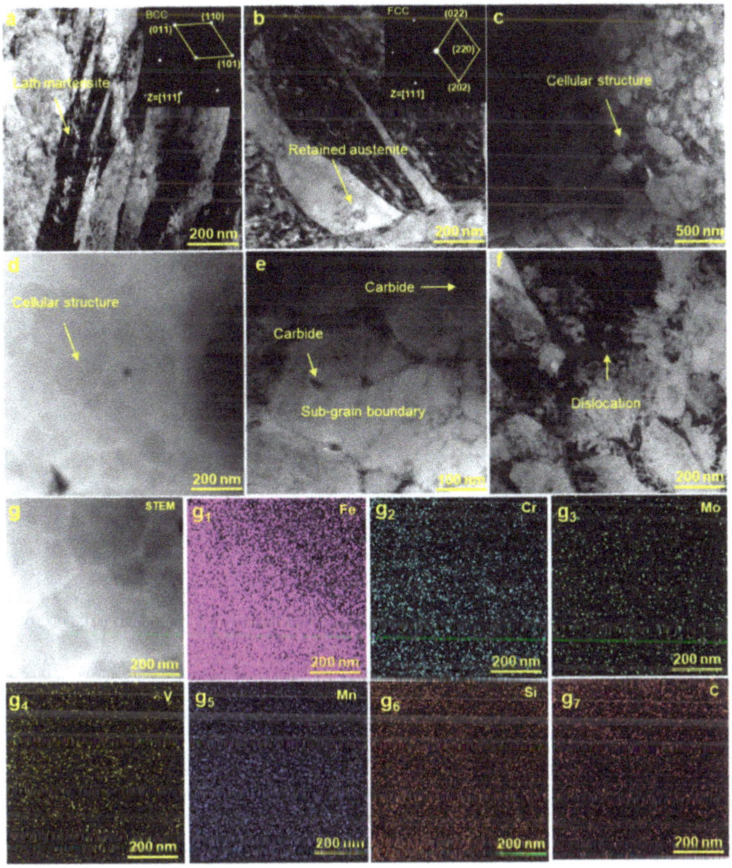

Figure 13. TEM micrographs of the as-SLMed H13 steel specimens, showing (**a**) bcc-structured lath martensite; (**b**) the formation of retained austenite with high cooling rate; (**c**) cellular structure; (**d**) STEM image of the cellular structure shown in (**c**); (**e**) carbides at and within sub-grain boundaries; (**f**) the density of dislocations; (**g**, $\mathbf{g_1}$–$\mathbf{g_7}$) the corresponding EDS maps of cellular structure, including elements of ($\mathbf{g_1}$) Fe, ($\mathbf{g_2}$) Cr, ($\mathbf{g_3}$) Mo, ($\mathbf{g_4}$) V, ($\mathbf{g_5}$) Mn, ($\mathbf{g_6}$) Si and ($\mathbf{g_7}$) C.

4.2. The Strengthening Mechanisms

The results showed that the as-SLMed H13 steel processed at the VED of 95.8 J/mm^3 exhibits the YS of 1468 MPa, UTS of 1837 MPa and fracture strain of 8.48%. Previous studies have shown that the cellular structure strengthening (σ_c), fine grain strengthening (σ_g) and dislocation strengthening (σ_{dis}) are the main contributing factors of the yield strength of the as-SLMed steel [5,49,52]. In this study, the cellular structure, the fine grain size (~1.67 μm) and the dislocations induced by the thermal contraction stress during the rapid solidification of SLM process are also the main contributors of high yield strength. Additionally, the precipitation strengthening induced by carbides could be negligible due to the very low volume fraction. Thus, the contribution of the main strengthening contributions of as-SLMed H13 can be calculated as

$$\sigma_y = \sigma_0 + \sigma_c + \sigma_g + \sigma_{dis} \tag{4}$$

where σ_0 is a constant (70 MPa for H13 alloy [53]). The cellular structure followed a Hall-Petch type of strengthening behaviour, where the yield strength contributed by the cellular structure (σ_c) with the average cell diameter (L_c).

$$\sigma c = 183.31 + 253.66/\sqrt{Lc} \tag{5}$$

The corresponding Lc was estimated as ~500 nm, as shown in Figures 10d and 13c. The strength contribution of the cellular structure to yield strength is estimated as 542 MPa. The strengthening contribution of fine grain strengthening (σg) can be estimated by the Hall-Petch equation:

$$\sigma_g = Kd^{-0.5} \tag{6}$$

where K is the Hall-Petch constant, and d is the average grain size. The k (572 MPa μm$^{1/2}$) is taken from literature [37,54]. From Figure 10e, the average grain size of as-SLMed H13 is 1.67 μm. Thus, the contribution of grain refinement to yield strength is calculated as 442.6 MPa. As shown in Figure 13e,f, the interaction between dislocation and wall of cellular structure impedes grain boundary motion and thus enhances strengthening. The contribution of σ_{dis} can be evaluated via Taylor's hardening law [55]:

$$\sigma_{dis} = M{\cdot}\alpha{\cdot}G{\cdot}b{\cdot}\rho^{0.5} \tag{7}$$

where M is the Taylor factor (3 for BCC crystal structure of iron [55]), α is a material constant (0.33 [55]), G is the shear modulus (80 GPa [55]), b is the Burgers vector (0.25 nm [55]), and ρ is the dislocation density (taken to the horizontal direction). The strength contribution of dislocation density to yield strength is estimated as 367.8 MPa. The estimated yield strength is ~1422.4 MPa ($\sigma_0 \approx 70$ MPa; $\sigma_c \approx 542$ MPa; $\sigma_g \approx 442.6$ MPa; $\sigma_{dis} \approx 367.8$ MPa), it is consistent well with the experimentally data of 1468 MPa. Furthermore, the high yield strength of H13 steel should also originate from other factors. For example, a small number of carbides distributed in and out of sub-grain boundaries were found for the as-SLMed H13 steel. Verifiably, the yield strength of the as-SLMed H13 specimens is mainly contributed by the strengthening from cellular structure and grain-refinement, followed by dislocation strengthening.

5. Conclusions

In this study, microstructure and mechanical properties of the as-SLMed H13 have been investigated. Our major conclusions are as follows:

(1) H13 steel could be fabricated successfully by SLM in this study. The maximum relative density reached 99.6% under the optimized process parameters, including P = 230 W, Vs = 800 mm/s and VED = 95.8 J/mm^3.

(2) The as-SLMed microstructure consists of lath martensite and 6~8.7% retained austenite. The as-SLMed structure with a typical cellular size of ~500 nm is predominantly

cellular morphology and Cr and Mo element were enriched toward the cell wall of each cellular structure.

(3) The as-SLMed H13 can offer a remarkable synergistic improvement in strengthen and ductility, in which the yield strength is 1468 MPa, the UTS is 1837 MPa and the fracture strain is 8.48%. Numerous dimples are found in the fractured surface, indicating the existence of ductile fractures.

(4) The excellent strength-ductility synergy can be attributed to the refined hierarchical microstructures with fine grains at an average size of 1.67 μm, cellular structures and dislocations. The calculated contribution to the yield is 542 MPa for the cellular structure strengthening, 442.6 MPa for the grain-refinement strengthening, and 367.8 MPa of dislocation strengthening.

Author Contributions: Data curation, Writing—original draft, F.L. and T.W.; Investigation, F.Y., J.W. (Jianying Wang) and J.F.; Conceptualization, Writing—review & editing, Supervision, J.W. (Jiong Wang); Conceptualization, Writing—review & editing, Supervision, Funding acquisition, Project administration, H.Y.; Conceptualization, Results review, Writing—review, J.R. and S.J. All authors have read and agreed to the published version of the manuscript.

Funding: This research was funded by National Key Research and Development Program of China (Grant No. 2020YFB0311300ZL), and National Natural Science Foundation of China (Grant No. 52071343).

Institutional Review Board Statement: Not applicable.

Informed Consent Statement: Not applicable.

Data Availability Statement: The data presented in this study are available on request from the corresponding author.

Conflicts of Interest: The authors declare no conflict of interest.

References

1. Herzog, D.; Seyda, V.; Wycisk, E.; Emmelmann, C. Additive manufacturing of metals. *Acta Mater.* **2016**, *117*, 371–392. [CrossRef]
2. Srinivas, M.; Babu, B.S. A critical review on recent research methodologies in additive manufacturing. *Mater. Today Proc.* **2017**, *4*, 9049–9059. [CrossRef]
3. Fette, M.; Sander, P.; Wulfsberg, J.; Zierk, H.; Herrmann, A.; Stoess, N. Optimized and cost-efficient compression molds manufactured by selective laser melting for the production of thermoset fiber reinforced plastic aircraft components. *Procedia CIRP* **2015**, *35*, 25–30. [CrossRef]
4. Lewandowski, J.; Seifi, M. Metal additive manufacturing: A review of mechanical properties (postprint). *Annu. Rev. Mater. Res.* **2016**, *46*, 151–186. [CrossRef]
5. Yan, J.J.; Song, H.; Dong, Y.P.; Quach, W.M.; Yan, M. High strength (~2000 MPa) or highly ductile (~11%) additively manufactured H13 by tempering at different conditions. *Mater. Sci. Eng. A* **2020**, *773*, 138845. [CrossRef]
6. Åsberg, M.; Fredriksson, G.; Hatami, S.; Fredriksson, W.; Krakhmalev, P. Influence of post treatment on microstructure, porosity and mechanical properties of additive manufactured H13 tool steel. *Mater. Sci. Eng. A* **2019**, *742*, 584–589. [CrossRef]
7. Mazur, M.; Brincat, P.; Leary, M.; Brandt, M. Numerical and experimental evaluation of a conformally cooled H13 steel injection mould manufactured with selective laser melting. *Int. J. Adv. Manuf. Technol.* **2017**, *93*, 881–900. [CrossRef]
8. Lee, J.; Choe, J.; Park, J.; Yu, J.H.; Kim, S.; Jung, I.D.; Sung, H. Microstructural effects on the tensile and fracture behavior of selective laser melted H13 tool steel under varying conditions. *Mater. Charact.* **2019**, *155*, 109817. [CrossRef]
9. Holzweissig, M.J.; Taube, A.; Brenne, F.; Schaper, M.; Niendorf, T. Microstructural characterization and me performance of hot work tool steel processed by selective laser melting. *Metall. Mater. Trans. B* **2015**, *46*, 545–549. [CrossRef]
10. Safka, J.; Ackermann, M.; Voleský, L. Structural properties of H13 tool steel parts produced with use of selective laser melting technology. *J. Phys. Conf. Ser.* **2016**, *709*, 012004. [CrossRef]
11. Wang, M.; Li, W.; Wu, Y.; Li, S.; Cai, C.; Wen, S.F.; Wei, Q.S.; Shi, Y.S.; Ye, F.Y., Chen, Z.P. High temperature properties and microstructural stability of the AISI H13 hot-work tool steel processed by selective laser melting. *Metall. Mater. Trans. B* **2019**, *50*, 531–542. [CrossRef]
12. Xue, L.; Chen, J.; Wang, S.H. Freeform laser consolidated H13 and CPM 9V tool steels. *Microsc. Microanal.* **2013**, *2*, 67–78. [CrossRef]
13. Ning, A.G.; Mao, W.W.; Chen, X.C.; Guo, H.J.; Guo, J. Precipitation behavior of carbides in H13 hot work die steel and its strengthening during tempering. *Metals* **2017**, *7*, 70. [CrossRef]

14. Lima, E.P.R.; Maurício, D.M.; Neves, D.; Nogueira, R.A.; Oliveira, L.G.C.; Filho, F.A. Effect of different tempering stages and temperatures on microstructure, tenacity and hardness of vacuum sintered HSS AISI T15. *Mater. Sci. Forum* **2008**, *591–593*, 68–73. [CrossRef]
15. Wang, M.; Wu, Y.; Wei, Q.; Shi, Y. Thermal fatigue properties of H13 Hot-work tool steels processed by selective laser melting. *Metals* **2020**, *10*, 116. [CrossRef]
16. Narvan, M.; Al-Rubaie, K.S.; Elbestawi, M. Process-structure-property relationships of AISI H13 tool steel processed with selective laser melting. *Materials* **2019**, *12*, 2284. [CrossRef]
17. Yan, J.J.; Zheng, D.L.; Li, H.X.; Jia, X.; Yan, M. Selective laser melting of H13: Microstructure and residual stress. *J. Mater. Sci.* **2017**, *52*, 12476–12485. [CrossRef]
18. Maziasz, P.J.; Payzant, E.A.; Schlienger, M.E.; Mchugh, K.M. Residual stresses and microstructure of H13 steel formed by combining two different direct fabrication methods. *Scr. Mater.* **1998**, *39*, 1471–1476. [CrossRef]
19. Cottam, R.; Wang, J.; Luzin, V. Characterization of microstructure and residual stress in a 3D H13 tool steel component produced by additive manufacturing. *J. Mater. Res.* **2014**, *29*, 1978–1986. [CrossRef]
20. Nguyen, V.L.; Kim, E.A.; Yun, J.; Choe, J.; Yang, D.Y.; Lee, H.S.; Lee, C.W.; Yu, J.H. Nano-mechanical behavior of H13 tool steel fabricated by a selective laser melting method. *Metall. Mater. Trans. A* **2019**, *50*, 523–528. [CrossRef]
21. Debroy, T.; Wei, H.L.; Zuback, J.; Mukherjee, T.; Zhang, W. Additive manufacturing of metallic components–process, structure and properties. *Prog. Mater. Sci.* **2018**, *92*, 112–224. [CrossRef]
22. Yadroitsev, I.; Gusarov, A.; Yadroitsav, I.; Smurov, I. Single track formation in selective laser melting of metal powders. *J. Mater. Process. Technol.* **2010**, *210*, 1624–1631. [CrossRef]
23. Wang, Z.; Palmer, T.A.; Beese, A.M. Effect of processing parameters on microstructure and tensile properties of austenitic stainless steel 304L made by directed energy deposition additive manufacturing. *Acta Mater.* **2016**, *110*, 226–235. [CrossRef]
24. Matilainen, V.P.; Piili, H.; Salminen, A.; Nyrhiläc, O. Preliminary investigation of keyhole phenomena during single layer fabrication in laser additive manufacturing of stainless steel. *Phys. Procedia* **2015**, *78*, 377–387. [CrossRef]
25. Król, M.; Kujawa, M.; Dobrzański, L.A.; Tański, T. Influence of technological parameters on additive manufacturing steel parts in Selective Laser Sintering. *Arch. Mater. Sci. Eng. A* **2014**, *67*, 84–92.
26. Merten, R.; Vrancken, B.; Holmstock, N.; Kinds, Y.; Kruth, J.P.; Humbeeck, J.V. Influence of powder bed preheating on microstructure and mechanical properties of H13 tool steel SLM parts. *Phys. Procedia* **2016**, *83*, 882–890. [CrossRef]
27. Buchbinder, D.; Meiners, W.; Pirch, N.; Wissenbach, K.; Schrage, J. Investigation on reducing distortion by preheating during manufacture of aluminum components using selective laser melting. *J. Laser Appl.* **2014**, *26*, 012004. [CrossRef]
28. Deirmina, F.; Peghini, N.; Almangour, B.; Grzesiak, D.; Pellizzari, M. Heat treatment and properties of a hot work tool steel fabricated by additive manufacturing. *Mater. Sci. Eng. A* **2019**, *753*, 109–121. [CrossRef]
29. Chen, C.J.; Yan, K.; Qin, L.L.; Zhang, M.; Wang, X.N.; Zou, T.; Hu, Z.R. Effect of heat treatment on microstructure and mechanical properties of laser additively manufactured AISI H13 tool steel. *J. Mater. Eng. Perform.* **2017**, *26*, 5577–5589. [CrossRef]
30. Cheng, L.; Zhao, Z.; Northwood, D.O.; Liu, Y. A new empirical formula for the calculation of Ms temperatures in pure iron and super-low carbon alloy steels. *J. Mater. Process. Technol.* **2001**, *113*, 556–562.
31. Zhong, Y.; Liu, L.F.; Wikman, S.; Cui, D.Q.; Shen, Z.J. Intragranular cellular segregation network structure strengthening 316L stainless steel prepared by selective laser melting. *J. Nucl. Mater.* **2016**, *470*, 170–178. [CrossRef]
32. Zhu, Z.G.; Nguyen, Q.B.; Ng, F.L.; An, X.H.; Liao, X.Z.; Liaw, P.K.; Nai, S.M.L.; Wei, J. Hierarchical microstructure and strengthening mechanisms of a CoCrFeNiMn high entropy alloy additively manufactured by selective laser melting. *Scr. Mater.* **2018**, *154*, 20–24. [CrossRef]
33. Wang, D.; Song, C.H.; Yang, Y.Q.; Bai, Y.C. Investigation of crystal growth mechanism during selective laser melting and mechanical property characterization of 316L stainless steel parts. *Mater. Des.* **2016**, *100*, 291–299. [CrossRef]
34. Saeidi, K.; Gao, X.; Zhong, Y.; Shen, Z.J. Hardened austenite steel with columnar sub-grain structure formed by laser melting. *Mater. Sci. Eng. A* **2015**, *625*, 221–229. [CrossRef]
35. Murr, L.E.; Gaytan, S.M.; Ramirez, D.A.; Ramirez, A.; Martinez, E.; Hernandez, J.; Amato, K.N.; Shindo, P.W.; Medina, F.R.; Wicker, R.B. Metal fabrication by additive manufacturing using laser and electron beam melting technologies. *J. Mater. Sci. Technol.* **2012**, *28*, 1–14. [CrossRef]
36. Saeidi, K.; Kvetkova, L.; Lofaj, F.; Shen, Z. Austenitic stainless steel strengthened by the in situ formation of oxide nanoinclusions. *Rsc. Adv.* **2015**, *5*, 20747–20750. [CrossRef]
37. Mao, W.W.; Ning, A.G.; Guo, H.J. Nanoscale precipitates and comprehensive strengthening mechanism in AISI H13 steel. *Int. J. Min. Metall. Mater.* **2016**, *23*, 1056–1064. [CrossRef]
38. Ma, M.; Wang, Z.; Zeng, X.A. A comparison on metallurgical behaviors of 316L stainless steel by selective laser melting and laser cladding deposition. *Mater. Sci. Eng. A* **2017**, *685*, 265–273. [CrossRef]
39. Xu, J.Y.; Ding, Y.T.; Gao, Y.B.; Wang, H.; Hu, Y.; Zhang, D. Grain refinement and crack inhibition of hard-to-weld Inconel 738 alloy by altering the scanning strategy during selective laser melting. *Mater. Des.* **2021**, *209*, 109940. [CrossRef]
40. Zhang, C.C.; Feng, K.; Kokawa, H.; Han, B.; Li, Z.G. Cracking mechanism and mechanical properties of selective laser melted CoCrFeMnNi high entropy alloy using different scanning strategies. *Mater. Sci. Eng. A* **2020**, *789*, 139672. [CrossRef]
41. Li, F.Y.; Ma, D.S.; Chen, Z.Z.; Liu, J.H.; Youg, Q.L.; Kang, Z. Structure and properties of high temperature diffused-superfining treated die Steel H13. *Spec. Steel* **2008**, *9*, 63–65.

42. Krell, J.; Röttger, A.; Geenen, K.; Theisen, W. General investiagtions on processing tool steel X40CrMoV5-1 with selective laser melting. *J. Mater. Process. Technol.* **2018**, *255*, 679–688. [CrossRef]
43. Katancik, M.; Mirzabaei, S.; Ghayoor, M.; Pasebani, S. Selective laser melting and tempering of H13 tool steel for rapid tooling applications. *J. Alloy. Compd.* **2018**, *849*, 156319. [CrossRef]
44. Yadollahi, A.; Shamsaei, N.; Thompson, S.M.; Elwany, A.; Bian, L. Effects of building orientation and heat treatment on fatigue behavior of selective laser melted 17-4 PH stainless steel. *Int. J. Fatigue* **2017**, *94*, 218–235. [CrossRef]
45. Zhu, J.; Zhang, Z.; Xie, J. Improving strength and ductility of H13 die steel by pre-tempering treatment and its mechanism. *Mater. Sci. Eng. A* **2019**, *752*, 101–114. [CrossRef]
46. Zhang, X.; Li, H.; Duan, S.; Yu, X.; Feng, J.; Wang, B.; Huang, Z. Modeling of residual stress field induced in Ti–6Al–4V alloy plate by two sided laser shock processing. *Surf. Coat. Technol.* **2015**, *280*, 163–173. [CrossRef]
47. Trelewicz, J.R.; Halada, G.P.; Donaldson, O.K.; Manogharan, G. Microstructure and corrosion resistance of laser additively manufactured 316L stainless steel. *JOM* **2016**, *68*, 850–859. [CrossRef]
48. Prashanth, K.G.; Eckert, J. Formation of metastable cellular microstructures in selective laser melted alloys. *J. Alloy. Compd.* **2016**, *707*, 27–34. [CrossRef]
49. Wang, Y.M.; Voisin, T.; McKeown, J.T.; Ye, J.C.; Calta, N.P.; Li, Z.; Zeng, Z.; Zhang, Y.; Chen, W.; Roehling, T.T.; et al. Additively manufactured hierarchical stainless steels with high strength and ductility. *Nat. Mater.* **2018**, *17*, 63–70. [CrossRef]
50. Zhang, J.; Yua, M.J.; Lia, Z.Y.; Liu, Y.; Zhang, Q.M.; Jiang, R.; Sun, S.F. The effect of laser energy density on the microstructure, residual stress and phase composition of H13 steel treated by laser surface melting. *J. Alloy. Compd.* **2021**, *856*, 158168. [CrossRef]
51. Fonseca, E.B.; Gabriel, A.H.G.; Araújo, L.C.; Santos, P.L.L.; Lopes, E.S.N. Assessment of laser power and scan speed influence on microstructural features and consolidation of AISI H13 tool steel processed by additive manufacturing. *Addit. Manuf.* **2020**, *34*, 101250. [CrossRef]
52. Bertsch, K.M.; Bellefon, G.M.D.; Kuehl, B.; Thoma, D.J. Origin of dislocation structures in an additively manufactured austenitic stainless steel 316L. *Acta Mater.* **2020**, *199*, 19–33. [CrossRef]
53. Qiang, L. Modeling the microstructure–mechanical property relationship for a 12Cr–2W–V–Mo–Ni power plant steel. *Mater. Sci. Eng. A* **2003**, *361*, 385–391.
54. Majta, J.; Lenard, J.G.; Pietrzyk, M. A study of the effect of the thermomechanical history on the mechanical properties of a high niobium steel. *Mater. Sci. Eng. A* **1996**, *208*, 249–259. [CrossRef]
55. Mazaheri, Y.; Kermanpur, A.; Najafizadeh, A. Strengthening mechanisms of ultrafine grained dual phase steels developed by new thermomechanical processing. *ISIJ Int.* **2015**, *55*, 218–226. [CrossRef]

materials

Article

An Effective Strategy to Maintain the CALPHAD Atomic Mobility Database of Multicomponent Systems and Its Application to Hcp Mg–Al–Zn–Sn Alloys

Ting Cheng, Jing Zhong and Lijun Zhang *

State Key Laboratory of Powder Metallurgy, Central South University, Changsha 410083, China; chengting@csu.edu.cn (T.C.); zhongjingjogy@csu.edu.cn (J.Z.)
* Correspondence: lijun.zhang@csu.edu.cn

Abstract: In this paper, a general and effective strategy was first developed to maintain the CALPHAD atomic mobility database of multicomponent systems, based on the pragmatic numerical method and freely accessible HitDIC software, and then applied to update the atomic mobility descriptions of the hcp Mg–Al–Zn, Mg–Al–Sn, and Mg–Al–Zn–Sn systems. A set of the self-consistent atomic mobility database of the hcp Mg–Al–Zn–Sn system was established following the new strategy presented. A comprehensive comparison between the model-predicted composition–distance profiles/inter-diffusivities in the hcp Mg–Al–Zn, Mg–Al–Sn, and Mg–Al–Zn–Sn systems from the presently updated atomic mobilities and those from the previous ones that used the traditional method indicated that significant improvement can be achieved utilizing the new strategy, especially in the cases with sufficient experimental composition–distance profiles and/or in higher-order systems. Furthermore, it is anticipated that the proposed strategy can serve as a standard for maintaining the CALPHAD atomic mobility database in different multicomponent systems.

Keywords: atomic mobility; CALPHAD; diffusion couple; HitDIC; Hcp Mg–Al–Zn–Sn alloys

Citation: Cheng, T.; Zhong, J.; Zhang, L. An Effective Strategy to Maintain the CALPHAD Atomic Mobility Database of Multicomponent Systems and Its Application to Hcp Mg–Al–Zn–Sn Alloys. *Materials* **2022**, *15*, 283. https://doi.org/10.3390/ma15010283

Academic Editor: Michele Bacciocchi

Received: 25 October 2021
Accepted: 24 December 2021
Published: 31 December 2021

1. Introduction

As is well known, the mechanical properties of metallic materials, such as strength, ductility, and hardness, are closely related to their microstructural formation during various preparation processes such as solidification, solid solution, and aging [1,2]. To achieve a comprehensive understanding of different preparation processes, accurate diffusion coefficients of composition and temperature dependence should be the prerequisite. For typical multicomponent technical alloys, direct experimental measurement of the complex diffusion coefficient matrices seems to be very difficult [3]. One alternative substitution in the CALPHAD (CALculation of PHAse Diagram) community is to predict a variety of composition- and temperature-dependent diffusion coefficients from the established atomic mobility database of the target alloys together with the corresponding thermodynamic database [4].

In terms of the CALPHAD framework, the traditional procedure for establishing the atomic mobility database of multicomponent alloys is referred to Figure 6.3 of a recent book chapter by Zhang and Chen [5], and it is also briefly described as follows: (*i*) Step 1: conduct a literature review of various diffusion properties in boundary unary, binary, ternary, and higher-order systems including the diffusion coefficients such as self/impurity diffusion coefficients for unary systems, interdiffusion coefficients for binary and ternary systems, and tracer coefficients for binary and higher-order systems as well as the experimental composition–distance profiles for quaternary and higher-order systems; (*ii*) Step 2: supplement the diffusion coefficients in boundary unary, binary, and ternary systems lacking the diffusion coefficients. The self/impurity diffusion coefficients of unary systems and tracer diffusion coefficients for binary and higher-order systems can be determined by the tracer method [6], first-principles [7], molecular dynamics [8], and also some indirect methods [9,10]. The interdiffusion coefficients in binary and ternary systems can be determined by either traditional Matano methods [11–15] or numerical inverse methods [16–19]. (*iii*) Step 3: Select the reasonable diffusion model(s) for the target phase(s). For detail information on this aspect, please also refer to the recent book chapter [5]. (*iv*) Step 4: Assess the atomic mobilities from the unary to binary and then ternary systems. (*v*) Extrapolate and validate the atomic mobilities of quaternary and higher-order systems, which can be directly extrapolated from those of boundary ternary systems, and then validated by comparing the model-predicted composition–distance profiles with the experimental ones. If most of the predicted results are inconsistent with the experimental data, Step 4 (maybe together with Step 3) should be repeated, until good agreement between the predicted and experimental observations of higher-order systems is achieved.

After the first version of the CALPHAD atomic mobility database is established, maintenance of the released database is essential, because some new experimental observations and theoretical calculations are likely to be produced from time to time. In general, for a technologically important multicomponent system, the atomic mobility descriptions in boundary unary and binary systems are typically reasonable, since sufficient and reliable diffusion coefficients are usually available [20]; thus, there is no need to update those atomic mobility descriptions frequently. For the boundary ternary systems, only scattered experimental interdiffusion coefficients are available in most cases due to the low efficiency of the Matano–Kirkaldy (M–K) method [12] with which only four independent interdiffusion coefficients can be obtained at the intersection point from the diffusion paths of two diffusion couples [21]. In order to improve the quality of the atomic mobility database, more interdiffusion coefficients covering wider compositions and temperature ranges are indispensable. Thus, continuous renewal of the corresponding atomic mobilities is necessary, but it is a really time- and cost-consuming process. While for the boundary quaternary and higher-order systems, some new experimental composition–distance profiles from the diffusion couples/multiples may appear and only be used to validate the established atomic mobility database, it cannot be directly employed to update the database [21] according to the traditional approach to CALPHAD database development. Therefore, there is an urgent need to improve the current situation.

One more superior approach is to utilize the numerical inverse method for maintaining the atomic mobility database of the target multicomponent system. Very recently, two of the present authors [22] developed a computational framework for the establishment of an atomic mobility database directly from the experimental composition–distance profiles based on the pragmatic numerical inverse method [16] and incorporated it into the freely accessible HitDIC (High-Throughput Determination of Interdiffusion Coefficients, https://hitdic.com/, accessed on 17 October 2021, version 2.3.0) software [23]. With this computational framework and HitDIC, the experimental composition–distance profiles, instead of interdiffusion coefficients, can be directly used as the input for the assessment of atomic mobilities and their related uncertainties. Then, for the ternary systems, the complex computational process of interdiffusion coefficients can be avoided, resulting in accuracy and efficiency improvements. One more important advantage lies in that the experimental composition–distance profiles in quaternary and higher-order systems can also be employed to assess the atomic mobility parameters in the target system with the computational framework and HitDIC.

Due to the fact of their good castability and low cost, Mg–Al–Zn (AZ) series alloys are widely used in various fields such as automobile, aerospace, and additive manufacturing [24–28]. Sn, as an important alloying element, is usually introduced to improve the mechanical properties of AZ series alloys [29–31]. In order to precisely design the optimal additional amount of Sn in AZ alloys, accurate diffusion coefficients in hcp (hexagonal close-packed) Mg–Al–Sn–Zn alloys are needed. Up to now, the atomic mobilities in the hcp Mg–Al–Zn–Sn quaternary system have only been assessed by Zhong et al. [32] according to the traditional approach. Moreover, the composition–distance profiles in one hcp Mg–Al–Zn–Sn quaternary diffusion couple measured by Bryan et al. [33] can only be used to validate the simulation results but cannot be employed in the optimization process of Zhong et al. [32] due to the limitations of the traditional approach. Furthermore, Zhang et al. [34] measured the new composition–distance profiles of the hcp Mg–Al–Sn ternary system at elevated temperatures recently, which should be considered during the update of the atomic mobility database of the hcp Mg–Al–Sn system.

Consequently, the major objectives of this paper were (*i*) to develop an efficient strategy for the maintenance of an atomic mobility database of multicomponent alloys based on the pragmatic numerical inverse method and HitDIC and (*ii*) to apply the developed efficient strategy to update the atomic mobility database of the hcp Mg–Al–Zn–Sn quaternary system and validate the reliability of the updated atomic mobility database.

2. An Effective Strategy to Maintain the CALPHAD Atomic Mobility Database of Multicomponent Systems

Starting from the pragmatic numerical inverse method and HitDIC software, an effective strategy to maintain the atomic mobility database of multicomponent systems is proposed in Figure 1, and it can be separated into the following steps:

1. The original atomic mobility descriptions of the target multicomponent system, together with the thermodynamic descriptions, should be ready or re-constructed according to the corresponding publication(s).
2. A critical review of all the composition–distance profiles of diffusion multiples/couples in ternary and higher-order systems available in the literature should be conducted both before and after the publication/release of the original atomic mobility database.
3. The atomic mobility descriptions in each boundary ternary system should be updated by means of the HitDIC software based on the reviewed composition–distance profiles. It should be noted that the atomic mobility descriptions in all the boundary binaries are fixed during the entire stage. Moreover, the reliability of the updated atomic mobilities should be validated by the experimental composition–distance profiles as well as the evaluated interdiffusion coefficients available in the literature.
4. Based on the updated atomic mobilities of boundary ternary systems, all the composition–distance profiles in the higher-order systems should be input into the HitDIC software to assess the possible interaction parameters in high-order systems. The interaction parameters in higher-order systems are introduced if their addition can really improve the fit to most of the experimental composition profiles. During this step, it should be noted that the interaction parameters of ternary atomic mobilities can be updated if a better fit to the experimental composition profiles in higher-order system can be achieved.
5. One needs to validate the updated atomic mobility database by comprehensively comparing the predicted diffusion properties with the experimental ones in all the related ternary, quaternary, and higher-order systems, verify the updated atomic mobility database by applying real applications if available, and finalize the documentation.

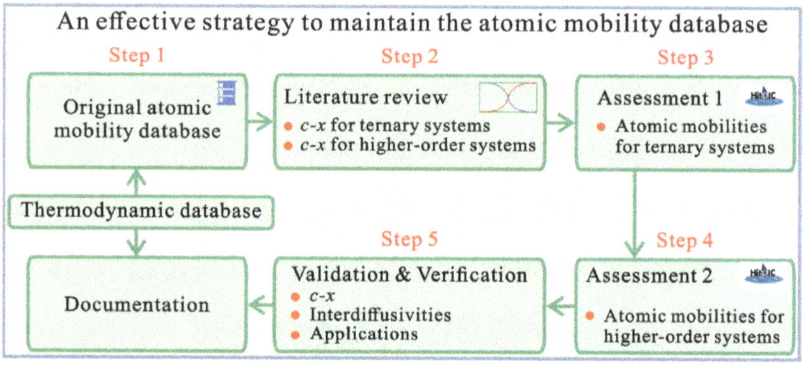

Figure 1. An effective strategy for maintaining the atomic mobility database of multicomponent systems proposed in this work. Here, the "*c*–*x*" represents the "composition–distance".

3. Literature Review on Diffusion Information in Hcp Mg–Al–Zn–Sn Alloys

In this paper, the atomic mobilities of three boundary binaries (i.e., hcp Mg–Al, Mg–Zn, and Mg–Sn) were directly taken from Zhong et al. [32] and fixed during the subsequent assessment of atomic mobilities in higher-order systems (i.e., hcp Mg–Al–Zn, Mg–Al–Sn, and Mg–Al–Zn–Sn); thus, there was no need to conduct the literature review for those boundary binaries. In the following, all the measured composition–distance profiles in the hcp Mg–Al–Zn, Mg–Al–Sn, and Mg–Al–Zn–Sn systems available in the literature are briefly introduced and are also summarized in Table 1. Moreover, in order to validate the reliability of the subsequently assessed mobilities, the experimental reports on different diffusivities in the hcp Mg–Al–Zn, Mg–Al–Sn, and Mg–Al–Zn–Sn systems were also briefly described as follows.

For the hcp Mg–Al–Zn ternary system, the interdiffusion behaviors in seven groups of diffusion couples at 673 and 723 K were investigated by Kammerer et al. [35]. Thereinto, the composition–distance profiles of four groups (i.e., Mg-9.08Al/Mg-2.55Zn, Mg-0.87Al/Mg-1.12Zn, Mg-9.10Al/Mg-2.03Zn, and Mg-2.27Al/Mg-1.06Zn, in at.%) were reported, while only the diffusion paths were given for the other three groups (i.e., Mg-3Al/Mg-1Zn, Mg-3Al/Mg-0.5Zn, and Mg/Mg-3Al-0.5Zn, in at.%). However, it should be noted that the composition–distance profiles of Mg-0.87Al/Mg-1.12Zn from Kammerer et al. [35] were not reasonable based on the analysis of Wang et al. [36]. In addition, the composition–distance profiles in the AZ91 (Mg-9Al-1Zn, in wt.%)/Mg diffusion couple at 663 and 708 K were also determined by Bryan et al. [33]. But Zhong et al. [32] pointed out that the existence of MgO on the surface of the diffusion couples showed a noticeable effect on the interdiffusion between pure Mg and AZ91 in the work of Bryan et al. [33]. Hence, the composition–distance profiles of three groups (i.e., Mg-9.08Al/Mg-2.55Zn, Mg-9.10Al/Mg-2.03Zn, and Mg-2.27Al/Mg-1.06Zn, in at.%) from Kammerer et al. [35] were employed in the present optimization, while the composition–distance profiles from Bryan et al. [33] as well as the composition–distance profiles of Mg-0.87Al/Mg-1.12Zn and diffusion paths from Kammerer et al. [35] were not.

For the hcp Mg–Al–Sn ternary system, Zhou et al. [37] determined the composition–distance profiles at 673 and 723 K based on the diffusion couple technique. Moreover, the composition–distance profiles for the Mg–Al–Sn ternary diffusion couples at 723, 773, and 823 K were determined by Zhang et al. [34], respectively. The experimental data from both Zhou et al. [37] and Zhang et al. [34] were employed in the present optimization.

Table 1. List of the composition–distance profiles of different hcp_A3 Mg–Al–Zn, Mg–Al–Sn, and Mg–Al–Zn–Sn alloys available in the literature.

Type of Diffusion Couple (in at.%)	Diffusion Temperature (K)	Diffusion Time (h)	References	Code
		Mg–Al–Zn ternary system		
Mg-9.08Al/Mg-2.55Zn Mg/Mg-0.87Al-1.12Zn Mg-3Al/Mg-1Zn Mg-3Al/Mg-0.5Zn	673	8 24 20 24	[35]	▲ △ △ △
Mg-9.10Al/Mg-2.03Zn Mg-2.77Al/Mg-1.06Zn Mg/Mg-3Al-0.5Zn	723	4 5 4	[35]	▲ ▲ △
Mg/Mg-8.41Al-0.45Zn	663	144	[33]	△
Mg/Mg-8.50Al-0.41Zn	708	144	[33]	△
		Mg–Al–Sn ternary system		
Mg-0.52Sn/Mg-7.81Al Mg-1.00Sn/Mg-7.37Al Mg-2.30Al-0.83Sn/Mg Mg-8.00Al-0.46Sn/Mg	673	216 216 216 216	[37]	▲ ▲ ▲ ▲
Mg-1.04Sn/Mg-3.59Al Mg-1.07Sn/Mg-7.63Al Mg/Mg-7.86Al-0.53Sn Mg-2.3Al-0.9Sn/Mg	723	216 216 216 216	[37]	▲ ▲ ▲ ▲
Mg-2.63Al-0.94Sn/Mg Mg-1.43Sn/Mg-3.80Al Mg-1.89Al/Mg-0.97Sn	723	9 9 9	[34]	▲ ▲ ▲
Mg/Mg-2.77Al-0.97Sn Mg-1.46Sn/Mg-3.81Al Mg-0.96Al-1.48Sn/Mg Mg-0.98Sn/Mg-1.92Al	773	6 6 6 6	[34]	▲ ▲ ▲ ▲
Mg/Mg-1.43Al-0.92Sn Mg-1.45Sn/Mg-3.74Al Mg-0.98Sn/Mg-1.83Al	823	3 3 3	[34]	▲ ▲ ▲
		Mg–Al–Zn–Sn quaternary system		
Mg-0.64Al-0.04Sn-0.59Zn/Mg-0.79Al-2.42Sn-0.66Zn	773	250	[33]	▲

▲, used in the optimization process; △, only used for comparison.

As for the hcp Mg–Al–Zn–Sn quaternary system, the composition–distance profiles in one quaternary diffusion couple annealed at 773 K for 250 h were measured by Bryan et al. [33] and thus were considered during the present assessment of the atomic mobilities.

Besides the above experimental information on the composition–distance profiles, there are also some reports on the inter-diffusivities available in the literature. Based on the experimental composition profiles in the hcp Mg–Al–Zn system by Kammerer et al. [35], Wang et al. [36] evaluated the main interdiffusion coefficients (i.e., $\tilde{D}^{Mg}_{AlAl}$ and $\tilde{D}^{Mg}_{ZnZn}$) and cross-interdiffusion coefficients (i.e., $\tilde{D}^{Mg}_{AlZn}$ and $\tilde{D}^{Mg}_{ZnAl}$) at common intersection points using the Whittle–Green (W–G) method [38]. For the hcp Mg–Al–Sn ternary system, the interdiffusion coefficients (i.e., $\tilde{D}^{Mg}_{AlAl}$, $\tilde{D}^{Mg}_{SnSn}$, $\tilde{D}^{Mg}_{AlSn}$, and $\tilde{D}^{Mg}_{SnAl}$) at the intersection compositions along diffusion paths were determined by Zhou et al. [37] also using the W–G method. Moreover, the inter-diffusivities of the hcp Mg–Al–Sn system were also determined by Zhang et al. [34] by means of the M–K method. As indicated above, all the related interdiffusion coefficients in ternary systems were not used in the assessment procedure but employed to validate the finally obtained atomic mobilities.

4. Results and Discussion

The thermodynamic descriptions for the Mg–Al–Zn–Sn quaternary system from our previous publications [39–41] were directly employed in the present work for providing accurate thermodynamic properties. In the following, the atomic mobilities in the hcp Mg–Al–Zn and hcp Mg–Al–Sn ternary systems were first updated by fixing the atomic mobilities in boundary binaries, from which the atomic mobility database in the hcp Mg–Al–Zn–Sn quaternary system was then established.

In the hcp Mg–Al–Zn ternary system, the composition–distance profiles measured by Kammerer [35] (except for those in the Mg-0.87Al/Mg-1.12Zn diffusion couple, in at.%) together with the atomic mobility descriptions of boundary binaries as well as the thermodynamic descriptions were first provided as input in HitDIC software. Subsequently, the initial values of the interaction parameters (i.e., $\Phi_{Al}^{Mg,Zn}$ and $\Phi_{Zn}^{Mg,Al}$) of the ternary system were automatically set, and the optimization of the two parameters was carried out automatically by the HitDIC software until the best fit between the model-predicted composition–distance profiles and the experimental data was achieved. Finally, the established atomic mobility database of the hcp Mg–Al–Zn ternary system was validated by comparing the predicted diffusion properties with the corresponding experimental data. Moreover, a similar strategy was adopted for the hcp Mg–Al–Sn ternary system.

As for the hcp Mg–Al–Zn–Sn quaternary system, the experimental composition–distance profiles by Bryan et al. [33] together with the updated atomic mobility descriptions of the hcp Mg–Al–Zn and Mg–Al–Sn as well as the thermodynamic descriptions of the hcp Mg–Al–Zn–Sn quaternary systems were first provided as the input in the HitDIC software. Subsequently, the assessment of the interaction parameters in the ternary and/or quaternary systems was automatically performed. It was found that introduction of an interaction parameter (i.e., $\Phi_{Zn}^{Mg,Sn}$) can result in the best fit to the experimental data. The finally obtained atomic mobility parameters of the hcp Mg–Al–Zn–Sn quaternary system are summarized in Table 2.

4.1. Hcp Mg–Al–Zn Ternary System

The model-predicted composition–distance profiles of four diffusion couples (i.e., Mg-9.08Al/Mg-2.55Zn at 673 K for 8 h, Mg/Mg-0.87Al-1.12Zn at 673 K for 24 h, Mg-9.10Al/Mg-2.03Zn at 723 K for 4 h, and Mg-2.77Al/Mg-1.06Zn at 723 K for 5 h, in at.%) according to the present atomic mobilities (solid lines) are displayed in Figure 2, compared with the corresponding experimental data (in symbols) by Kammerer et al. [35]. The model-predicted results by Zhong et al. [32] are also superimposed as dashed lines in the figure for direct comparison with the present results. Without specification, all the model-predicted results of Zhong et al. [32] are taken exactly from their original publication. As can be seen in Figure 2, the predicted results from the present work are consistent with those from Zhong et al. [32], and both predicted results are in good agreement with the experimental composition–distance profiles [35], expect for Figure 2b. As shown in Figure 2b, a large deviation between the model-predicted composition–distance profile of Zn and the experimental ones can be observed. This fact is quite normal because the composition–distance profiles of Mg/Mg-0.87Al-1.12Zn at 673 K for 24 h are not reasonable based on the suggestion by Wang et al. [36] and, thus, were not employed in the present optimization. Furthermore, the model-predicted diffusion paths at 673 and 723 K, based on the present atomic mobilities together with those by Zhong et al. [32], are shown in Figure 3 compared with the experimental data [35]. The diffusion paths predicted according to the present atomic mobilities are in very good agreement with the experimental data [35] and also the ones by Zhong et al. [32]. Moreover, the comparison between the model-predicted composition–distance profiles due to the present atomic mobilities and the experimental data by Bryan et al. [33] as well as those by Zhong et al. [32] are displayed in the Supplementary Materials for readers' reference. As can be seen Figur S1, certain deviations exist between the simulated composition profiles of Al/Zn and the experimental data. This is because MgO exits on the surface of the diffusion couples, as pointed out by Zhong et al. [32], and hinders the diffusion of both Al and Zn. Thus, the composition–distance profiles of Bryan et al. [33] were not considered in the present update of atomic mobilities.

Table 2. List of the atomic mobility parameters of hcp the Mg–Al–Zn–Sn system assessed in the present work together with those taken in the literature [32].

Mobility	Parameters	References
Mobility of Mg	$\Phi_{Mg}^{Mg} = -125{,}748.3 - 86.924 \times T$	[32]
	$\Phi_{Mg}^{Al} = -105{,}022.4 - 100.826 \times T$	[32]
	$\Phi_{Mg}^{Zn} = -97{,}239.0 - 87.338 \times T$	[32]
	$\Phi_{Mg}^{Sn} = -76{,}913.9 - 71.922 \times T$	[32]
	$\Phi_{Mg}^{Mg,Al} = 154{,}978.2$	[32]
Mobility of Al	$\Phi_{Al}^{Al} = -115{,}705.9 - 104.143 \times T$	[32]
	$\Phi_{Al}^{Mg} = -133{,}378.9 - 86.232 \times T$	[32]
	$\Phi_{Al}^{Zn} = -97{,}239.0 - 87.338 \times T$	[32]
	$\Phi_{Al}^{Sn} = -76{,}913.9 - 71.922 \times T$	[32]
	$\Phi_{Al}^{Mg,Al} = 125{,}172.6$	[32]
	$\Phi_{Al}^{Mg,Zn} = 313{,}977.051$	This work
	$\Phi_{Al}^{Mg,Sn} = 214{,}599.609$	This work
Mobility of Zn	$\Phi_{Zn}^{Zn} = -97{,}239.0 - 87.338 \times T$	[32]
	$\Phi_{Zn}^{Mg} = -125{,}731.0 - 76.734 \times T$	[32]
	$\Phi_{Zn}^{Al} = -115{,}705.9 - 104.143 \times T$	[32]
	$\Phi_{Zn}^{Sn} = -76{,}913.9 - 71.922 \times T$	[32]
	$\Phi_{Zn}^{Mg,Zn} = 80{,}988.7$	[32]
	$\Phi_{Zn}^{Mg,Al} = 90{,}957.031$	This work
	$\Phi_{Zn}^{Mg,Sn} = -11{,}209.270$	This work
Mobility of Sn	$\Phi_{Sn}^{Sn} = -76{,}913.9 - 71.922 \times T$	[32]
	$\Phi_{Sn}^{Mg} = -143{,}787.3 - 72.615 \times T$	[32]
	$\Phi_{Sn}^{Al} = -115{,}705.9 - 104.143 \times T$	[32]
	$\Phi_{Sn}^{Zn} = -97{,}239.0 - 87.338 \times T$	[32]
	$\Phi_{Sn}^{Mg,Sn} = -162{,}023.5$	[32]
	$\Phi_{Sn}^{Mg,Al} = 191{,}345.215$	This work

According to the presently updated atomic mobility descriptions together with the thermodynamic descriptions [42], the interdiffusion coefficients of the hcp Mg–Al–Zn system over the composition range of 0–5.0 at.% Al and 0–3.0 at.% Zn at 623, 673, and 723 K are predicted in Figure 4. Figure 4a,b show the calculated main interdiffusion coefficients, $\widetilde{D}_{AlAl}^{Mg}$ and $\widetilde{D}_{ZnZn}^{Mg}$, in three-dimensional space, respectively. As shown in Figure 4a,b, $\widetilde{D}_{ZnZn}^{Mg}$ was larger than $\widetilde{D}_{AlAl}^{Mg}$ at the same temperature by approximately one order of magnitude, which means that the diffusion rate of Zn in hcp Mg–Al–Zn alloys is faster than that of Al. Moreover, it can be observed that both $\widetilde{D}_{AlAl}^{Mg}$ and $\widetilde{D}_{ZnZn}^{Mg}$ increased with the increase in temperature and concentrations of both Al and Zn. Figure 4c displays variations in the cross interdiffusion coefficient $\widetilde{D}_{AlZn}^{Mg}$ along with the concentrations of Al and Zn. It should be noted that the predicted $\widetilde{D}_{AlZn}^{Mg}$ over wide composition and temperature range is negative. Hence, the $\log_{10}(-\widetilde{D}_{AlZn}^{Mg})$ was adopted for the label of ordinate of Figure 4c in order to facilitate the analysis. The sign of cross-interdiffusion coefficients had been analyzed in detail by Liu et al. [43] in terms of thermodynamics. According to Liu et al. [43], the cross-interdiffusion coefficient $\widetilde{D}_{AlZn}^{Mg}$ in the hcp Mg–Al–Zn ternary system can be expressed as follows:

$$\widetilde{D}_{AlZn}^{Mg} = \left[(1-x_{Al})^2 x_{Al} M_{Al} + x_{Al}^2 x_{Zn} M_{Zn} + x_{Al}^2 x_{Mg} M_{Mg}\right] \frac{\partial(\mu_{Al} - \mu_{Mg})}{\partial x_{Zn}} \quad (1)$$

where M_{Al}, M_{Zn}, and M_{Mg} are the atomic mobilities for Al, Zn, and Mg, respectively. x_{Al}, x_{Zn}, and x_{Mg} are the mole fractions for Al, Zn, and Mg, respectively. μ_{Al} and μ_{Mg} represent the chemical potentials of Al and Mg, respectively. Because the term before $\partial(\mu_{Al} - \mu_{Mg})/\partial x_{Zn}$ in Equation (1) is positive, the negative sign of $\widetilde{D}_{AlZn}^{Mg}$ is determined by the $\partial(\mu_{Al} - \mu_{Mg})/\partial x_{Zn}$. As can be seen in Figure 4c, the $\widetilde{D}_{AlZn}^{Mg}$ was lower than the main interdiffusion coefficients, $\widetilde{D}_{AlAl}^{Mg}$ and $\widetilde{D}_{ZnZn}^{Mg}$, at the same temperature by approximately one to two orders of magnitude. Moreover, the $\widetilde{D}_{AlZn}^{Mg}$ increased with the increase in temperature and Zn concentration. While $\widetilde{D}_{AlZn}^{Mg}$ increased rapidly as the Al concentration increased in the region where the Al concentration was close to zero, but then increased slowly with the further increase in Al concentration. Furthermore, it is interesting to see in Figure 4c that the cross-interdiffusion coefficients $\widetilde{D}_{AlZn}^{Mg}$ at 623, 673, and 723 K were all approaching zero as the concentration of Al approaches zero. It should be noted that such an interesting phenomenon is reasonable and can be obviously proved by Equation (1). The relationship between the cross-interdiffusion coefficient $\widetilde{D}_{ZnAl}^{Mg}$ and concentrations of both Al and Zn is displayed in Figure 4d. Different from $\widetilde{D}_{AlZn}^{Mg}$, the presently predicted $\widetilde{D}_{ZnAl}^{Mg}$ is positive, and can be expressed as the following equation similar to Equation (1):

$$\widetilde{D}_{ZnAl}^{Mg} = \left[(1-x_{Zn})^2 x_{Zn} M_{Zn} + x_{Zn}^2 x_{Al} M_{Al} + x_{Zn}^2 x_{Mg} M_{Mg}\right] \frac{\partial(\mu_{Zn} - \mu_{Mg})}{\partial x_{Al}} \quad (2)$$

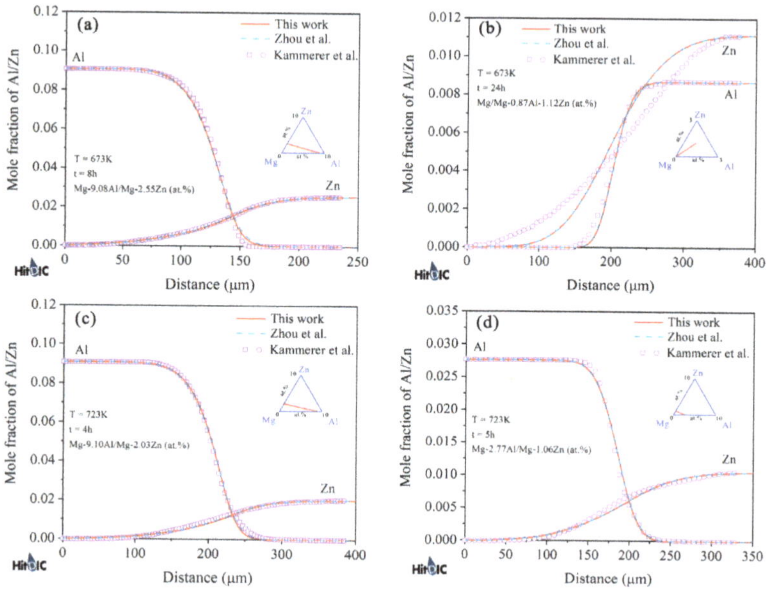

Figure 2. Model-predicted composition–distance profiles of different hcp Mg-Al–Zn diffusion couples annealed at (**a**) 673 K for 8 h, (**b**) 673 K for 24 h, (**c**) 723 K for 4 h, and (**d**) 723 K for 5 h, due to the present atomic mobilities (solid lines), compared with those of Zhong et al. [32] (dashed lines) and the experimental data [35] (in symbols).

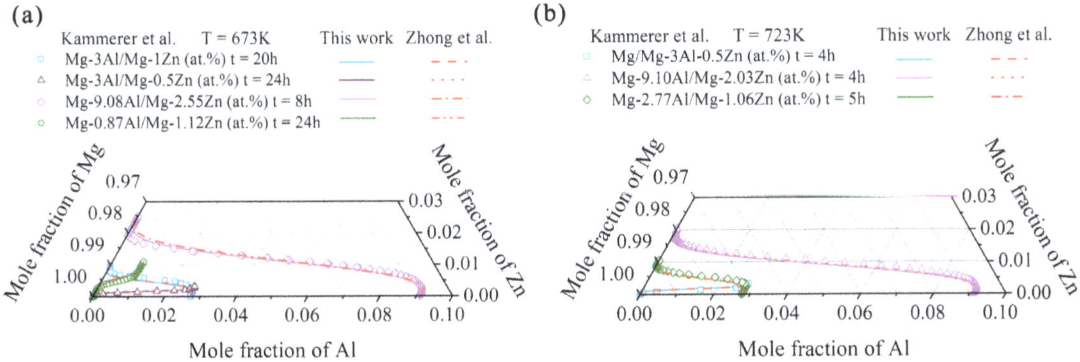

Figure 3. Model-predicted diffusion paths in the hcp Mg–Al–Zn system at (**a**) 673 and (**b**) 723 K due to the present atomic mobilities (solid lines) compared with those of Zhong et al. [32] (dashed lines) and the experimental data [35] (in symbols).

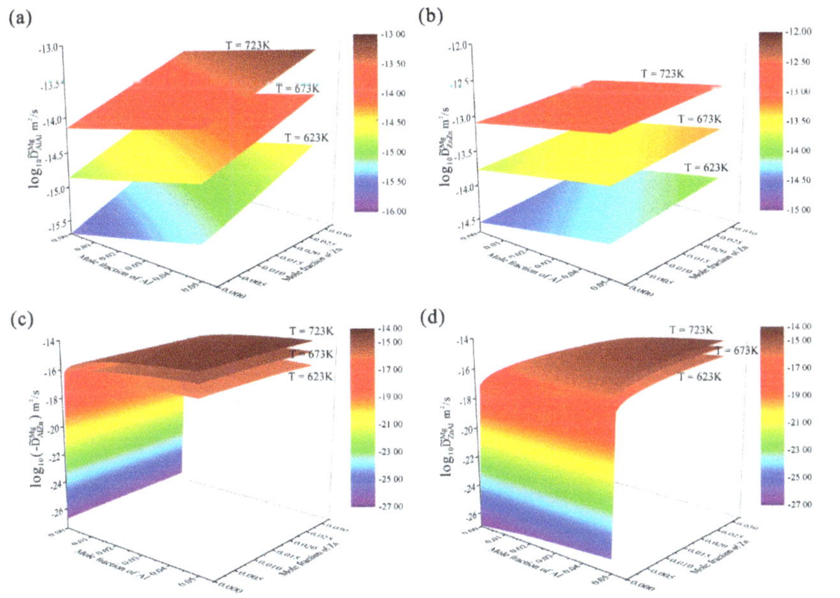

Figure 4. Model-predicted composition-dependent inter-diffusivities of (**a**) $\widetilde{D}^{Mg}_{AlAl}$, (**b**) $\widetilde{D}^{Mg}_{ZnZn}$, (**c**) $\widetilde{D}^{Mg}_{AlZn}$, and (**d**) $\widetilde{D}^{Mg}_{ZnAl}$ over the composition range of 0–5.0 at.% Al and 0–3.0 at.% Zn at 623, 673, and 723 K according to the present atomic mobilities together with the thermodynamic descriptions [42].

According to Equation (2), the positive sign of $\widetilde{D}^{Mg}_{ZnAl}$ is due to the positive $\partial(\mu_{Zn} - \mu_{Mg})/\partial x_{Al}$. As shown in Figure 4d, the value of $\widetilde{D}^{Mg}_{ZnAl}$ is in the same order of the absolute one of $\widetilde{D}^{Mg}_{AlZn}$, but lower than the main interdiffusion coefficients, $\widetilde{D}^{Mg}_{AlAl}$ and $\widetilde{D}^{Mg}_{ZnZn}$. In addition, the $\widetilde{D}^{Mg}_{ZnAl}$ increased with the increase in temperature and Al concentration, while the $\widetilde{D}^{Mg}_{ZnAl}$ increased rapidly in the region where the Zn concentration was close to zero, and then increased slowly with the further increase of Zn. Moreover, an interesting phenomenon can also be found with the cross-interdiffusion coefficients $\widetilde{D}^{Mg}_{ZnAl}$ at 623, 673, and 723 K all approaching zero as the concentration of Zn approached zero.

To further illustrate the reliability of the presently updated atomic mobilities, the calculated main inter-diffusivities according to the present work are compared with the determined ones by Wang et al. [36] in Figure 5a. Along the diagonal lines, the model-predicted values are exactly equal to the experimental ones. The region of empirical errors for inter-diffusivities is constructed by the two dashed lines that represent the interdiffusion coefficients multiplied with a pre-factor of 2 or 0.5, respectively, according to the suggestion in [44]. A similar plot was also made in Figure 5b between the calculated main inter-diffusivities by Zhong et al. [32] and the ones determined by Wang et al. [36]. Based on the comparison in Figure 5a,b, it can be found that the calculated main interdiffusion coefficients from the present work are consistent with those of Zhong et al. [32], and the calculated main interdiffusion coefficients in both the present work and Zhong et al. [32] agree well with all the experimental data (within the dashed lines), expected for 6 values marked by black circles in the figure. It should be noted that those 6 points marked by black circles were determined by Wang et al. [36] based on three diffusion couples (i.e., Mg-3Al/Mg-1Zn, Mg-3Al/Mg-0.5Zn, and Mg-0.87Al/Mg-1.12Zn) from Kammerer et al. [35] of which the composition–distance profiles were not employed in the present optimization because the original experimental data were either unreasonable (i.e., Mg-0.87Al/Mg-1.12Zn) or not provided (i.e., Mg-3Al/Mg-1Zn and Mg-3Al/Mg-0.5Zn) according to the original publications.

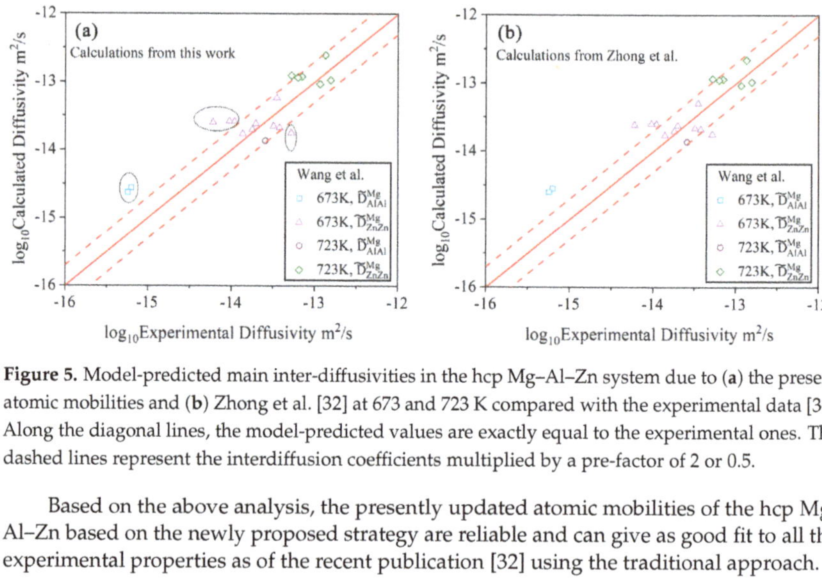

Figure 5. Model-predicted main inter-diffusivities in the hcp Mg–Al–Zn system due to (**a**) the present atomic mobilities and (**b**) Zhong et al. [32] at 673 and 723 K compared with the experimental data [36]. Along the diagonal lines, the model-predicted values are exactly equal to the experimental ones. The dashed lines represent the interdiffusion coefficients multiplied by a pre-factor of 2 or 0.5.

Based on the above analysis, the presently updated atomic mobilities of the hcp Mg–Al–Zn based on the newly proposed strategy are reliable and can give as good fit to all the experimental properties as of the recent publication [32] using the traditional approach.

4.2. Hcp Mg–Al–Sn Ternary System

Figures 6 and 7 display the model-predicted composition–distance profiles of eight diffusion couples (i.e., Mg-0.52Sn/Mg-7.81Al, Mg-1.00Sn/Mg-7.37Al, Mg-2.30Al-0.83Sn/Mg, and Mg-8.00Al-0.46Sn/Mg, annealed at 673 K for 216 h, in at.%; Mg-1.04Sn/Mg-3.59Al, Mg-1.07Sn/Mg-7.63Al, Mg/Mg-7.86Al-0.53Sn, and Mg-2.3Al-0.9Sn/Mg, annealed at 723 K for 216 h, in at.%) from the present work (solid lines) compared with the experimental data (in symbols) by Zhou et al. [37]. The model-predicted results according to Zhong et al. [32] (dashed lines) are also superimposed in the figure for direct comparison. Figures 8 and 9 also show the model-predicted composition–distance profiles of 10 diffusion couples (i.e., Mg/Mg-2.77Al-0.97Sn, Mg-1.46Sn/Mg-3.81Al, Mg-0.96Al-1.48Sn/Mg, and Mg-0.98Sn/Mg-1.92Al, annealed at 773 K for 6 h, in at.%; Mg-2.63Al-0.94Sn/Mg, Mg-1.43Sn/Mg-3.80Al, and Mg-1.89Al/Mg-0.97Sn, annealed at 723 K for 9 h, in at.%; Mg/Mg-1.43Al-0.92Sn, Mg-1.45Sn/Mg-3.74Al, and Mg-0.98Sn/Mg-1.83Al, annealed at 823 K for 3 h, in at.%) according to the present atomic mobilities and also those from Zhong et al. [32] compared with the experimental data by Zhang et al. [34]. As can be seen in Figures 6–9, the model-predicted composition–distance profiles according to the present work are in better agreement with the experimental data by Zhou et al. [37] and Zhang et al. [34] than the model-predicted ones due from Zhong et al. [32], especially in the figures, i.e., Figure 6a,b, Figure 7a,b, Figure 8b,d and Figure 9b,c,e,f. Furthermore, the model-predicted diffusion paths at 673, 723, 773, and 823 K, based on the presently updated atomic mobilities and also those by Zhong et al. [32], are displayed in Figure 10 compared with the experimental data from Zhou et al. [37] and Zhang et al. [34]. As can be seen in Figure 10, the model-predicted diffusion paths by the present work again agree better with the experimental data [34,37] than the model-predicted ones by Zhong et al. [32].

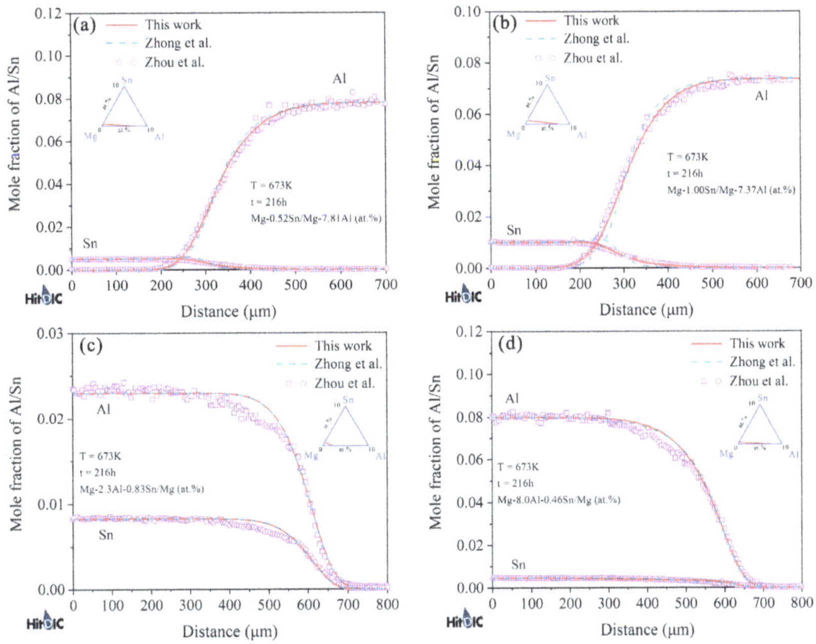

Figure 6. (**a–d**) Model-predicted composition–distance profiles of the different hcp Mg–Al–Sn diffusion couples annealed at 673 K for 216 h from the present atomic mobilities (solid lines) compared with those of Zhong et al. [32] (dashed lines) and the experimental data [37] (in symbols).

Based on the updated atomic mobility descriptions by the present work together with the thermodynamic descriptions [39], the inter-diffusivities of the hcp Mg–Al–Sn system over the composition range of 0–5.0 at.% Al and 0–2.0 at.% Sn at 723, 773, and 823 K are predicted in Figure 11. Similar to the hcp Mg–Al–Zn system, the interdiffusion coefficients of the hcp Mg–Al–Sn system were also processed with a logarithm. As shown in Figure 11a,b, the main interdiffusion coefficient $\widetilde{D}^{\mathrm{Mg}}_{\mathrm{AlAl}}$ was in the same order of magnitude as the $\widetilde{D}^{\mathrm{Mg}}_{\mathrm{SnSn}}$ at the same temperature. Besides, both $\widetilde{D}^{\mathrm{Mg}}_{\mathrm{AlAl}}$ and $\widetilde{D}^{\mathrm{Mg}}_{\mathrm{SnSn}}$ increased with the increase in temperature and concentrations of Al and Sn. Figure 11c,d show that the cross-interdiffusion coefficients, $\widetilde{D}^{\mathrm{Mg}}_{\mathrm{AlSn}}$ and $\widetilde{D}^{\mathrm{Mg}}_{\mathrm{SnAl}}$, varied apparently along with the concentrations of Al and Sn. Similar to Equations (1) and (2), $\widetilde{D}^{\mathrm{Mg}}_{\mathrm{AlSn}}$ and $\widetilde{D}^{\mathrm{Mg}}_{\mathrm{SnAl}}$ can be expressed as:

$$\widetilde{D}^{\mathrm{Mg}}_{\mathrm{AlSn}} = \left[(1-x_{Al})^2 x_{Al} M_{Al} + x_{Al}^2 x_{Sn} M_{Sn} + x_{Al}^2 x_{Mg} M_{Mg}\right] \frac{\partial(\mu_{Al}-\mu_{Mg})}{\partial x_{Sn}} \quad (3)$$

$$\widetilde{D}_{\mathrm{SnAl}}^{\mathrm{Mg}} = \left[(1 - x_{Sn})^2 x_{Sn} M_{Sn} + x_{Sn}^2 x_{Al} M_{Al} + x_{Sn}^2 x_{Mg} M_{Mg}\right] \frac{\partial(\mu_{Sn} - \mu_{Mg})}{\partial x_{Al}} \tag{4}$$

where M_{Al}, M_{Sn}, and M_{Mg} are the atomic mobilities for Al, Sn, and Mg, respectively. x_{Al}, x_{Sn}, and x_{Mg} are the mole fractions for Al, Sn, and Mg, respectively. μ_{Al}, μ_{Sn}, and μ_{Mg} represent the chemical potentials of Al, Sn, and Mg, respectively. Here, it should be noted that the signs of $\widetilde{D}_{\mathrm{AlSn}}^{\mathrm{Mg}}$ and $\widetilde{D}_{\mathrm{SnAl}}^{\mathrm{Mg}}$ are positive, which are determined by the terms $\partial(\mu_{Al} - \mu_{Mg})/\partial x_{Sn}$ and $\partial(\mu_{Sn} - \mu_{Mg})/\partial x_{Al}$, respectively. As can be seen in Figure 11c,d, the cross-interdiffusion coefficient $\widetilde{D}_{\mathrm{AlSn}}^{\mathrm{Mg}}$ was in the same order of the main interdiffusion coefficients, $\widetilde{D}_{\mathrm{AlAl}}^{\mathrm{Mg}}$ and $\widetilde{D}_{\mathrm{SnSn}}^{\mathrm{Mg}}$, at the same temperature, while the cross-interdiffusion coefficient $\widetilde{D}_{\mathrm{SnAl}}^{\mathrm{Mg}}$ was lower than the main interdiffusion coefficients, $\widetilde{D}_{\mathrm{AlAl}}^{\mathrm{Mg}}$ and $\widetilde{D}_{\mathrm{SnSn}}^{\mathrm{Mg}}$, by approximately one order of magnitude. Moreover, $\widetilde{D}_{\mathrm{AlSn}}^{\mathrm{Mg}}$ increased with the increase in temperature and Sn concentration, while $\widetilde{D}_{\mathrm{AlSn}}^{\mathrm{Mg}}$ increased rapidly in the region where the Al concentration was close to zero, and then increased slowly with the further increase in Al. As the concentration of Al (Sn) approached zero, the cross-interdiffusion coefficient, $\widetilde{D}_{\mathrm{AlSn}}^{\mathrm{Mg}}$ ($\widetilde{D}_{\mathrm{SnAl}}^{\mathrm{Mg}}$) at 723, 773, and 823 K were all approaching zero, which can be reasonably explained by Equations (3) and (4). Figure 12a,b respectively show the calculated main interdiffusion coefficients according to the present atomic mobilities and those of Zhong et al. [32], compared with the experimental data [34,37]. Along the diagonal lines, the model-predicted values are exactly equal to the experimental ones. The two dashed lines represent the interdiffusion coefficients multiplied by a pre-factor of 2 or 0.5, respectively. A comparison between Figure 12a,b clearly indicates that the calculated interdiffusion coefficients from the present work can reproduce more experimental data than those by Zhong et al. [32].

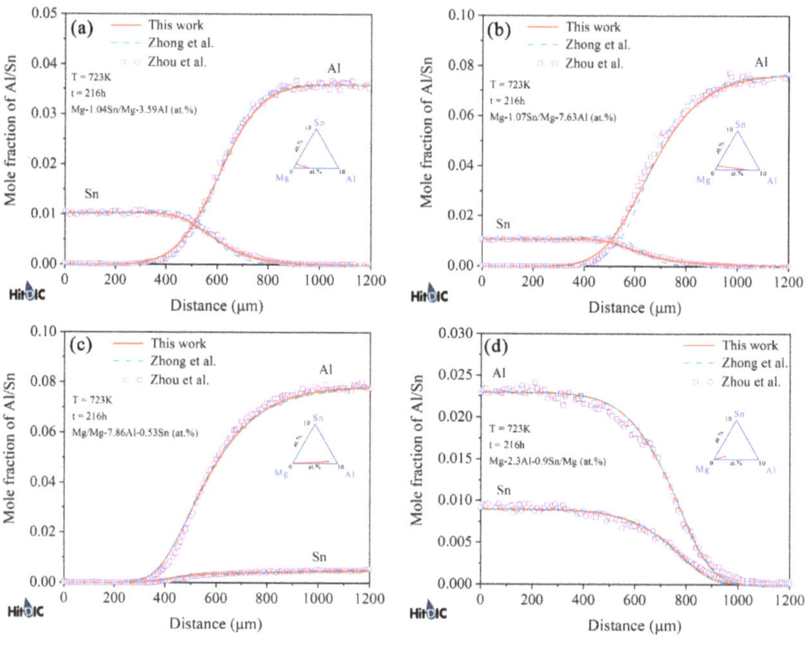

Figure 7. (**a**–**d**) Model-predicted composition–distance profiles of the different hcp Mg-Al-Sn diffusion couples annealed at 723 K for 216 h from the present atomic mobilities (solid lines) compared with those of Zhong et al. [32] (dashed lines) and the experimental data [37] (in symbols).

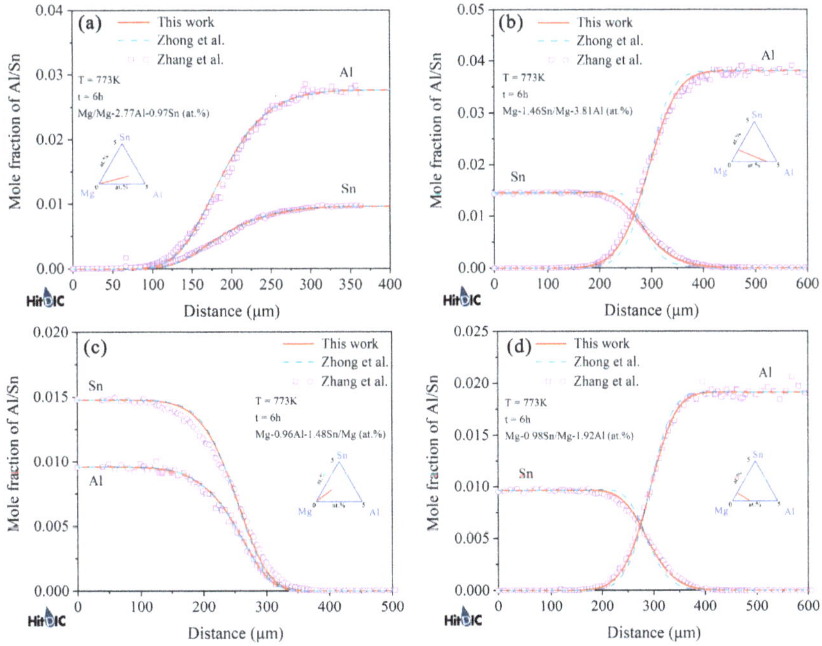

Figure 8. (**a**–**d**) Model-predicted composition–distance profiles of the different hcp Mg–Al–Sn diffusion couples annealed at 773 K for 6 h from the present atomic mobilities (solid lines) compared with those of Zhong et al. [32] (dashed lines) and the experimental data [34] (in symbols).

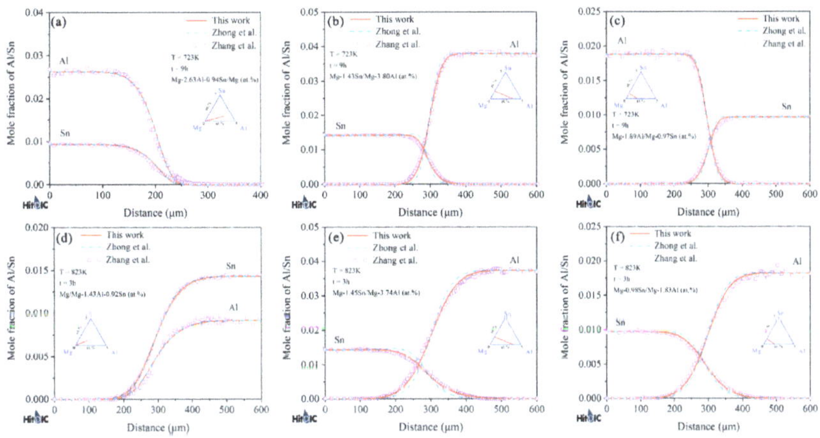

Figure 9. Model-predicted composition–distance profiles of the different hcp Mg–Al–Sn diffusion couples annealed at (**a**–**c**) 723 K for 9 h and (**d**–**f**) 823 K for 3 h from the present atomic mobilities (solid lines) compared with these of Zhong et al. [32] (dashed lines) and the experimental data [34] (in symbols).

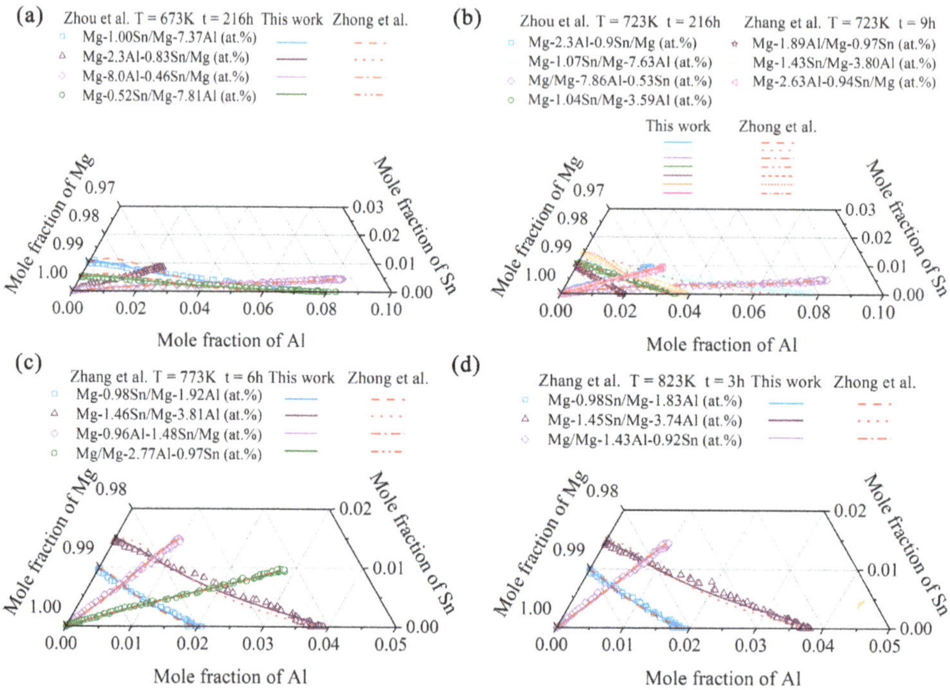

Figure 10. Model-predicted diffusion paths in the hcp Mg–Al–Sn system at (**a**) 673 K for 216 h, (**b**) 723 K for 216 and 9 h, (**c**) 773 K for 6 h, and (**d**) 823 K for 3 h from the present mobilities (solid lines) compared with those of Zhong et al. [32] (dashed lines) and the experimental data [34,37] (in symbols).

Based on the above comprehensive comparison among the model-predicted results from the present work, the ones by Zhong et al. [32], and the experimental data [34,37], the reliability of the atomic mobility descriptions of the hcp Mg–Al–Sn system was significantly improved by using the newly proposed strategy compared with the traditional approach. The major reason lies in that although 18 groups of diffusion couples were investigated by Zhou et al. [37] and Zhang et al. [34], only very scattered experimental interdiffusion coefficients at the intersection compositions of diffusion paths can be determined by the traditional methods and then utilized in the traditional optimization process, which may lead to the lower accuracy of the obtained atomic mobility descriptions. By contrast, all 18 groups of composition–distance profiles can be employed in the optimization process using the new strategy based on HitDIC, which can largely improve the reliability of the atomic mobility descriptions.

4.3. Hcp Mg–Al–Zn–Sn Quaternary System

Figure 13 displays the comparison between the model-predicted composition–distance profiles of the only quaternary diffusion couple, Mg-0.64Al-0.04Sn-0.59Zn/Mg-0.79Al-2.42Sn-0.66Zn, annealed at 773 K for 250 h due to the present atomic mobilities (solid lines) and the experimental data (in symbols) by Bryan et al. [33]. The model-predicted composition–distance profiles from Zhong et al. [32] (dashed lines) are also superimposed in the figure for direct comparison. As can be seen in Figure 13, the model-predicted composition–distance profiles of Sn and Al in the present work show much better agreement with the experimental data of Bryan et al. [33], compared with the results from Zhong et al. [32]. The model-predicted composition–distance curve of Zn from both the present work and Zhong et al. [32] slightly deviate from the experimental data of

Bryan et al. [33]. It should be noted that the difference in the Zn concentration in both end alloys was only 0.07 at.%, which may cause large difficulties in the accurate experimental measurement of Zn concentration.

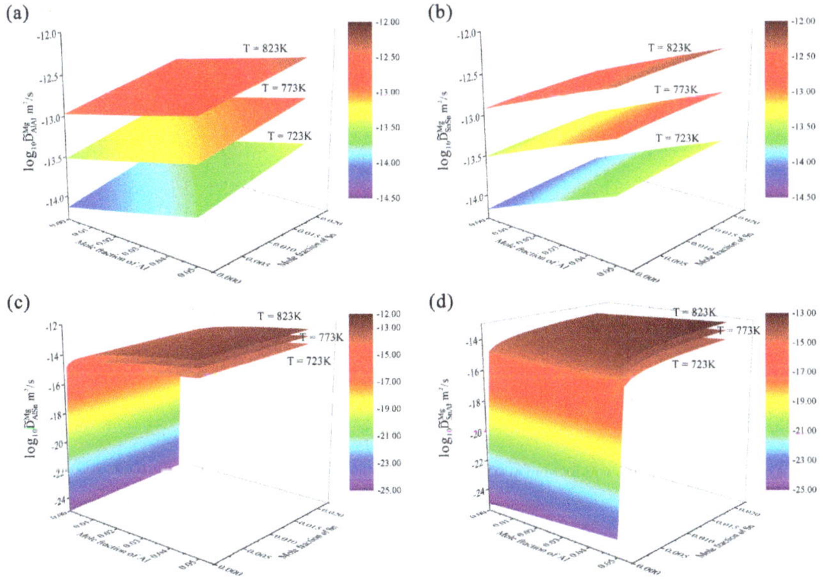

Figure 11. Model-predicted composition-dependent inter-diffusivities of (**a**) $\widetilde{D}^{Mg}_{AlAl}$, (**b**) $\widetilde{D}^{Mg}_{SnSn}$, (**c**) $\widetilde{D}^{Mg}_{AlSn}$, and (**d**) $\widetilde{D}^{Mg}_{SnAl}$ over the composition range of 0–5.0 at.% Al and 0–2.0 at.% Sn at 723, 773, and 823 K according to the present atomic mobilities together with the thermodynamic descriptions [39].

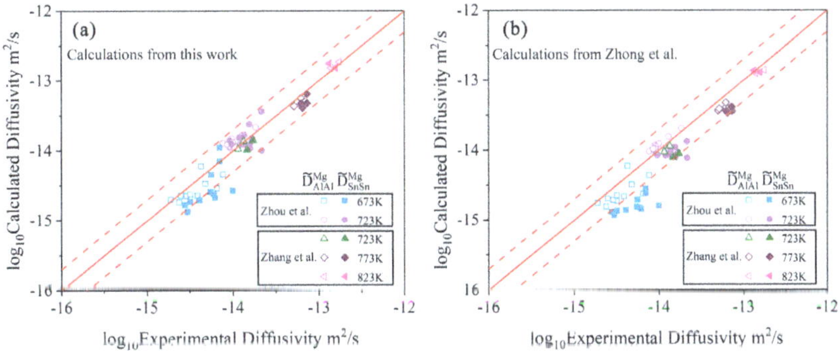

Figure 12. Model-predicted main inter-diffusivities in the hcp Mg–Al–Sn system due to (**a**) the present atomic mobilities and (**b**) Zhong et al. [32] at 673, 723, 773, and 823 K compared with the experimental data [34,37]. Along the diagonal lines, the model-predicted values are exactly equal to the experimental ones. The dashed lines represent the interdiffusion coefficients multiplied with a pre-factor of 2 or 0.5.

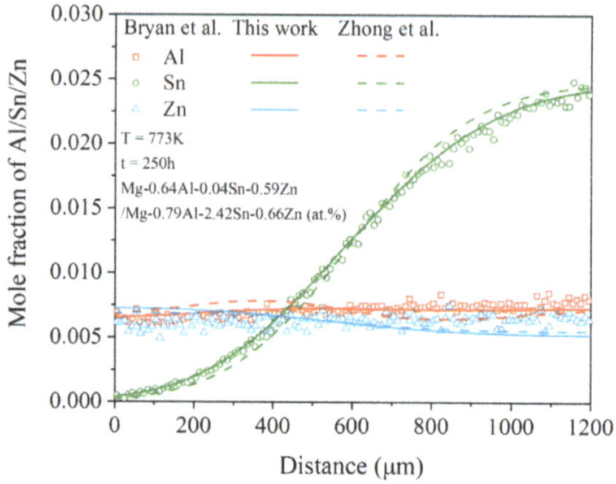

Figure 13. Model-predicted composition–distance profiles of the only quaternary diffusion couple in quaternary system, Mg-0.64Al-0.04Sn-0.59Zn/Mg-0.79Al-2.42Sn-0.66Zn, annealed at 773 K for 250 h from the present atomic mobilities (solid lines) compared with these of Zhong et al. [32] (dashed lines) and the experimental data [33] (in symbols).

Based on the above analysis, a real improvement in the reproduction of the experimental data was achieved by the present work compared with the results of Zhong et al. [32], even though only one more quaternary diffusion couple was included in the present work. It is anticipated that the reliability of the atomic mobilities in the hcp Mg–Al–Zn–Sn quaternary system can be further improved by using the newly proposed strategy if more experimental composition profiles in the quaternary Mg–Al–Zn–Sn system are available. By contrast, the reliability of atomic mobilities in the hcp Mg–Al–Zn–Sn quaternary system cannot be improved based on the traditional method, no matter whether the experimental data for Mg–Al–Zn–Sn system are sufficient.

To illustrate the influence of Sn concentration and temperature on the inter0diffusivities of the hcp Mg–Al–Zn–Sn quaternary system, the matrix (α-Mg) phase with an average composition of 2.4 at.% Al and 0.56 at.% Zn in as-cast AZT640 (Mg-6Al-4Zn-0.6Sn, in wt.%), according to Dong et al. [45], was chosen as the target in the present work. According to the presently updated atomic mobility descriptions together with the thermodynamic descriptions [40], the interdiffusion coefficients of the Mg–Al–Zn–Sn quaternary system over a compositions range of 2.4 at.% Al, 0.56 at.% Zn, and 0–0.6 at.% Sn at 623, 673, and 723 K were predicted in Figure 14. As can be seen in Figure 14a, the $\tilde{D}^{Mg}_{AlAl}$ and $\tilde{D}^{Mg}_{SnSn}$ were quite close to each other and lower than $\tilde{D}^{Mg}_{ZnZn}$ by approximately one order of magnitude. Moreover, the $\tilde{D}^{Mg}_{AlAl}$ and $\tilde{D}^{Mg}_{SnSn}$ increased with the increase in temperature and Sn concentration, while $\tilde{D}^{Mg}_{ZnZn}$ increased with the increase in temperature and kept nearly constant with the increment in Sn concentration. Figure 14b displays the variations in cross-interdiffusion coefficients, $\tilde{D}^{Mg}_{AlSn}$ and $\tilde{D}^{Mg}_{ZnSn}$, with Sn concentration. $\tilde{D}^{Mg}_{ZnSn}$ was larger than $\tilde{D}^{Mg}_{AlSn}$ but lower than the main interdiffusion coefficients, $\tilde{D}^{Mg}_{AlAl}$ and $\tilde{D}^{Mg}_{SnSn}$. In addition, $\tilde{D}^{Mg}_{ZnSn}$ increased with the increase in temperature but kept nearly constant with the increment in Sn concentration, while $\tilde{D}^{Mg}_{AlSn}$ rose with the increase in temperature and Sn concentration.

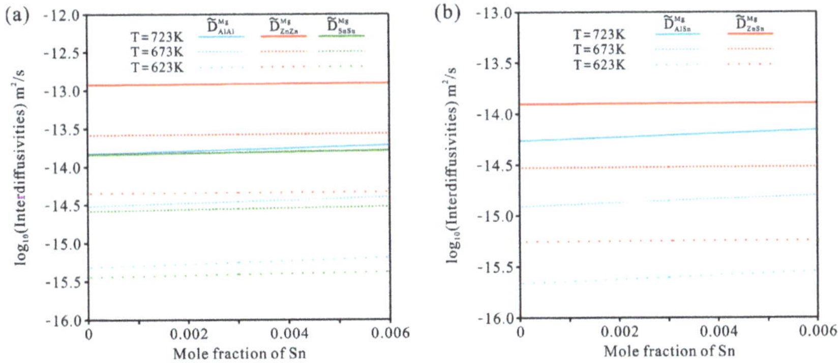

Figure 14. Model-predicted component-dependent inter-diffusivities of (**a**) $\widetilde{D}_{AlAl}^{Mg}$, $\widetilde{D}_{ZnZn}^{Mg}$, and $\widetilde{D}_{SnSn}^{Mg}$ and (**b**) $\widetilde{D}_{AlSn}^{Mg}$ and $\widetilde{D}_{ZnSn}^{Mg}$ over a composition range of 2.4 at.% Al, 0.56 at.% Zn, and 0–0.3 at.% Sn at 623, 673, and 723 K according to the present atomic mobilities together with the thermodynamic descriptions [40].

5. Conclusions

- A general and effective strategy for the maintenance of the CALPHAD atomic mobility database of multicomponent systems was developed based on the pragmatic numerical inverse method and HitDIC software;
- Following the newly proposed strategy, the atomic mobility descriptions of the hcp Mg–Al–Zn and Mg–Al–Sn ternary systems were updated based on the experimental composition profiles in the respective ternary systems. It was found that the presently updated atomic mobilities of the hcp Mg–Al–Zn system provided a good fit for all of the experimental diffusion properties as did the previous assessment [32] using the traditional approach, while the presently updated atomic mobilities of the hcp Mg–Al–Sn system showed better agreement with the experimental diffusion properties than the previous assessment [32] using the traditional approach. Moreover, the variation trend of inter-diffusivities of the hcp Mg–Al–Zn and Mg–Al–Sn systems with the temperature and solute (i.e., Al, Zn, and Sn) concentrations was also fully analyzed;
- Based on the updated atomic mobility descriptions of the hcp Mg–Al–Zn and Mg–Al–Sn systems, together with only one set of composition–distance profiles, the atomic mobility descriptions of the hcp Mg–Al–Zn–Sn quaternary system were further updated following the newly proposed strategy. A real improvement in the reproduction of experimental data was achieved by the present work compared with the previous assessment. Furthermore, the influence of Sn concentration and temperature on the inter-diffusivities of the hcp Mg–Al–Zn–Sn quaternary alloys was also illustrated;
- It is anticipated that the presently proposed strategy can serve as a standard for maintaining the CALPHAD atomic mobility database of different multicomponent systems.

Supplementary Materials: The following are available online at https://www.mdpi.com/article/10.3390/ma15010283/s1, Figure S1: Model-predicted composition–distance profiles of different hcp Mg–Al–Zn diffusion couples annealed at (a) 663 K for 144 h and (b) 708 K for 144 h as well as an enlarged composition–distance curve of Zn annealed at (c) 663 K for 144 h and (d) 708 K for 144 h from the present atomic mobilities (solid lines) compared with those of Zhong et al. [32] (dashed lines) and the experimental data [33] (symbols).

Author Contributions: L.Z., J.Z. and T.C. conceived the prototype of the approach and designed the practical cases; T.C. performed the studies with the practical alloys; T.C., J.Z. and L.Z. wrote the paper. All authors have read and agreed to the published version of the manuscript.

Funding: This research was funded by the National Key Research and Development Program of China (Grant No. 2016YFB0301101), the Youth Talent Project of Innovation-Driven Plan at Central South University (Grant No. 2282019SYLB026), and with financial support from the Hunan Province Scientific Research and Innovation Project for Postgraduates (Grant No. 2020zzts077).

Institutional Review Board Statement: Not applicable.

Informed Consent Statement: Not applicable.

Data Availability Statement: The data presented in this study are available upon reasonable request from the corresponding author.

Conflicts of Interest: The authors declare no conflict of interest.

References

1. Luo, A.A.; Fu, P.; Peng, L.; Kang, X.; Li, Z.; Zhu, T. Solidification Microstructure and Mechanical Properties of Cast Magnesium-Aluminum-Tin Alloys. *Metall. Mater. Trans. A* **2012**, *43*, 360–368. [CrossRef]
2. Lu, Z.; Zhang, L. Thermodynamic description of the quaternary Al-Si-Mg-Sc system and its application to the design of novel Sc-additional A356 alloys. *Mater. Des.* **2017**, *116*, 427–437. [CrossRef]
3. Cheng, K.; Chen, W.; Liu, D.; Zhang, L.; Du, Y. Analysis of the Cermak–Rothova method for determining the concentration dependence of ternary interdiffusion coefficients with a single diffusion couple. *Scr. Mater.* **2014**, *76*, 5–8. [CrossRef]
4. Andersson, J.; Ågren, J. Models for numerical treatment of multicomponent diffusion in simple phases. *J. Appl. Phys.* **1992**, *72*, 1350–1355. [CrossRef]
5. Zhang, L.; Chen, Q. Chapter 6—CALPHAD-Type Modeling of Diffusion Kinetics in Multicomponent Alloys. In *Handbook of Solid State Diffusion*; Paul, A., Divinski, S., Eds.; Elsevier: Amsterdam, The Netherlands, 2017; Volume 1, pp. 321–362.
6. Vaidya, M.; Trubel, S.; Murty, B.S.; Wilde, G.; Divinski, S.V. Ni tracer diffusion in CoCrFeNi and CoCrFeMnNi high entropy alloys. *J. Alloys Compd.* **2016**, *688*, 994–1001. [CrossRef]
7. Mantina, M.; Wang, Y.; Arroyave, R.; Chen, L.; Liu, Z.; Wolverton, C. First-principles calculation of self-diffusion coefficients. *Phys. Rev. Lett.* **2008**, *100*, 215901. [CrossRef]
8. Wang, J.; Hou, T. Application of molecular dynamics simulations in molecular property prediction II: Diffusion coefficient. *J. Comput. Chem.* **2011**, *32*, 3505–3519. [CrossRef]
9. Hall, L.D. An analytical method of calculating variable diffusion coefficients. *J. Chem. Phys.* **1953**, *21*, 87–89. [CrossRef]
10. Zhang, Y.; Zou, J.; Wu, X.; Deng, C.; Zhang, L. An Effective Approach to Acquire the Impurity Diffusion Coefficients in Binary Alloys with Quantified Uncertainties. *Metals* **2021**, *11*, 809. [CrossRef]
11. Matano, C. On the relation between the diffusion-coefficients and concentrations of solid metals. *Jpn. J. Phys.* **1933**, *8*, 109–113.
12. Kirkaldy, J.S.; Young, D.J. *Diffusion in the Condensed State*; The Institute of Metals: London, UK, 1987.
13. Wu, X.; Zhong, J.; Zhang, L. A general approach to quantify the uncertainty of interdiffusion coefficients in binary, ternary and multicomponent systems evaluated using Matano-based methods. *Acta Mater.* **2020**, *188*, 665–676. [CrossRef]
14. Paul, A. A pseudobinary approach to study interdiffusion and the Kirkendall effect in multicomponent systems. *Philos. Mag.* **2013**, *93*, 2297–2315. [CrossRef]
15. Esakkiraja, N.; Paul, A. A novel concept of pseudo ternary diffusion couple for the estimation of diffusion coefficients in multicomponent systems. *Scr. Mater.* **2018**, *147*, 79–82. [CrossRef]
16. Chen, W.; Zhang, L.; Du, Y.; Tang, C.; Huang, B. A pragmatic method to determine the composition-dependent interdiffusivities in ternary systems by using a single diffusion couple. *Scr. Mater.* **2014**, *90–91*, 53–56. [CrossRef]
17. Chen, W.; Zhong, J.; Zhang, L. An augmented numerical inverse method for determining the composition-dependent interdiffusivities in alloy systems by using a single diffusion couple. *MRS Commun.* **2016**, *6*, 295–300. [CrossRef]
18. Xu, H.; Chen, W.; Zhang, L.; Du, Y.; Tang, C. High-throughput determination of the composition-dependent interdiffusivities in Cu-rich fcc Cu–Ag–Sn alloys at 1073 K. *J. Alloys Compd.* **2015**, *644*, 687–693. [CrossRef]
19. Xu, H.; Cheng, K.; Zhong, J.; Wu, X.; Wei, M.; Zhang, L. Determination of accurate interdiffusion coefficients in fcc Ag-In and Ag-Cu-In alloys: A comparative study on the Matano method with distribution function and the numerical inverse method with HitDIC. *J. Alloys Compd.* **2019**, *798*, 26–34. [CrossRef]
20. McCall, D.W.; Douglass, D.C. Diffusion in binary solutions. *J. Chem. Phys.* **1967**, *71*, 987–997. [CrossRef]
21. DeHoff, R.; Kulkarni, N. The trouble with diffusion. *Mater. Res.* **2002**, *5*, 209–229. [CrossRef]
22. Zhong, J.; Chen, L.; Zhang, L. Automation of diffusion database development in multicomponent alloys from large number of experimental composition profiles. *NPJ Comput. Mater.* **2021**, *7*, 35. [CrossRef]
23. Zhong, J.; Zhang, L.; Wu, X.; Chen, L.; Deng, C. A novel computational framework for establishment of atomic mobility database directly from composition profiles and its uncertainty quantification. *J. Mater. Sci. Technol.* **2020**, *48*, 163–174. [CrossRef]
24. Luo, A.A. Magnesium casting technology for structural applications. *J. Magnes. Alloys* **2013**, *1*, 2–22. [CrossRef]
25. Luo, A.A. Recent magnesium alloy development for elevated temperature applications. *Int. Mater. Rev.* **2004**, *49*, 13–30. [CrossRef]
26. Gontarz, A.; Drozdowski, K.; Michalczyk, J.; Wiewiórowska, S.; Pater, Z.; Tomczak, J.; Samołyk, G.; Winiarski, G.; Surdacki, P. Forging of Mg–Al–Zn Magnesium Alloys on Screw Press and Forging Hammer. *Materials* **2021**, *14*, 32. [CrossRef]

27. Aldalur, E.; Suárez, A.; Veiga, F. Metal transfer modes for Wire Arc Additive Manufacturing Al-Mg alloys: Influence of heat input in microstructure and porosity. *J. Mater. Process. Technol.* **2021**, *297*, 117271. [CrossRef]
28. Wang, Z.; Lin, X.; Kang, N.; Hu, Y.; Chen, J.; Huang, W. Strength-ductility synergy of selective laser melted Al-Mg-Sc-Zr alloy with a heterogeneous grain structure. *Addit. Manuf.* **2020**, *34*, 101260. [CrossRef]
29. Liang, J.Y. Effects of Sn on Mechanical Properties of Magnesium Alloy AZ61. *Appl. Mech. Mater.* **2012**, *204–208*, 4161–4164. [CrossRef]
30. Mahmudi, R.; Moeendarbari, S. Effects of Sn additions on the microstructure and impression creep behavior of AZ91 magnesium alloy. *Mater. Sci. Eng. A* **2013**, *566*, 30–39. [CrossRef]
31. Jung, I.C.; Kim, Y.K.; Cho, T.H.; Oh, S.H.; Kim, T.E.; Shon, S.W.; Kim, W.T.; Kim, D.H. Suppression of discontinuous precipitation in AZ91 by addition of Sn. *Met. Mater. Int.* **2014**, *20*, 99–103. [CrossRef]
32. Zhong, W.; Zhao, J.-C. A comprehensive diffusion mobility database comprising 23 elements for magnesium alloys. *Acta Mater.* **2020**, *201*, 191–208. [CrossRef]
33. Bryan, Z.L.; Alieninov, P.; Berglund, I.S.; Manuel, M.V. A diffusion mobility database for magnesium alloy development. *Calphad* **2015**, *48*, 123–130. [CrossRef]
34. Zhang, Y.; Du, C.; Liu, Y.; Wen, S.; Liu, S.; Huang, Y.; Hort, N.; Du, Y. Interdiffusion and atomic mobility in hcp Mg–Al–Sn alloys. *J. Alloys Compd.* **2021**, *871*, 159517. [CrossRef]
35. Kammerer, C.C.; Kulkarni, N.S.; Warmack, B.; Sohn, Y.H. Interdiffusion in Ternary Magnesium Solid Solutions of Aluminum and Zinc. *J. Phase Equilib. Diffus.* **2016**, *37*, 65–74. [CrossRef]
36. Wang, J.; Li, N.; Wang, C.; Beltran, J.I.; Llorca, J.; Cui, Y. Computational study of atomic mobility in hcp Mg–Al–Zn ternary alloys. *Calphad* **2016**, *54*, 134–143. [CrossRef]
37. Zhou, Z.; Gu, Y.; Xu, G.; Guo, Y.; Cui, Y. Diffusion research in HCP Mg–Al–Sn ternary alloys. *Calphad* **2020**, *68*, 101710. [CrossRef]
38. Whittle, D.P.; Green, A. The measurement of diffusion coefficient in ternary systems. *Scr. Metall.* **1974**, *8*, 883–884. [CrossRef]
39. Cheng, T.; Tang, Y.; Zhang, L. Update of thermodynamic descriptions of the binary Al-Sn and ternary Mg–Al–Sn systems. *Calphad* **2019**, *64*, 354–363. [CrossRef]
40. Cheng, T.; Zhang, L. Thermodynamic Descriptions of the Quaternary Mg–Al–Zn–Sn System and Their Experimental Validation. In *Magnesium Technology 2020*; Springer: Berlin/Heidelberg, Germany, 2020; pp. 269–279.
41. Cheng, T.; Zhang, L.-J. Thermodynamic re-assessment of the Al-Sn-Zn ternary system. *J. Min. Metall. Sect. B-Metall.* **2019**, *55*, 439–449. [CrossRef]
42. Liang, P.; Tarfa, T.; Robinson, J.; Wagner, S.; Ochin, P.; Harmelin, M.; Seifert, H.; Lukas, H.; Aldinger, F. Experimental investigation and thermodynamic calculation of the Al–Mg–Zn system. *Thermochim. Acta* **1998**, *314*, 87–110. [CrossRef]
43. Liu, D.; Zhang, L.; Du, Y.; Xu, H.; Jin, Z. Ternary diffusion in Cu-rich fcc Cu–Al–Si alloys at 1073 K. *J. Alloys Compd.* **2013**, *566*, 156–163. [CrossRef]
44. Du, Y.; Chang, Y.; Huang, B.; Gong, W.; Jin, Z.; Xu, H.; Yuan, Z.; Liu, Y.; He, Y.; Xie, F.-Y. Diffusion coefficients of some solutes in fcc and liquid Al: Critical evaluation and correlation. *Mater. Sci. Eng. A* **2003**, *363*, 140–151. [CrossRef]
45. Dong, X.; Fu, J.; Wang, J.; Yang, Y. Microstructure and tensile properties of as-cast and as-aged Mg-6Al-4Zn alloys with Sn addition. *Mater. Des.* **2013**, *51*, 567–574. [CrossRef]

materials

MDPI

Article

Experimental Investigation and Thermodynamic Verification for the Phase Relation around the ε-Mg_{23} (Al, Zn)$_{30}$ Intermetallic Compound in the Mg-Zn-Al System

Yan Zheng [1], Jiaxing Sun [1,2], Kaiming Cheng [1,*], Jin Wang [1], Chengwei Zhan [1], Jingrui Zhao [3], Xitao Wang [1], Shouqiu Tang [1], Jixue Zhou [1,*], Lijun Zhang [4] and Yong Du [1,4]

1 Shandong Provincial Key Laboratory of High Strength Lightweight Metallic Materials, Advanced Materials Institute, Qilu University of Technology (Shandong Academy of Sciences), Jinan 250014, China; zhengyan08212013@163.com (Y.Z.); jiaxingsung@foxmail.com (J.S.); wangjin@sdas.org (J.W.); chengwei.zhan@sdas.org (C.Z.); xtwang@ustb.edu.cn (X.W.); tangshq@sdas.org (S.T.); yong-du@csu.edu.cn (Y.D.)
2 Laboratory of Materials Phase Equilibria and New Materials Design, School of Materials Science and Engineering, University of Science and Technology Beijing, Beijing 100083, China
3 School of Materials Science and Engineering, Shandong Jianzhu University, Jinan 250101, China; jingr_zhao@126.com
4 State Key Laboratory of Powder Metallurgy, Central South University, Changsha 410083, China; lijun.zhang@csu.edu.cn
* Correspondence: chengkm@sdas.org (K.C.); zhoujx@sdas.org (J.Z.)

Abstract: The appearance of the ε phase during the welding process can severely weaken the welding strength of dissimilar metals of Mg-Zn-Al alloy systems. An understanding of the accurate phase diagram, especially the equilibrium phase relation around the ε phase, is thus of particular importance. However, the phase interrelation near the ε-Mg_{23}(Al, Zn)$_{30}$ phase has not yet been fully studied. In this work, the local phase diagrams of the ε phase and its surrounding phases in the Mg-Zn-Al system are systematically determined by experimental investigation and thermodynamic verification. Five Mg-Zn-Al alloys and one diffusion couple were fabricated and analyzed to get accurate phase constituents and relationships adjacent to ε phase. The current experimental data obtained from Scanning Electron Microscope (SEM), X-ray diffraction (XRD), Differential Scanning Calorimetry (DSC), and Electron Probe Micro Analysis (EPMA) were further compared with the thermodynamically computed phase relations around ε phase for verification, showing good agreements. Several important conclusions are drawn based on current experimental work, which can provide supporting information for the follow-up studies on ε phase in the Mg-Zn-Al alloy systems.

Keywords: Mg-Zn-Al alloy; ε phase; phase equilibrium; calculation of phase diagram (CALPHAD); diffusion

Citation: Zheng, Y.; Sun, J.; Cheng, K.; Wang, J.; Zhan, C.; Zhao, J.; Wang, X.; Tang, S.; Zhou, J.; Zhang, L.; et al. Experimental Investigation and Thermodynamic Verification for the Phase Relation around the ε-Mg_{23} (Al, Zn)$_{30}$ Intermetallic Compound in the Mg-Zn-Al System. *Materials* **2021**, *14*, 6892. https://doi.org/10.3390/ma14226892

Academic Editor: Frank Czerwinski

Received: 17 September 2021
Accepted: 13 November 2021
Published: 15 November 2021

1. Introduction

Due to their relatively low density, good specific stiffness, specific strength and electromagnetic shielding, biocompatibility, recyclability, large hydrogen storage capacity, and high theoretical specific capacity for battery, magnesium alloys have attracted more and more attention for their application in the automotive, aerospace, biomedical, and energy industries [1–6]. For decades, efforts have been made to improve the mechanical properties, creep resistance and electrochemical stability of magnesium alloys, in order to broaden their practical applications. One of the most common methods is to design the alloy composition so as to modify the microstructure [7], and controlling over the formation of various intermetallic compounds (IMCs) is essential. In Mg-Zn-Al alloy systems, the existence of binary (γ-$Mg_{17}Al_{12}$, ε-$Mg_{23}Al_{30}$, and β-Al_3Mg_2) and ternary (φ-$Mg_5Al_2Zn_2$ and τ-Mg_{32}(Al, Zn)$_{49}$) intermetallic compounds can be manipulated to adjust the alloy performance.

For instance, the presence or absence of $Mg_{17}Al_{12}$ precipitates has a great influence on the grain structure upon hot deformation, thereby affecting the strength of the deformed magnesium alloys [8]. The discontinuous precipitation of coarse γ-$Mg_{17}Al_{12}$ phase in AZ alloys causes softening of the grain boundary at high temperatures, reducing the strength and creep resistance, thus limiting the service temperature to below 120 °C [9–12]. It has been found that the high temperature creep resistance of ZA series alloy is much better than that of AZ91 [13]. The unfavorable $Mg_{17}Al_{12}$ phase can be replaced by the thermally stable Mg-Zn-Al ternary phases, such as the τ-Mg_{32}(Al, Zn)$_{49}$ phase and φ-$Mg_5Al_2Zn_2$ phase, at high temperatures, making the creep behavior of ZA magnesium alloys with high Zn content better than that of AZ alloys [14,15]. Shi et al. [16] reported that the age hardening response of Al in Mg-6Zn-5Al alloys can be further improved due to the existence of τ-Mg_{32}(Al, Zn)$_{49}$ phase. Other intermetallic compounds, such as ε-$Mg_{23}Al_{30}$, can be detected in the welding process. For example, the ε phase was detected in the central region of the friction stir welding of AZ31 and 6061 aluminum alloys [17,18]. The formation of $Mg_{23}Al_{30}$ phase should be avoided, since it is the main weakness in the welding strength of Mg-Al dissimilar metals, as stated by Sun et al. [19]. The formation of ε-Mg_{23}(Al, Zn)$_{30}$ phase, where Zn partially occupies the sublattice of Al in the binary $Mg_{23}Al_{30}$ phase, was also detected by Wang et al. [20] when studying the welding coating of Mg-Al_xZn_{1-x} alloys using the diffusion couple technique. An understanding of the accurate phase diagram, especially the equilibrium phase relation among different intermetallic compounds for different alloy composition, is of particular importance during the alloy design for the Mg-Zn-Al system.

There are several experimental studies on the phase diagram of Mg-Zn-Al alloys. Eger et al. [21] presented the first systematic investigation on the liquidus surface. Bergman et al. [22,23] and Clark et al. [24,25] then studied the ternary intermetallic compounds of τ phase and φ phase, respectively. Later, the first summarizing review was presented by Willey et al. [26], based on which Liang et al. [27] and Liang et al. [28] presented detailed thermodynamic models of the Mg-Zn-Al system, respectively. More recent experimental studies include the isothermal section of the Mg-rich corner in the Mg-Zn-Al system determined at 300 °C, 320 °C, and 335 °C by Ren et al. [29–31], and the Mg-rich phase equilibria and solidification behaviors studied by Ohno et al. [32]. Although there is a large volume of experimental data of the Mg-Zn-Al phase diagram, the phase interrelation near the ε-Mg_{23} (Al, Zn)$_{30}$ phase has not yet been fully studied. In this work, we present an experimental investigation and thermodynamic calculation of the Mg-Zn-Al system, with a particular focus on the ε-Mg_{23} (Al, Zn)$_{30}$ phase and its equilibrium phase relations. Alloys with different compositions and diffusion couples were fabricated and analyzed by Scanning Electron Microscope (SEM), X-ray diffraction (XRD), Differential Scanning Calorimetry (DSC), and Electron Probe Micro Analysis (EPMA). The current experimental results, together with the data from previous research, were then compared with the available thermodynamic description of the Mg-Zn-Al system for further discussion.

2. Experimental Procedure

In this study, five alloys with different compositions were prepared from high-purity Mg (99.9%), Al (99.99%), and Zn (99.99%), which were melted in an argon atmosphere by resistance furnace. The nominal compositions of the samples were selected according to the isothermal section at 350 °C, which was calculated based on the previous thermodynamic description [27] containing the ε-Mg_{23} (Al, Zn)$_{30}$ phase, including the ε, $\varepsilon + \gamma$, $\varepsilon + \beta$, $\varepsilon + \gamma + \tau$, and $\varepsilon + \beta + \tau$ phase regions, as shown in Table 1. During alloying, the resistance furnace was first heated up to 710 °C, and the Al blocks were placed in a stainless-steel crucible in the controlled atmosphere furnace before the addition of Mg and Zn. The samples were then heated up to 720 °C and held for 25 min before being cooled down to 690 °C before casting. The heating process was carefully controlled to avoid massive volatilization. The melts poured in graphite crucibles for casting were then annealed at 350 °C for 305 h in an argon atmosphere to reach homogenization.

Table 1. The nominal compositions of current Mg-Zn-Al alloys and the phases constitutions and concentrations determined by XRD and EPMA.

Samples	Nominal Compositions (at.%)			Phases and Composition (at.%) at 350 °C			
	Mg	Zn	Al	Phase	Mg	Zn	Al
1	43.35	3.04	53.61	ε	42.92	3.90	53.18
2	47.73	2.87	49.40	ε	44.72	4.05	51.23
				γ	51.80	2.89	45.31
3	41.21	2.50	56.29	ε	43.72	3.27	53.01
				β	39.64	2.64	57.72
4	47.32	7.60	45.08	ε	46.38	7.70	45.93
				γ	54.04	4.09	41.87
				τ	45.03	11.28	43.69
5	41.33	7.00	51.68	ε	43.77	7.03	49.20
				β	39.29	5.39	55.32
				τ	41.06	10.50	48.44

Two end members, i.e., $MgZn_2$ and $Mg_{17}Al_{12}$, were selected in this work to form a diffusion couple, in order to observe the diffusion path across ε-Mg_{23} (Al, Zn)$_{30}$ phase. The end-member alloys of intermetallic compounds were fabricated according to the stoichiometric ratio using an identical processing route to that detailed in our previous work [14]. The alloys were annealed at 450 °C for 12 h in an argon atmosphere, and then cut into 8 mm × 5 mm × 3 mm blocks before being polished in ethanol. The treated surface was bonded with a tantalum clamp before it was sealed in the vacuum quartz tube to avoid oxidation during the diffusion annealing process. The packaged diffusion couple was annealed in a 410 °C furnace for 4 h to form intermetallic compound layers with a temperature error within ± 1 °C.

The annealed alloy samples were investigated by XRD (Rigaku D-Max/2550VB+, Rigaku, Japan) and SEM (Zeiss EVO M10, Zeiss, Germany) for phase identification. The compositions of the annealed samples and the phase constitutions were determined by EPMA (JXA-8230, JEOL, Japan). These samples were further subjected to DSC analysis (NETZSCH STA449F3A, NETZSCH, Germany) to detect the phase transformation temperature, where a heating rate of 10 K/min and a temperature range of 32 °C to 600 °C were applied. The annealed diffusion couples were unclamped from the Ta jigs and subjected to wire-cutting parallel to the diffusion direction. The cutting surfaces were metallographically prepared for analysis of the microstructure and diffusion path by SEM and EPMA.

3. Results and Discussion

3.1. *Phase Equilibrium Containing ε Intermetallic Compound*

Figure 1 shows the SEM images of the five alloy samples after heat treatment, and the corresponding XRD patterns are summarized in Figure 2. Table 1 lists the composition of phases that appeared in each sample measured by EPMA. The phase constitutions of each alloy can be determined by analyzing all the SEM images, XRD patterns, and EPMA results. It can be seen from Figure 2a that sample 1 contained a single ε phase, which was consistent with the uniform microstructure of Figure 1a. The EPMA results in Table 1 indicate a solid dissolution of 3.9% Zn in the ε single phase. From Figure 1b, it can be seen that the light gray and dark gray phases distributed evenly in sample 2, which was selected to be within a two-phase region of ε and γ. According to the XRD and EPMA analysis, as shown in Figure 2b and Table 1, the two phases were determined as γ phase and ε phase, respectively. The content of Zn in ε phase was 4.05% when ε phase is in equilibrium with γ phase. In Figure 1c, it can be observed that sample 3 contains bright and dark phase regions with clear phase boundaries, which are ε and β phase according to the XRD patterns shown in Figure 2c. The ε phase is brighter than the β phase due to the higher content of Zn (3.27% in ε phase and 2.64% in β phase). The SEM image of Figure 1d shows a three-phase region with different colors of bright, light gray, and dark gray. Upon analysis of the EPMA result and the XRD pattern shown in Figure 2d, the bright, light gray, and dark gray phase regions corresponded to τ, ε, and γ phase, respectively. That is, sample 4 is a ternary alloy within the three-phase equilibrium of ε + γ + τ, showing a large solid solubility of 7.7% Zn in ε phase. Figure 1e exhibits an obvious three-phase region in sample 5. According to Figure 2e and Table 1, the light, dark, and light gray phase regions correspond to τ, β, and ε phase, respectively, and the solid solubility of Zn in ε phase is determined as 7.03%.

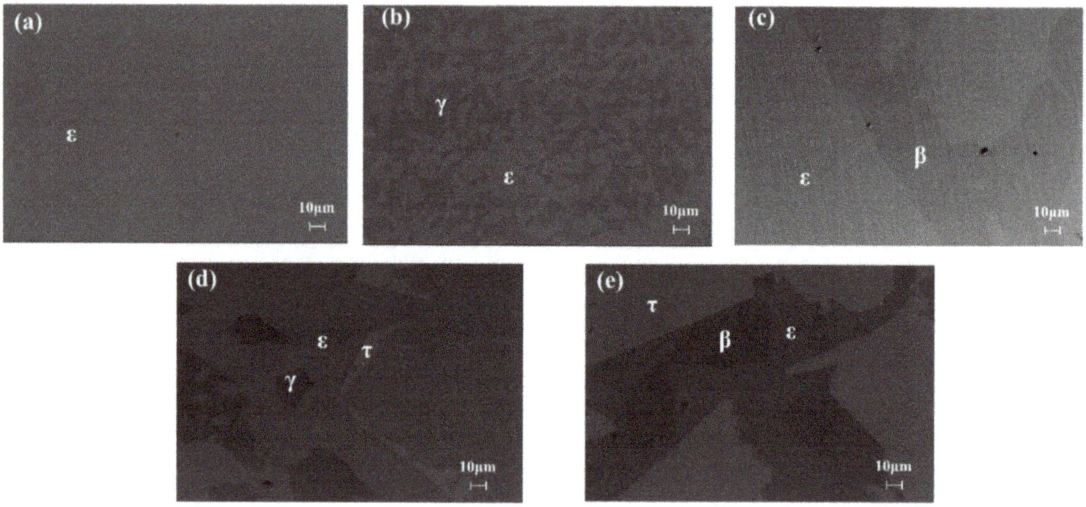

Figure 1. SEM images of samples annealed at 350 °C for 305 h: (**a**–**e**) represent samples 1, 2, 3, 4, and 5, respectively.

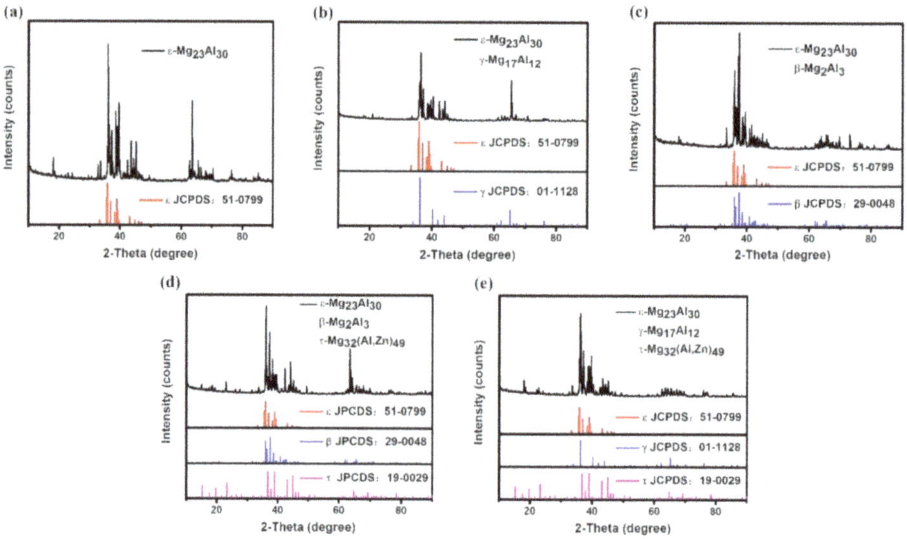

Figure 2. XRD patterns of samples annealed at 350 °C for 305 h: (**a**–**e**) represent samples 1, 2, 3, 4, and 5, respectively.

The experimental data points obtained by analyzing the XRD patterns and EPMA results are shown in Figure 3a, along with the isothermal section calculated at 350 °C. The nominal compositions of samples 1–5 are within the corresponding phase regions of ε, ε + γ, ε + β, ε + γ + τ, and ε + β + τ, which confirms that the phase constitutions of the currently prepared samples are located in the desired phase region of the Mg-Zn-Al alloy system. Tie lines for the phase regions of each alloy sample are also plotted, showing reasonable agreements with the calculated isothermal section. There are two three-phase equilibrium regions directedly connected with ε phase, which can be seen from samples 4 and 5 in the phase regions of ε + γ + τ and ε + β + τ. The maximum solid solubility of Zn in ε phase was determined to be 7.7% at 350 °C.

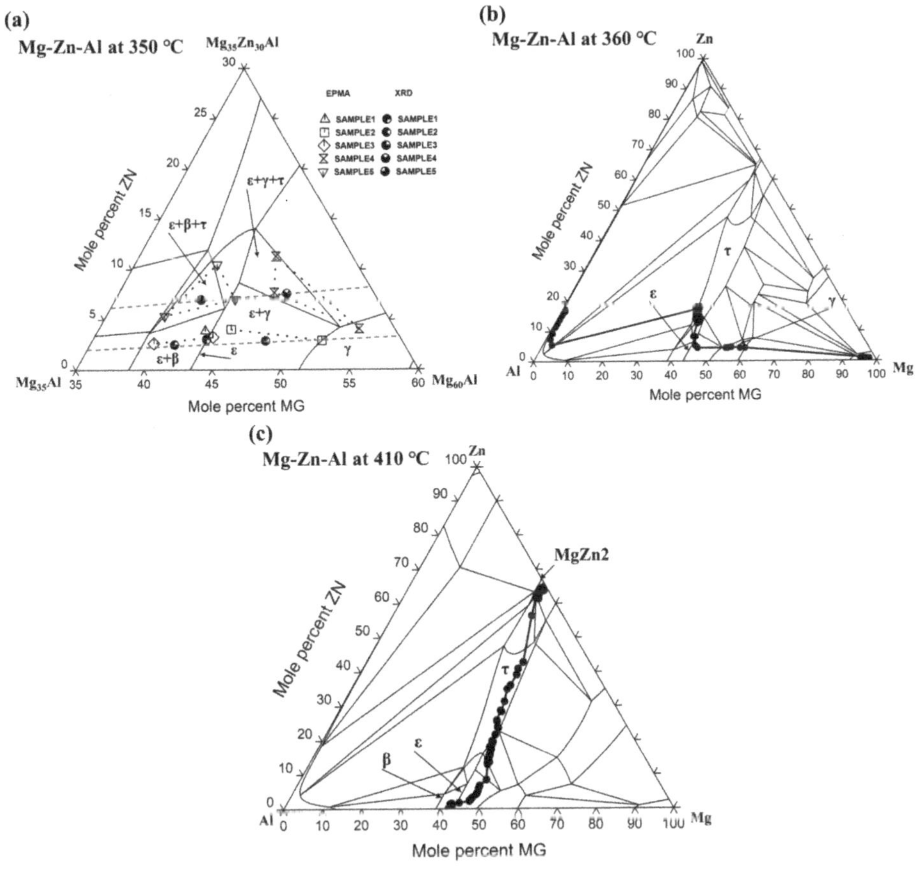

Figure 3. Isothermal section of the Mg-Zn-Al system (**a**) close to the ε phase calculated at 350 °C along with the data from XRD and EPMA; (**b**) calculated at 360 °C along with the diffusion profile of Al-Mg20Zn from literature; (**c**) calculated at 410 °C along with the diffusion profile of $MgZn_2$-Al_3Mg_2 diffusion couple.

3.2. Vertical Sections around ε Intermetallic Compound

In order to study the extension of ε phase relation in the direction of temperature, the phase transformation temperatures in samples 1–5 were then measured by DSC. The experimental heating curves of all five samples determined by DSC are shown in Figure 4. The original DSC data are presented in the Supplementary Materials. The phase transformation temperatures were obtained by acquiring the intersection point of the tangent lines at the initial onset temperature, while the peak values were determined to be the liquidus temperatures. These data are also shown in Figure 5, where the vertical sections across samples 1–3 and 4–5 are calculated respectively for comparison. These two vertical sections are marked using red dashed lines in Figure 3a.

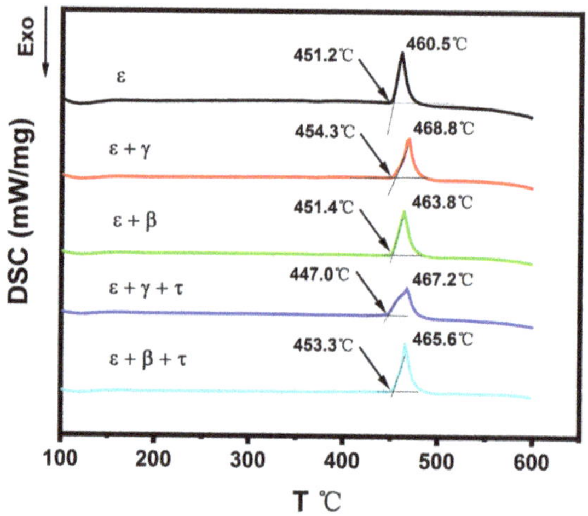

Figure 4. Current DSC curves of samples 1–5 with a heating rate of 10 K min^{-1}.

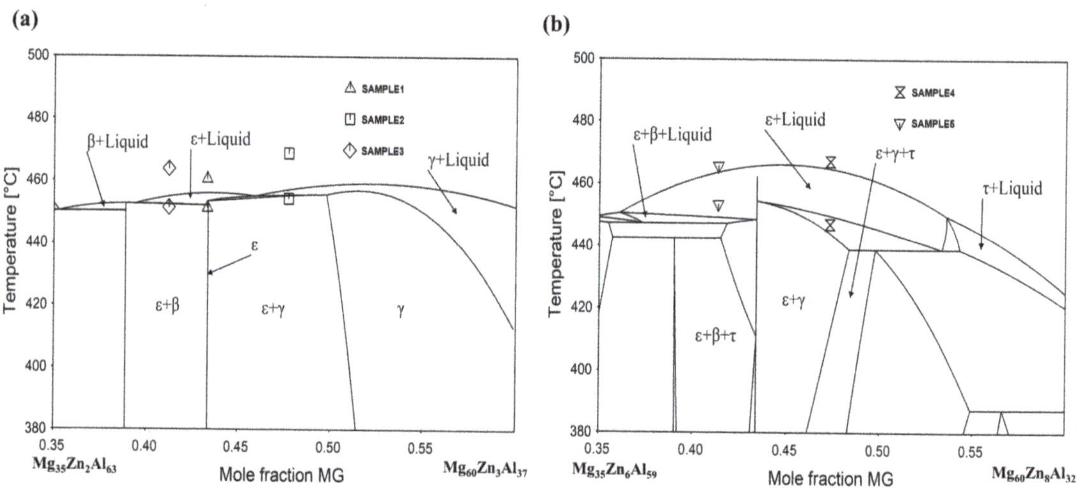

Figure 5. Calculated vertical section of the Mg-Zn-Al system (**a**) along with the DSC data of samples 1–3 and (**b**) along with the DSC data of samples 4 and 5.

It can be seen that the invariant equilibria temperature and liquidus temperature for sample 1, i.e., single ε phase, are 451.2 °C and 460.5 °C. These two temperatures correspond to the calculated vertical section in Figure 5a, where the phase should first enter into a two-phase region of liquid + ε at 451 °C before being completely melted at 461 °C. For samples 2 and 3, the temperature for eutectic reactions L = ε + γ and L = ε + β were measured to be 454.3 °C and 451.4 °C, showing a good agreement with the calculated vertical section in Figure 5a. However, the measured melting temperatures for samples 1–3 were about 10 K higher than the calculation results. The deviation can be attributed to the relatively fast heating rate of 10 K/min during the DSC analysis, where signals of the liquidus overlapped with the invariant reaction temperature. In Figure 5b, the observed thermal signals at 447 °C and 453.3 °C correspond to the temperatures when samples 4 and 5 entering the ε + L two phase region, respectively. The measured melting temperatures of 467.2 °C and 465.6 °C are in good agreement with the calculation results. Based on the calculated vertical section of Figure 5, it can be further noted that the ε phase with the dissolved 3.9% Zn (sample 1) persisted up to 456 °C, and this temperature range extended to 465 °C when the 7.7% Zn (sample 4) dissolved in ε phase, indicating that Zn is a stabilizer for ε phase within its solid solubility.

3.3. *Diffusion Path Related to ε Intermetallic Compound*

In this study, the composition profiles of two diffusion couples, one from previous experimental work by Wang et al. [20] and the other from our experiment, were presented, along with the calculated isothermal section of the Mg-Zn-Al system at certain temperatures, to further verify the phase relations around ε intermetallic compounds.

Figure 3b shows the diffusion path of the diffusion couple Mg-Al20Zn determined at 360 °C in previous research [20], where the diffusion profile moves across the hcp-Mg, γ, τ, and ε single-phase region in sequence. According to [20], Zn was found in all three IMC layers of γ, τ, and ε, and its content increased discontinuously towards the Al-Zn substrate. The average composition of the thickest layer formed adjacent to the Al-Zn side of the diffusion couple was $Al_{45}Mg_{40}Zn_{15}$, which is further verified by TEM characterization to be the τ-$(Al, Zn)_{49}Mg_{32}$ phase. The β-Al_3Mg_2 phase usually seen in Al-Mg binary systems was replaced by τ phase in this diffusion, indicating that high Zn content can be sufficient to change the diffusion path to fully suppress the formation of the undesirable β phase. The addition of Zn can significantly retard the thickening rate of the γ-$Al_{12}Mg_{17}$ phase, while it is weak at inhibiting the thickening rate of the overall IMC reaction layer in an Al-Mg diffusion couple. It can be further noted from the diffusion path that after entering the ε single-phase region from γ phase, the content of Zn increases clearly from 5% Zn to 8% Zn before entering the τ phase region, which is in good agreement with the current maximum solid solubility of 7.7% Zn determined form the alloy sample 4, as shown in Table 1.

Figure 6 shows the EPMA and SEM results of the $MgZn_2$-Al_3Mg_2 diffusion couples used in this study after annealing at 410 °C for 4 h, where the composition profile corresponding to the metallographic structure can be detected in detail. Figure 6a presents the EPMA results of a series of points selected on a line perpendicular to the phase interface along the diffusion direction, where the two interlayers of τ and ε phase can be detected. In Figure 6b, layers of the intermetallic compounds τ phase can be easily observed between $MgZn_2$ and Al_3Mg_2, while ε phase is difficult to distinguish without noticing the composition jump between τ and Al_3Mg_2 phase, as shown in Figure 6a. Comparing the two figures, it can be seen that the thickness of τ phase is greater than that of ε phase, indicating a faster elemental diffusion within the τ phase. This situation agrees well with the experimental detection by Wang et al. [20]. It can be further noticed that the composition profiles within τ phase and ε phase exhibit clear nonlinear distributions, implying that the diffusivities of the elements are strongly composition-dependent.

The composition profile in Figure 6a is further plotted in the isothermal section of the Mg-Zn-Al system computed at 410 °C to detect the diffusion path, as shown in Figure 3c. It can be seen that ε phase lies between τ and Al_3Mg_2 phase, exhibiting a wide solid solubility range of Zn. The maximum homogeneity content of Zn in ε phase is 8.5% on the diffusion path, which is similar to the measured value of the current alloy sample 4 in Table 1. By comparing sample 4 with sample 1, the increment of Zn content from 3.9% to 7.7% leads to a clear decrease in Al content from 53.18% to 41.38%, indicating the Zn atoms dissolved in ε phase preferred to occupy the sites of the Al atoms. This result is consistent with the previous experimental findings by Wang et al. [20]. However, the diffusion path of the experimental data in this study deviates from the calculated single ε phase region to exhibit a higher Mg content, which was also detected in sample 4, where the Mg content ε phase (46.38%) is higher than that of sample 1 with a single ε phase (42.92%). As a result, there should be a homogenization range of Mg in the ε phase from 43.77 % (sample 5) to 46.38 % (sample 4) within the isothermal section, which may have been due to the Zn dissolution. Further modification of the thermodynamic description of Mg-Zn-Al is required to take these experimental data into consideration.

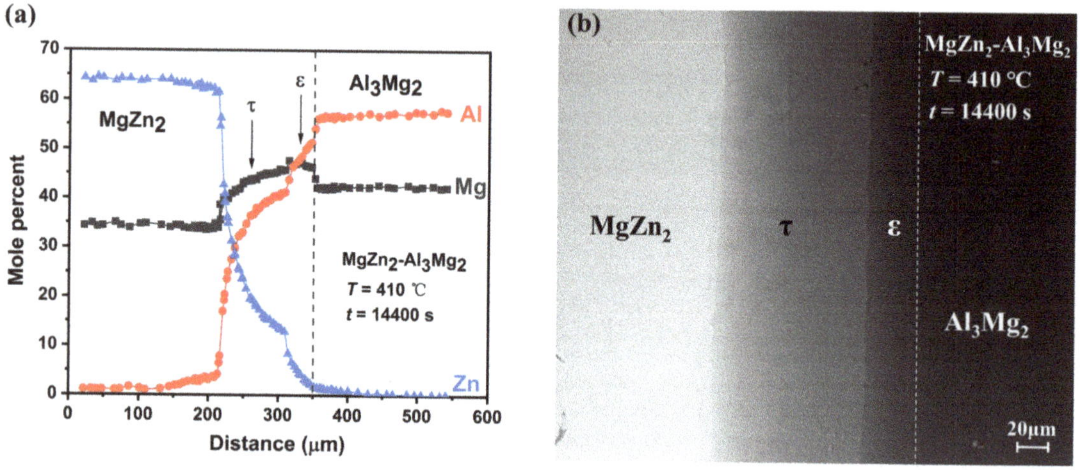

Figure 6. (**a**) EPMA data of $MgZn_2$-Al_3Mg_2 diffusion couple annealed at 410 °C for 4 h. (**b**) SEM image of $MgZn_2$-Al_3Mg_2 diffusion couple annealed at 410 °C for 4 h.

4. Conclusions

The phase diagram of the Mg-Zn-Al system focusing on the ε intermetallic compound and its surrounding phase relationships was obtained through experimental study and thermodynamic calculation. Some important conclusions are highlighted as follows.

- The existence of ε, ε + γ, ε + β, ε + γ + τ, and ε + β + τ phase regions in the isothermal section of the Mg-Zn-Al alloy system were confirmed by the analysis of the currently prepared samples. The maximum solid solubility of Zn in ε phase was determined as 7.7% at 350 °C.
- The melting temperature of ε phase with dissolved 3.9% Zn was 456 °C, which increased to 465 °C as the Zn content rose to 7.7%, indicating that Zn can improve the high temperature stability of ε phase within its solid solubility.
- Zn atoms dissolved in ε phase preferred to replace the sites of the Al atoms to decrease the total content of Al, and the homogenization range of Mg in the ε phase was from 43.77% to 46.38%. Further modification of the thermodynamic description of Mg-Zn-Al in this region is required.
- The $MgZn_2$-Al_3Mg_2 diffusion couple at 410 °C in this study showed a thicker interlayer of τ phase than of ε phase, indicating a faster elemental diffusion within τ phase. Moreover, the nonlinear elemental distributions within τ and ε IMCs imply the composition dependence of elemental diffusivities.

Supplementary Materials: The following are available online at https://www.mdpi.com/article/10.3390/ma14226892/s1, Figure S1: The DSC primary data of sample 1 (ε), Figure S2: The DSC primary data of sample 2 (ε + γ), Figure S3: The DSC primary data of sample 3 (ε + β), Figure S4: The DSC primary data of sample 4 (ε + γ + τ), Figure S5: The DSC primary data of sample 5 (ε + β + τ).

Author Contributions: Data curation, Y.Z.; formal analysis, J.S., C.Z. and J.Z. (Jingrui Zhao); funding acquisition, K.C. and J.Z. (Jixue Zhou); methodology, K.C., J.Z. (Jixue Zhou), L.Z. and Y.D.; project administration, K.C., J.W., X.W., S.T. and J.Z. (Jixue Zhou); writing—original draft, Y.Z.; writing—review and editing, Y.Z., J.S., K.C., J.W., C.Z., X.W., S.T., J.Z. (Jingrui Zhao), J.Z. (Jixue Zhou), L.Z. and Y.D. All authors will be informed about each step of manuscript processing including submission, revision, revision reminder, etc. via emails from our system or the assigned Assistant Editor. All authors have read and agreed to the published version of the manuscript.

Funding: This work was funded by the National Natural Science Foundation of China (No. 51801116, 51901117, 52001176 and 52001187), the Shandong Province Key Research and Development Plan (No. 2019GHZ019, 2019JZZY010364, 2019JZZY020329), the Youth Innovation and Technology Support Program of Shandong Provincial Colleges and Universities (No. 2020KJA002), and The Innovation Pilot Project for Fusion of Science, Education and Industry (International Cooperation) from Qilu University of Technology (Shandong Academy of Sciences) (No. 2020KJC-GH03).

Institutional Review Board Statement: Not applicable.

Informed Consent Statement: Not applicable.

Data Availability Statement: All data are available from the corresponding author on reasonable request.

Acknowledgments: This work is supported by the National Natural Science Foundation of China (No. 51801116, 51901117, 52001176 and 52001187), the Shandong Province Key Research and Development Plan (No. 2019GHZ019, 2019JZZY010364, 2019JZZY020329), and the Youth Innovation and Technology Support Program of Shandong Provincial Colleges and Universities (No. 2020KJA002). The Innovation Pilot Project for Fusion of Science, Education and Industry (International Cooperation) from Qilu University of Technology (Shandong Academy of Sciences) (No. 2020KJC-GH03) is also acknowledged.

Conflicts of Interest: The authors declare no conflict of interest.

References

1. Mordike, B.L.; Ebert, T. Magnesium—Properties—Applications—Potential. *Mater. Sci. Eng. A* **2001**, *302*, 37–45. [CrossRef]
2. Zeng, R.C.; Zhang, J.; Huang, W.J.; Dietzel, W.; Kainer, K.U.; Blawert, C.; Ke, W. Review of studies on corrosion of magnesium alloys. *Trans. Nonferrous Met. Soc. China* **2006**, *16*, 763–771. [CrossRef]
3. Zhang, J.; Guo, Z.X.; Pan, F.; Li, Z.; Luo, X. Effect of composition on the microstructure and mechanical properties of Mg–Zn–Al alloys. *Mater. Sci. Eng. A* **2007**, *456*, 43–51. [CrossRef]
4. Van Phuong, N.; Moon, S.; Chang, D.; Lee, K.H. Effect of microstructure on the zinc phosphate conversion coatings on magnesium alloy AZ91. *Appl. Surf. Sci.* **2013**, *264*, 70–78. [CrossRef]
5. Yeganeh, M.; Mohammadi, N. Superhydrophobic surface of Mg alloys: A review. *J. Magnes. Alloys* **2018**, *6*, 59–70. [CrossRef]
6. Song, J.; She, J.; Chen, D.; Pan, F. Latest research advances on magnesium and magnesium alloys worldwide. *J. Magnes. Alloys* **2020**, *8*, 1–41. [CrossRef]
7. Aryshenskii, E.; Hirsch, J.; Konovalov, S.; Aryshenskii, V.; Drits, A. Influence of Mg Content on Texture Development during Hot Plain-Strain Deformation of Aluminum Alloys. *Metals* **2021**, *11*, 865. [CrossRef]
8. Shi, R. Nonisothermal dissolution kinetics on $Mg_{17}Al_{12}$ intermetallic in Mg-Al alloys. *J. Magnes. Alloys* **2021**. [CrossRef]
9. Luo, A.A.; Balogh, M.P.; Powell, B.R. Creep and microstructure of magnesium-aluminum-calcium based alloys. *Metall. Mater. Trans. A* **2002**, *33*, 567–574. [CrossRef]
10. Kabirian, F.; Mahmudi, R. Effects of Zr Additions on the Microstructure and Impression Creep Behavior of AZ91 Magnesium Alloy. *Metall. Mater. Trans. A* **2010**, *41*, 3488–3498. [CrossRef]
11. Karakulak, E. A review: Past, present and future of grain refining of magnesium castings. *J. Magnes. Alloys* **2019**, *7*, 355–369. [CrossRef]
12. Tang, Y.; Li, Y.; Zhao, W.; Roslyakova, I.; Zhang, L. Thermodynamic descriptions of quaternary Mg–Al–Zn–Bi system supported by experiments and their application in descriptions of solidification behavior in Bi-additional AZ casting alloys. *J. Magnes. Alloys* **2020**, *8*, 1238–1252. [CrossRef]
13. Zhang, Z.; Couture, A. An investigation of the properties of Mg-Zn-Al alloys. *Scr. Mater.* **1998**, *39*, 45–53. [CrossRef]
14. Cheng, K.; Sun, J.; Xu, H.; Wang, J.; Zhou, J.; Tang, S.; Wang, X.; Zhang, L.; Du, Y. On the temperature-dependent diffusion growth of ϕ $Mg_5Al_2Zn_2$ ternary intermetallic compound in the Mg-Al-Zn system. *J. Mater. Sci.* **2020**, *56*, 3488–3497. [CrossRef]
15. Cheng, K.; Sun, J.; Xu, H.; Wang, J.; Zhan, C.; Ghomashchi, R.; Zhou, J.; Tang, S.; Zhang, L.; Du, Y. Diffusion growth of ϕ ternary intermetallic compound in the Mg-Al-Zn alloy system: In-situ observation and modeling. *J. Mater. Sci. Technol.* **2021**, *60*, 222–229. [CrossRef]
16. Shi, Z.Z.; Zhang, W.Z. Investigation on the microstructure of a τ-$Mg_{32}(Al, Zn)_{49}$ strengthened Mg-Zn-Al alloy with relatively low Zn content. *Phase Transit.* **2012**, *85*, 41–51. [CrossRef]
17. Wang, D.; Liu, J.; Xiao, B.; Ma, Z. Mg/Al reaction and Mechanical properties of Al alloy/Mg alloy fricton stir weding joints. *Acta Metall. Sin.* **2010**, *46*, 589–594. [CrossRef]

18. Mohammadi, J.; Behnamian, Y.; Mostafaei, A.; Izadi, H.; Saeid, T.; Kokabi, A.H.; Gerlich, A.P. Friction stir welding joint of dissimilar materials between AZ31B magnesium and 6061 aluminum alloys: Microstructure studies and mechanical characterizations. *Mater. Charact.* **2015**, *101*, 189–207. [CrossRef]
19. Sun, L.; Hui, W.; Zhou, Y.; Zhai, W.; Dong, H.; Liu, Y.; Gao, Q.; Dang, M.; Peng, J. The Electronic Structural and Elastic Properties of $Mg_{23}Al_{30}$ Intermediate Phase under High Pressure. *Crystals* **2020**, *10*, 642. [CrossRef]
20. Wang, Y.; Prangnell, P.B. Evaluation of Zn-rich coatings for IMC reaction control in aluminum-magnesium dissimilar welds. *Mater. Charact.* **2018**, *139*, 100–110. [CrossRef]
21. Eger, G. Study of the Constitution of Ternary Magnesium-Aluminum-Zinc Alloys. *Z. Fur Met.* **1913**, *4*, 29–128.
22. Bergman, G.; Waugh, J.L.T.; Pauling, L. Crystal Structure of the Intermetallic Compound $Mg_{32}(AI,Zn)_{49}$ and Related Phases. *Nature* **1952**, *169*, 1057–1058. [CrossRef]
23. Bergman, G.; Waugh, J.L.; Pauling, L. The crystal structure of the metallic phase $Mg_{32}(Al,Zn)_{49}$. *Acta Crystallogr.* **1957**, *10*, 254–259. [CrossRef]
24. Clark, J.B.; Rhines, F.N. Diffusion layer formation in the ternary system aluminum-magnesium-zinc. *Trans. Am. Math. Soc.* **1959**, *51*, 199–221.
25. Clark, J.B. Phase relations in the magnesium-rich region of the Mg-Al-Zn phase diagram. *Trans. Am. Math. Soc.* **1961**, *53*, 295–306.
26. Willey, L.D. *Metals Handbook*; American Society for Metals: Novelty, OH, USA, 1973; Volume 8, pp. 397–399.
27. Liang, P.; Tarfa, T.; Robinson, J.A.; Wagner, S.; Ochin, P.; Harmelin, M.G.; Seifert, H.J.; Lukas, H.L.; Aldinger, F. Experimental investigation and thermodynamic calculation of the Al-Mg-Zn system. *Thermochim. Acta* **1998**, *314*, 87–110. [CrossRef]
28. Liang, H.; Chen, S.L.; Chang, Y.A. A Thermodynamic Description of the Al-Mg-Zn System. *Metall. Mater. Trans. A* **1997**, *28*, 1725–1734. [CrossRef]
29. Ren, Y.P.; Qin, G.W.; Pei, W.L.; Guo, Y.; Zhao, H.D.; Li, H.X.; Jiang, M.; Hao, S.M. The α-Mg solvus and isothermal section of Mg-rich corner in the Mg-Zn-Al ternary system at 320 °C. *J. Alloys Compd.* **2009**, *481*, 176–181. [CrossRef]
30. Ren, Y.P.; Qin, G.W.; Pei, W.L.; Li, S.; Guo, Y.; Zhao, H.D. Phase equilibria of Mg-rich corner in Mg-Zn-Al ternary system at 300 °C. *Trans. Nonferrous Met. Soc. China* **2012**, *22*, 241–245. [CrossRef]
31. Ren, Y.P.; Sun, S.N.; Wang, L.Q.; Guo, Y.; Li, H.X.; Li, S.; Qin, G.W. Isothermal section of Mg-rich corner in Mg-Zn-Al ternary system at 335 °C. *Trans. Nonferrous Met. Soc. China* **2014**, *24*, 3405–3412. [CrossRef]
32. Ohno, M.; Mirkovic, D.; Schmid-Fetzer, R. Phase equilibria and solidification of Mg-rich Mg-Al-Zn alloys. *Mater. Sci. Eng. A* **2006**, *421*, 328–337. [CrossRef]

materials

Article

Effect of High Strain Rate on Adiabatic Shearing of α+β Dual-Phase Ti Alloy

Fang Hao [1,2], Yuxuan Du [2,*], Peixuan Li [1], Youchuan Mao [1,2], Deye Lin [3], Jun Wang [1], Xingyu Gao [4], Kaixuan Wang [2], Xianghong Liu [2], Haifeng Song [4], Yong Feng [1,2], Jinshan Li [1] and William Yi Wang [1,*]

1 State Key Laboratory of Solidification Processing, Northwestern Polytechnical University, Xi'an 710072, China; haofang85@163.com (F.H.); peixuan_li@mail.nwpu.edu.cn (P.L.); myc2525@163.com (Y.M.); nwpuwj@nwpu.edu.cn (J.W.); yfeng@c-wst.com (Y.F.); ljsh@nwpu.edu.cn (J.L.)
2 Western Superconducting Technologies Co., Ltd., Xi'an 710018, China; kingsin@c-wst.com (K.W.); xhliu@c-wst.com (X.L.)
3 CAEP Software Center for High Performance Numerical Simulation, Institute of Applied Physics and Computational Mathematics, Beijing 100088, China; lin_deye@iapcm.ac.cn
4 Laboratory of Computational Physics, Institute of Applied Physics and Computational Mathematics, Beijing 100088, China; gao_xingyu@iapcm.ac.cn (X.G.); song_haifeng@iapcm.ac.cn (H.S.)
* Correspondence: Eason@c-wst.com (Y.D.); wywang@nwpu.edu.cn (W.Y.W.)

Abstract: In the present work, the localized features of adiabatic shear bands (ASBs) of our recently designed damage tolerance α+β dual-phase Ti alloy are investigated by the integration of electron backscattering diffraction and experimental and theoretical Schmid factor analysis. At the strain rate of 1.8×10^4 s^{-1} induced by a split Hopkinson pressure bar, the shear stress reaches a maximum of 1951 MPa with the shear strain of 1.27. It is found that the α+β dual-phase colony structures mediate the extensive plastic deformations along α/β phase boundaries, contributing to the formations of ASBs, microvoids, and cracks, and resulting in stable and unstable softening behaviors. Moreover, the dynamic recrystallization yields the dispersion of a great amount of fine α grains along the shearing paths and in the ASBs, promoting the softening and shear localization. On the contrary, low-angle grain boundaries present good resistance to the formation of cracks and the thermal softening, while the non-basal slipping dramatically contributes to the strain hardening, supporting the promising approaches to fabricate the advanced damage tolerance dual-phase Ti alloy.

Keywords: deformation and fracture; microstructure; adiabatic shear bands; Schmid factor; dual-phase

Citation: Hao, F.; Du, Y.; Li, P.; Mao, Y.; Lin, D.; Wang, J.; Gao, X.; Wang, K.; Liu, X.; Song, H.; et al. Effect of High Strain Rate on Adiabatic Shearing of α+β Dual-Phase Ti Alloy. *Materials* **2021**, *14*, 2044. https://doi.org/10.3390/ma14082044

Academic Editor: Elena Pereloma

Received: 1 April 2021
Accepted: 15 April 2021
Published: 19 April 2021

1. Introduction

Titanium (Ti) alloys are widely used in aerospace and biomedicine, attributed to their excellent high specific strength, good corrosion resistance, and high heat resistance [1–3]. In the development of advanced damage-tolerant Ti alloys, it is a challenge to dramatically enhance their mechanical properties under extreme conditions, such as high temperature and pressure, severe corrosion environments, or high loading rates. Recently, defect engineering has been considered as an effective strategy for modifying the local microstructures, properties, and performance of advanced metal materials [2,4–8]. Without changing the chemical composition, the introduction of defects provides more freedom to optimize microstructures and mechanical properties [4,5,8]. In particular, dual-phase materials [9–11] present ultra-strong and ductile behaviors through solid solution strengthening, the grain refinement effect, and precipitation hardening, which excellently deal with the planar defects, including stacking faults, grain boundaries, and phase boundaries. Through combining a high volume fraction of pyramidally arranged non-shearable super refined α precipitates in the constrained β matrix, the ultimate strength of α+β dual-phase Ti-15Mo-3Nb-2.7Al-0.2Si (wt %) alloy (the β-21S or TB8 alloy) can be optimized in the range of 1–2 GPa [12]. Similarly, the α+β dual-phase Ti-3Mo-3Cr-2Fe-2Al alloy presents an excellent combination of ultimate tensile strength and ductility at 1324 MPa and 0.37, respectively [13]. Attributed to the twinning induced plasticity (TWIP) and transformation-induced plasticity (TRIP) effects, the strain-transformable Ti alloys display a superior combination of strength, ductility, and strain-hardening [14]. The sliding of α/β interfacial and colony boundaries plays the dominant role in the deformation of a newly developed Ti-0.85Al-4V-0.25Fe-0.25Si-0.15O alloy [15]. Moreover, in hexagonal close-packed (HCP) structures, the selection of key slip systems is generally controlled by the stacking fault energies in the basal and the prismatic planes, which dominate the slip process to form either twins or dislocations [2,7,16]. The Schmid law [1,16–18] provides an efficient approach to estimate the most possible activated slip variants for the given orientation relationships, the tensile direction and the crystallographic orientations, and thus to estimate the corresponding plastic deformation behaviors.

The strain-rate-dependent mechanical properties and structural evolutions should be revealed comprehensively [19–23]. Adiabatic shear bands (ASBs) are obtained at a high strain rate, yielding from the competition between thermal softening and strain/stress-rate hardening within the narrow planar region [20,24,25]. Although twinning is the most dominant plastic deformation behavior for HCP structures at ambient temperature, it is suppressed by dislocation gliding at high temperatures, which is for colony microstructures [17,26,27]. For instance, in the commercial $\alpha+\beta$ dual-phase Ti-10V-2Fe-3Al alloy with TWIP/TRIP properties, the strain-induced martensite α'' is treated as a relaxation mechanism at the α/β interface [14]. With the enhancement in strain rate from 500 to 1000 s^{-1}, the stress-induced martensitic transformation was observed in the TB8 alloy [28]. At the strain rate of 1000 s^{-1}, multiple deformation mechanisms were comprehensively investigated in Ti-25Nb-3Zr-3Mo-2Sn, presenting the dynamic deformation sequences as {332}<113> and {112}<111> mechanical twining + stress-induced α'' and ω phase transformations + dislocation slip at 293K $\rightarrow$ {332}<113> and {112}<111> mechanical twining + dislocation slip at 573K $\rightarrow$ dislocation slip only at 873 K [29]. The softening behavior of dual-phase Ti17 alloy (Ti-5Al-4Cr-4Mo-2Sn-2Zr) at a high temperature is attributed to the combinations of dynamic recrystallization, dynamic transformation, adiabatic heating, and morphological texture evolution [30]. Instead of thermal softening mechanisms, the microstructure evolutions or transformations play an important role in the initiation of ASB [21,24,31–33]. It is essential to reveal the microstructure–property relationship under dynamic loading in order to develop advanced damage-tolerant Ti alloys. Therefore, the $\alpha+\beta$ dual-phase Ti alloy (Ti-6Al-2Cr-2Mo-2Nb-2Sn-2Zr) fabricated by Western Superconducting Technologies Co, Ltd (Xi'an, China) was utilized to comprehensively reveal the localized features of ASBs in the present work.

2. Materials and Methods

In the present work, the hat-shaped specimen [34] of the annealed forging $\alpha+\beta$ Ti dual-phase alloy was deformed at ambient temperature using split Hopkinson pressure bar (SHPB), which conventionally captures the well-controlled and designed localized shear in the study of large strain and high strain rate deformation. In order to show earlier behaviors of shear localization, a lower shear strain rate of 1.8×10^4 s^{-1} was deliberately selected. Three samples were utilized in the SHPB tests, but the one yielding the best results is reported and was analyzed further. Afterward, the standard metallographic procedure was utilized to prepare all samples, which were etched in a solution of 8% tetrafluoroboric acid for 90 s for the final characterizations. For the reference state, the static tensile properties were tested at the strain rate of 0.01 based on the national standard of GB/T 228.1-2010. Electron backscattered diffraction (EBSD) analysis was executed on a JEOL 7800F field emission SEM equipped with SymmetryTM (Oxford Instruments, Oxfordshire, UK). The accelerating voltage was 20 kV, the probe current was 14 nA, the scanning speed was 815 Hz, and the scanning step sizes of 0.15 and 1 µm were used for areas of 0.16 and 2.88 mm^2, respectively.

In line with the Schmid law ($\tau = \sigma \cos\varphi \cos\lambda$), the slip system, with either a maximum position or minimum negative SF ($m = \cos\varphi \cos\lambda$), was activated by the largest shear stress

(τ). Here, λ is the angle between the slip direction and external force (σ) and φ is the angle between the normal of the slip plane and center axis. Based on the geometry of the hat-shaped specimen [34], the slip plane was fixed by the weakest regions, as presented by a group of planes paralleling to the designed shear region. Therefore, the global <phi> depended on the sample's geometry structure, and cosφ was fixed at $1/\sqrt{5}$.

3. Results and Discussion

3.1. Microstructure Characterizations

Based on the EBSD analysis of the as-received specimen in Figure 1, we found that the α+β dual-phase Ti alloy consisted of equiaxed α_p with an average size of about 10 μm and lamellar secondary α phases (α_s) with a thickness of about 0.9 μm. The amounts of α and β phases were 80.3% and 18.9%, respectively, which are identified in red and blue, respectively. The high-angle grain boundaries among the colony structures of the as-received specimen were constructed by α and β phases. The amount of these equiaxed α_p grains were about 85% without forming texture.

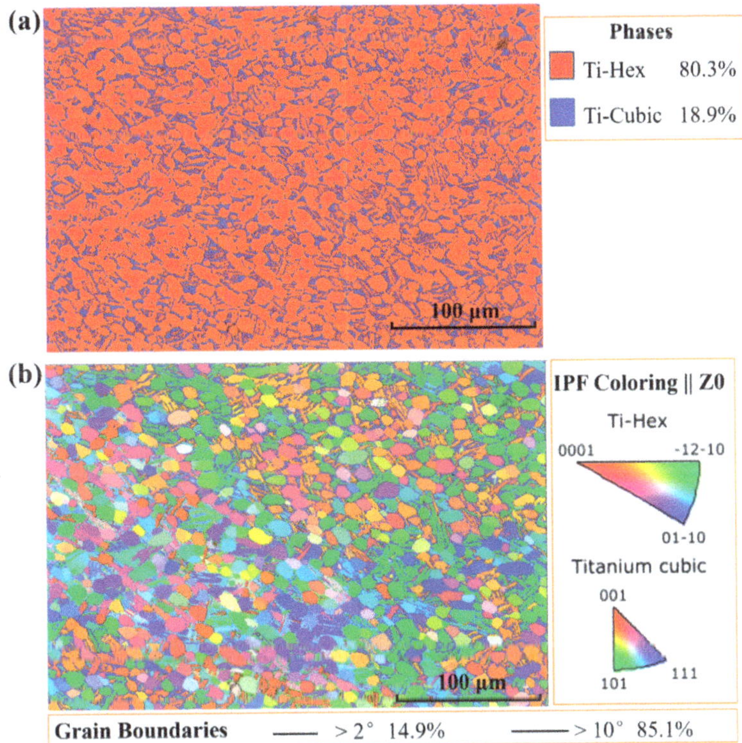

Figure 1. The electron backscattered diffraction (EBSD) analysis of the as-received dual-phase Ti alloy, (**a**) phase map; (**b**) the coupling analysis of inverse pole figure and the grain boundary map.

3.2. Mechanical Response at High Strain Rate

Based on the geometry of the hat-shaped specimen, the force applied to the shear region of the hat-shaped specimen in the SHPB test was obtained from the strain gauges on the incident and transmitted bars. The following equations [35] were used to calculate the shear stress τ_s, shear strain γ_s, and shear strain rate $\dot{\gamma}_s$ within the shear region:

$$\tau_s(t) = \frac{E_0 A_b}{A_s}\varepsilon_t \tag{1}$$

$$\gamma_s = -\frac{C_0}{d_1 - d_0} \int_0^t (\varepsilon_i - \varepsilon_t) dt \quad (2)$$

$$\dot{\gamma}_s = \frac{C_0}{d_1 - d_0} (\varepsilon_i - \varepsilon_t) \quad (3)$$

where d_0 and d_1 are the outside diameter of the hat and the inner diameter of the brim ring of specimen, respectively; C_0 and E_0 are the elastic wave speed and elastic modulus of Hopkinson bars, respectively; A_s and A_b are the area of shear section in the hat-shaped specimen and the area of cross-section in the transmitted bar, respectively; and ε_i and ε_t are the elastic strains of incident and transmitted bars, respectively.

At the strain rate of $1.8 \times 10^4\ s^{-1}$, this alloy presented an excellent damage tolerance capability, which had a maximum shear stress as high as 2 GPa at the strain-hardening stage, as shown in Figure 2. It can also be seen that this stress–strain curve consisted of four stages, including an elastic region, strain hardening, stable softening, and unstable softening. In particular, the shear stress increased with increased shear strain, and reached a maximum as high as 1951 MPa with a shear strain of 1.27, presenting an ultra-strong behavior compared with the classical high-strength Ti alloys. In the range of shear strain from 1.27 to 1.60, the shear stress slowly decreased by 104 MPa from the peak, attributed to the thermal softening during the adiabatic shear. The dynamic stored deformation energy until the work-hardening stage can be characterized as the integration of the stress–strain curve [36], which is much larger than those obtained from the static test. Correspondingly, it was expected that dynamic recrystallization would occur and result in the following softening. Moreover, the adiabatic shear behavior related to the microstructure evolution was revealed comprehensively, which is considered to be driven by the aforementioned stored energies.

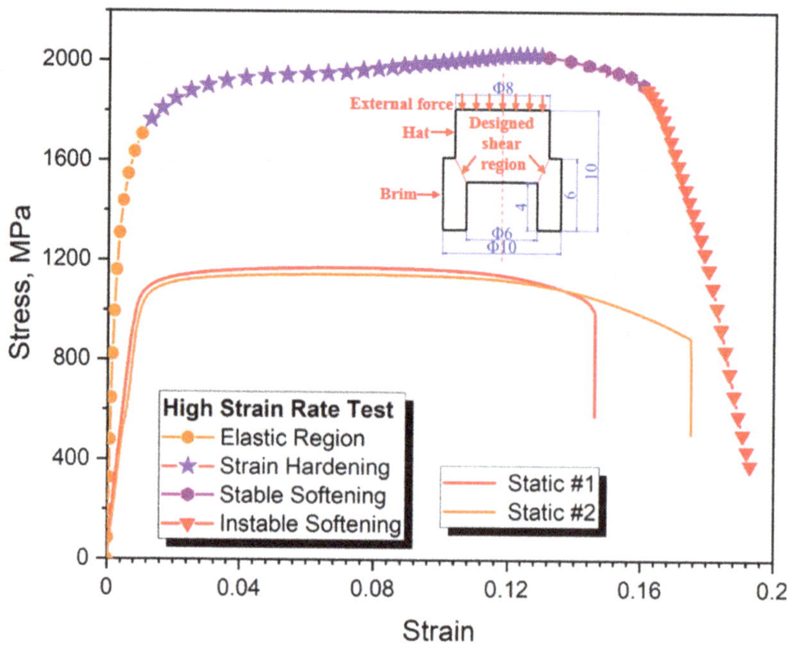

Figure 2. The stress–strain of dual-phase Ti alloy at the strain rate of $1.8 \times 10^4\ s^{-1}$ and the static one ($0.001\ s^{-1}$) together with the geometry of the hat-shaped specimen.

3.3. Local Features of Adiabatic Shear Bands

In order to comprehensively reveal these stable and unstable softening mechanisms, multi-scale characterizations of the localized ASBs were essential, the typical features of which are displayed in Figure 3. The ASBs were mainly located at the α/β dual-phase colony structures, yielding severe plastic deformations along the α/β phase boundaries. This contributed to the formations of microvoids and cracks, thus resulting in stable and unstable softening behaviors. Moreover, these ASBs caused by the thermal softening and the stress concentrations were the preferred sites for nucleation, growth, and coalescence of microvoids, contributing to the formation of cracks in the ASBs after reaching the critical length [37]. Interestingly, these microvoids were generally located at the center of the ASBs, which are attributed to the gradients of temperature and stress from the center of the ASBs to the boundary of matrix [37]. Furthermore, there were several well-aligned parallel ASBs, constructing the primary and the subordinate shear paths/zones. Even within the primary shear path, it seemed as if these shear bands formed periodically were caused by the thermoplastic instability/softening, which highlights the significance of geometry and microstructures dominated by the shear angle, shear strain, and shear rate [38].

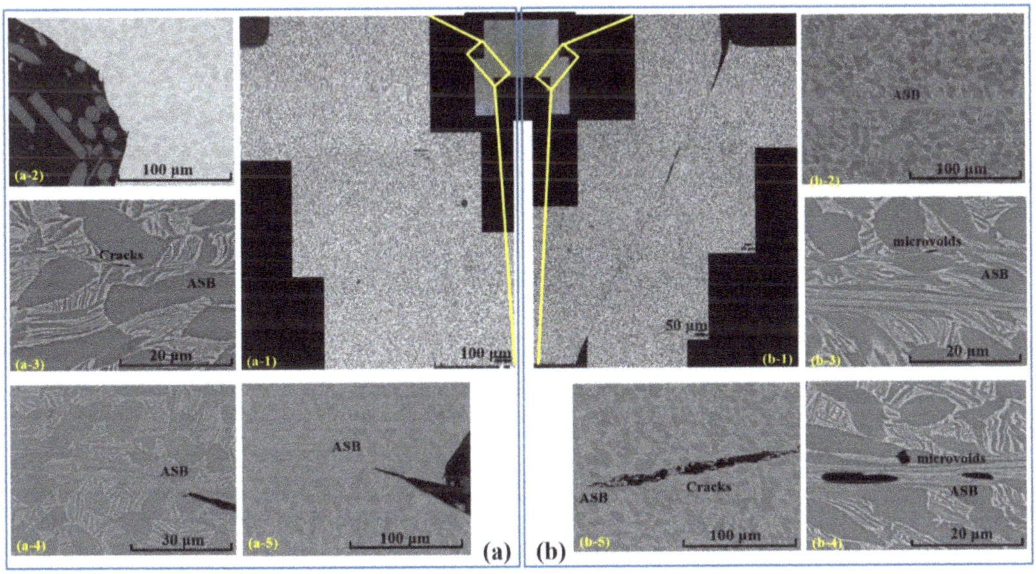

Figure 3. The multi-scale SEM characterizations of adiabatic shear bands in the hat-shaped specimen. The left part and the right part of the weak areas in the hat-shaped specimen are marked by yellow frames in (**a-1**) and (**b-1**), respectively. (**a-2**)~(**a-5**) reflect the typical features of the selected right region, such as ASB, microvoids and $\alpha+\beta$ phases, while the selected left region is identified in (**b-2**)~(**b-5**).

Figure 4 shows the full scans of the ASBs and the corresponding affected zones in terms of the phase map, the geometrically necessary dislocation (GND)map, and the coupling analysis of inverse pole figure and grain boundary map (IPF+GB). Different plastic strains between the left and the right sections may also yield different microstructure evolutions. In the phase maps, the estimated amounts of β phase decreased to less than 10%, which also indicates that the dominated shearing path was along the α/β phase boundaries, which caused the β grains within the elongated and the recrystallized extremely fine α grains to be indistinguishable. Dynamic recrystallization yielded a large amount of fine α grains dispersed along the shearing paths and in the ASBs, which could contribute to the formation of textures. It is understood that recrystallization is dominated by the entropic effect arising from the competition between the formation of dislocation and that of grain boundaries while the shear band instability emerges, which is attributed to the thermal heating occurring faster than heat dissipation [39]. Moreover, the thickness of the ASB-affected regions resulting in the collaborating plastic deformation was clearly captured by GND mapping. It seems as if the low-angle grain boundaries of the left side presented a good resistance to the formation of cracks and the thermal softening. The grain boundaries with high angles within the ASBs were also geometrical necessary boundaries, which is the same as previous observations in pure Ti [40]. In line with the grain boundary complexions [41–43], it is understood that free volumes exist at the boundaries, constructing the weakly bonded regions and contributing to the initiation of ASBs. Furthermore, the $\{0001\}\langle 11\bar{2}0\rangle$ slip systems highlighted in red in the IPF maps play an important role in the formation of ASBs, which is validated by the following theoretical analysis.

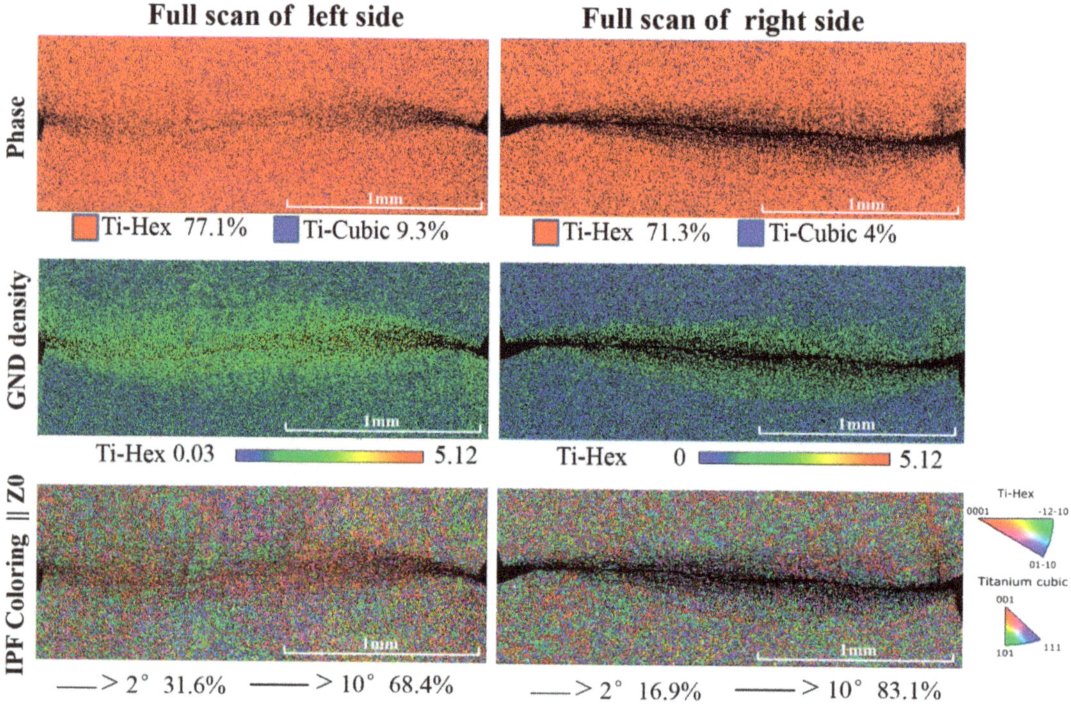

Figure 4. The EBSD features of the hat-shaped specimens in terms of phase map (Ph), geometrically necessary dislocation (GND) map, and the coupling analysis of inverse pole figure and grain boundary map (IPF+GB).

As displayed in Figure 5, the α grains were elongated while the β grains were difficult to capture in the colony structures. The solid and dashed arrows identify the cracks and the elongated α phase, respectively. It is understood that rotational dynamic recrystallization (RDR) yields a large amount of fine equiaxed α grains dispersed along the shearing paths and in ASBs, promoting the aforementioned stable softening and shear localization [20,36]. The equiaxed grains could be re-elongated by further deformation. In line with the progressive subgrain misorientation recrystallization, a mechanical subgrain rotation model to account for the recrystallized grains has been proposed and observed in ASBs in a number of materials [32,33,44]. The mechanical subgrain rotations at a high strain rate would assist the mechanism of recrystallizations. Recently, it was reported that the initiation of ASBs will occur ahead of the apparent temperature rise in Ti [23,24], motivating further investigation of the traditional well-accepted thermal-softening mechanism of ASBs. Thus, the local microstructure evolutions caused by shearing lead to dynamic recrystallizations and play an important role of softening during severe plastic deformation.

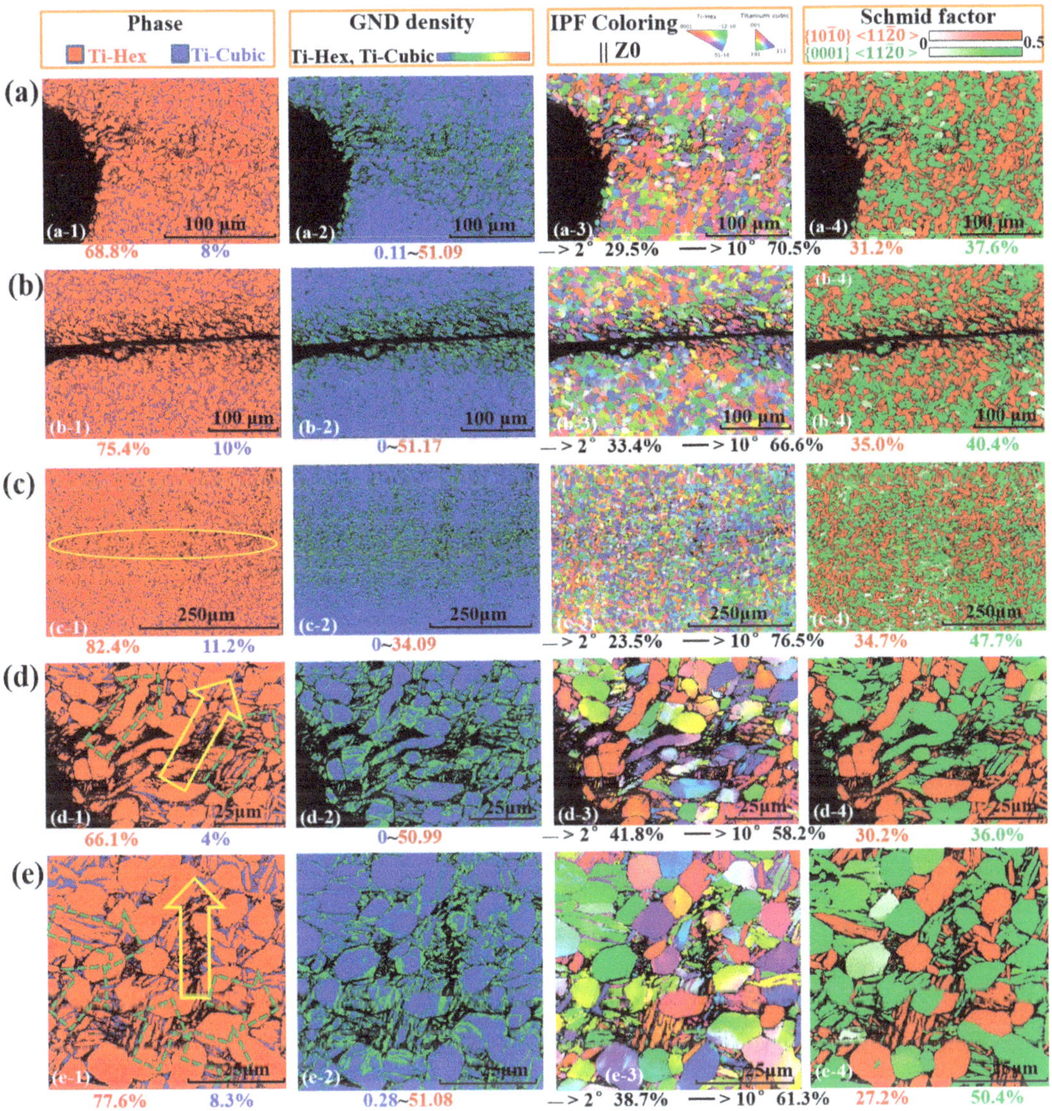

Figure 5. The typical features of ASBs and the affected zones in views of the phase map, GND map, and IPF+GB: (**a**–**c**) the cracks and ASB shear paths; (**d**,**e**) the localized shear deformation areas together with solid and dashed arrows identifying the cracks and the elongated α phase, respectively.

3.4. Discussion

During shock loading, uncommon slip systems might be activated. In general, plastic deformation initiates on the crystallographic plane with the highest Schmid factor when the stress resolved on that plane in the slip direction reaches a critical value. Based on the experimental analysis of Schmid factor, as shown in Figures 5 and 6a,b, the basal and the prismatic slips mediated by $\{0001\}\langle 11\bar{2}0\rangle$ and $\{10\bar{1}0\}\langle 11\bar{2}0\rangle$ are the dominated slip systems. Therefore, the physical reasons for the selection of these two slip systems are essential, which can also verify whether the Schmid law breaks down. Under the condition of $|\cos \lambda(a,b,c)| \leq 1$, the SF of a given slip direction <a,b,c> can be screened comprehensively. As displayed in Figure 6c,d, although the pyramidal slip systems have the largest SFs compared with those of basal and prismatic ones within the low angle ranges, they are difficult to activate because of the requirements of the largest critical resolved shear stress and the rotations of deformed grains. On the contrary, in the range of 25~35°, the basal and prismatic slips are the dominant approaches since both of them have the larger SFs, matching well with the present experimental observations. Since non-basal slipping enhances both the strength and ductility of an HCP structure [45], the $\{10\bar{1}0\}\langle 11\bar{2}0\rangle$ prismatic slips dramatically contribute to the aforementioned strain hardening. Finally, it is highlighted that the Schmid or the non-Schmid behaviors of the materials should be comprehensively investigated, revealing the variations in critical resolved shear stress and the selections of deformation model [18,26,46], providing the promising strategies to enhance damage tolerance properties.

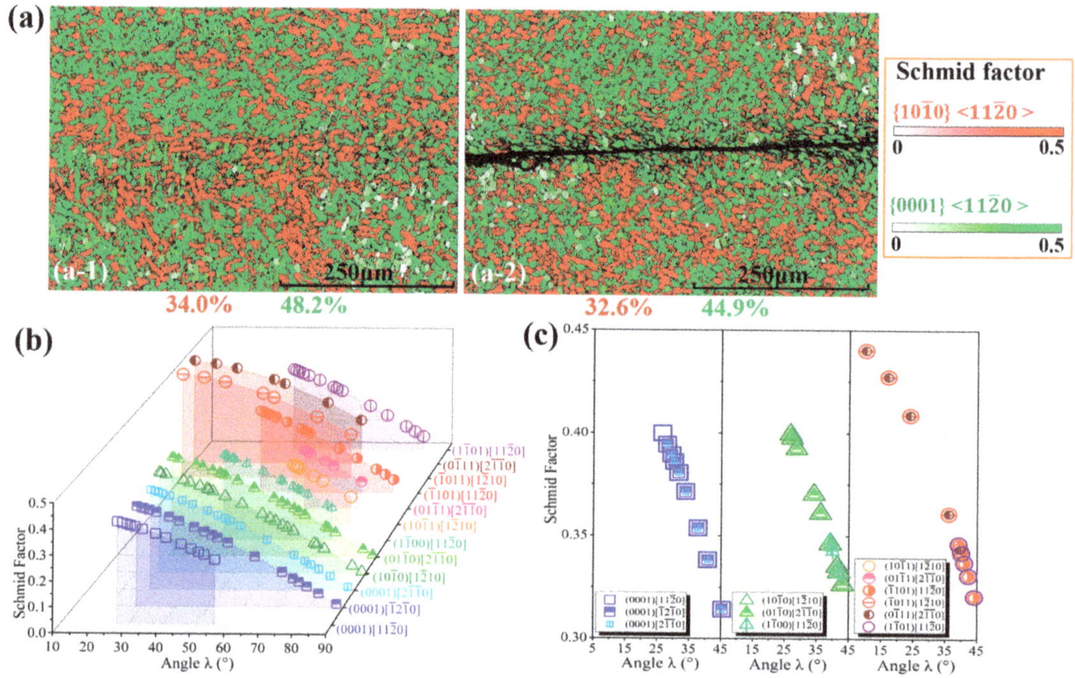

Figure 6. The theoretical Schmid factors (SFs) referring to experimental observations: (**a**) experimental characterizations of SFs of the left and the right side of the hat-shaped specimen; (**b**,**c**) 3D and 2D views of the theoretical SFs.

4. Conclusions

In summary, the α+β dual-phase colony structures mediate the extensive plastic deformations along α/β phase boundaries, which contribute to the formations of ASBs, microvoids, and cracks, thus resulting in stable and unstable softening behaviors. Dynamic recrystallization yields a large amount of fine α grains dispersed along the shearing paths and in ASBs, promoting the softening and shear localization. On the contrary, the low-angle grain boundaries present a good resistance to the formation of cracks and thermal softening, while the non-basal slipping dramatically contributes to strain hardening, supporting the promising approaches to fabricating advanced damage-tolerant dual-phase Ti alloy.

Author Contributions: Conceptualization, F.H., K.W., X.L., H.S., Y.F. and J.L.; methodology, F.H., Y.M., D.L., X.G. and W.Y.W.; validation, Y.D., Y.M., D.L., X.G. and W.Y.W.; formal analysis, F.H., P.L., Y.M., J.W., Y.D. and W.Y.W.; investigation, F.H., P.L., Y.M., J.W., Y.D. and W.Y.W.; data curation, F.H.; Y.D., Y.M., W.Y.W.; writing—original draft preparation, F.H., P.L., Y.M., J.W., Y.D. and W.Y.W.; writing—review and editing, K.W., X.L., H.S., Y.F. and J.L.; visualization, F.H., P.L., Y.M., Y.D. and W.Y.W.; supervision, K.W., X.L., H.S., Y.F. and J.L.; project administration, W.Y.W., H.S. and F.H.; funding acquisition, W.Y.W., H.S. and F.H. All authors have read and agreed to the published version of the manuscript.

Funding: This research was funded by the Science Challenge Project, grant number TZ2018002, and the Western Superconducting Technologies Co., Ltd., EPRCTC project 41422010505. The APC was funded by the Western Superconducting Technologies Co., Ltd., EPRCTC project 41422010505.

Data Availability Statement: Correspondence and requests for data and materials should be addressed to W.Y.W. (wywang@nwpu.edu.cn) and Y.X.D. (Eason@c-wst.com).

Acknowledgments: The authors acknowledge the financial supports by the Science Challenge Project (TZ2018002) and the Western Superconducting Technologies Co., Ltd. (EPRCTC project 41422010505).

Conflicts of Interest: The authors declare no conflict of interest.

References

1. Yu, Q.; Shan, Z.-W.; Li, J.; Huang, X.; Xiao, L.; Sun, J.; Ma, E. Strong crystal size effect on deformation twinning. *Nature* **2010**, *463*, 335–338. [CrossRef]
2. Clouet, E.; Caillard, D.; Chaari, N.; Onimus, F.; Rodney, D. Dislocation locking versus easy glide in titanium and zirconium. *Nat. Mater.* **2015**, *14*, 931–936. [CrossRef] [PubMed]
3. Polmear, I.; StJohn, D.; Nie, J.-F.; Qian, M. Chapter 7—Titanium Alloys. In *Light Alloys*, 5th ed.; Butterworth-Heinemann: Boston, MA, USA, 2017; pp. 369–460. [CrossRef]
4. Li, X.; Lu, K. Playing with defects in metals. *Nat. Mater.* **2017**, *16*, 700–701. [CrossRef] [PubMed]
5. Kaplan, W.D. The mechanism of crystal deformation. *Science* **2015**, *349*, 1059–1060. [CrossRef] [PubMed]
6. Zhang, Y.; Li, J.; Wang, W.Y.; Li, P.; Tang, B.; Wang, J.; Kou, H.; Shang, S.; Wang, Y.; Kecskes, L.J.; et al. When a defect is a pathway to improve stability: A case study of the $L1_2$ Co_3TM superlattice intrinsic stacking fault. *J. Mater. Sci.* **2019**, *54*, 13609–13618. [CrossRef]
7. Wang, W.Y.; Tang, B.; Shang, S.-L.; Wang, J.; Li, S.; Wang, Y.; Zhu, J.; Wei, S.; Wang, J.; Darling, K.A.; et al. Local lattice distortion mediated formation of stacking faults in Mg alloys. *Acta Mater.* **2019**, *170*, 231–239. [CrossRef]
8. Bae, J.W.; Lee, J.; Zargaran, A.; Kim, H.S. Enhanced cryogenic tensile properties with multi-stage strain hardening through partial recrystallization in a ferrous medium-entropy alloy. *Scr. Mater.* **2021**, *194*, 113653. [CrossRef]
9. Wu, G.; Chan, K.-C.; Zhu, L.; Sun, L.; Lu, J. Dual-phase nanostructuring as a route to high-strength magnesium alloys. *Nature* **2017**, *545*, 80–87. [CrossRef]
10. Basu, S.; Li, Z.; Pradeep, K.G.; Raabe, D. Strain rate sensitivity of a TRIP-assisted dual-phase high-entropy alloy. *Front. Mater.* **2018**, *5*, 30. [CrossRef]
11. Zou, C.; Li, J.; Wang, W.Y.; Zhang, Y.; Lin, D.; Yuan, R.; Wang, X.; Tang, B.; Wang, J.; Gao, X.; et al. Integrating data mining and machine learning to discover high-strength ductile Titanium alloys. *Acta Mater.* **2021**, *202*, 211–221. [CrossRef]
12. Mantri, S.A.; Choudhuri, D.; Alam, T.; Viswanathan, G.B.; Sosa, J.M.; Fraser, H.L.; Banerjee, R. Tuning the scale of α precipitates in β-titanium alloys for achieving high strength. *Scr. Mater.* **2018**, *154*, 139–144. [CrossRef]
13. Lee, S.W.; Park, C.H.; Hong, J.-K.; Yeom, J.-T. Development of sub-grained $\alpha+\beta$ Ti alloy with high yield strength showing twinning- and transformation-induced plasticity. *J. Alloys Compd.* **2020**, *813*, 152102. [CrossRef]
14. Danard, Y.; Poulain, R.; Garcia, M.; Guillou, R.; Thiaudiere, D.; Mantri, S.; Banerjee, R.; Sun, F.; Prima, F. Microstructure design and in-situ investigation of TRIP/TWIP effects in a forged dual-phase Ti-10V-2Fe-3Al alloy. *Materialia* **2019**, *8*, 100507. [CrossRef]
15. Kloenne, Z.; Viswanathan, G.; Fox, S.; Loretto, M.; Fraser, H.L. Interface and colony boundary sliding as a deformation mechanism in a novel titanium alloy. *Scr. Mater.* **2020**, *178*, 418–421. [CrossRef]
16. Bertrand, E.; Castany, P.; Péron, I.; Gloriant, T. Twinning system selection in a metastable β-titanium alloy by Schmid factor analysis. *Scr. Mater.* **2011**, *64*, 1110–1113. [CrossRef]
17. Wang, L.; Eisenlohr, P.; Yang, Y.; Bieler, T.R.; Crimp, M.A. Nucleation of paired twins at grain boundaries in titanium. *Scr. Mater.* **2010**, *63*, 827–830. [CrossRef]
18. Qin, H.; Jonas, J.J. Variant selection during secondary and tertiary twinning in pure titanium. *Acta Mater.* **2014**, *75*, 198–211. [CrossRef]
19. Reddy, V.S.; Nath, P.; Horbach, J.; Sollich, P.; Sengupta, S. Nucleation theory for yielding of nearly defect-free crystals: Understanding rate dependent yield points. *Phys. Rev. Lett.* **2020**, *124*, 025503. [CrossRef] [PubMed]
20. Rittel, D.; Landau, P.; Venkert, A. Dynamic recrystallization as a potential cause for adiabatic shear failure. *Phys. Rev. Lett.* **2008**, *101*, 165501. [CrossRef]

21. Ma, X.; Zhao, D.; Yadav, S.; Sagapuram, D.; Xie, K.Y. Grain-subdivision-dominated microstructure evolution in shear bands at high rates. *Mater. Res. Lett.* **2020**, *8*, 328–334. [CrossRef]
22. Bisht, A.; Kumar, S.; Pang, K.H.; Zhou, R.X.; Roy, A.; Silberschmidt, V.V.; Suwas, S. Shear band widening mechanism in Ti-6Al-4V under high strain rate deformation. *J. Mater. Res.* **2020**, *35*, 1623–1634. [CrossRef]
23. Guo, Y.; Ruan, Q.; Zhu, S.; Wei, Q.; Lu, J.; Hu, B.; Wu, X.; Li, Y. Dynamic failure of titanium: Temperature rise and adiabatic shear band formation. *J. Mech. Phys. Solids* **2020**, *135*, 103811. [CrossRef]
24. Guo, Y.; Ruan, Q.; Zhu, S.; Wei, Q.; Chen, H.; Lu, J.; Hu, B.; Wu, X.; Li, Y.; Fang, D. Temperature rise associated with adiabatic shear band: Causality clarified. *Phys. Rev. Lett.* **2019**, *122*, 015503. [CrossRef] [PubMed]
25. El-Azab, A. The statistical mechanics of strain-hardened metals. *Science* **2008**, *320*, 1729–1730. [CrossRef] [PubMed]
26. Arul Kumar, M.; Wronski, M.; McCabe, R.J.; Capolungo, L.; Wierzbanowski, K.; Tome, C.N. Role of microstructure on twin nucleation and growth in HCP titanium: A statistical study. *Acta Mater.* **2018**, *148*, 123–132. [CrossRef]
27. Fan, X.G.; Jiang, X.Q.; Zeng, X.; Shi, Y.G.; Gao, P.F.; Zhan, M. Modeling the anisotropy of hot plastic deformation of two-phase titanium alloys with a colony microstructure. *Int. J. Plast.* **2018**, *104*, 173–195. [CrossRef]
28. Tang, B.; Tang, B.; Han, F.; Yang, G.; Li, J. Influence of strain rate on stress induced martensitic transformation in beta solution treated TB8 alloy. *J. Alloys Compd.* **2013**, *565*, 1–5. [CrossRef]
29. Zhan, H.; Wang, G.; Kent, D.; Dargusch, M. The dynamic response of a metastable β Ti-Nb alloy to high strain rates at room and elevated temperatures. *Acta Mater.* **2016**, *105*, 104–113. [CrossRef]
30. Ben Boubaker, H.; Mareau, C.; Ayed, Y.; Germain, G.; Tidu, A. Impact of the initial microstructure and the loading conditions on the deformation behavior of the Ti17 titanium alloy. *J. Mater. Sci.* **2020**, *55*, 1765–1778. [CrossRef]
31. Landau, P.; Osovski, S.; Venkert, A.; Gärtnerová, V.; Rittel, D. The genesis of adiabatic shear bands. *Sci. Rep.* **2016**, *6*, 37226. [CrossRef]
32. Baik, S.-I.; Gupta, R.K.; Kumar, K.S.; Seidman, D.N. Temperature increases and thermoplastic microstructural evolution in adiabatic shear bands in a high-strength and high-toughness 10 wt.% Ni steel. *Acta Mater.* **2021**, *205*, 116568. [CrossRef]
33. Lins, J.F.C.; Sandim, H.R.Z.; Kestenbach, H.J.; Raabe, D.; Vecchio, K.S. A microstructural investigation of adiabatic shear bands in an interstitial free steel. *Mater. Sci. Eng. A* **2007**, *457*, 205–218. [CrossRef]
34. Ran, C.; Chen, P.; Li, L.; Zhang, W. Dynamic shear deformation and failure of Ti-5Al-5Mo-5V-1Cr-1Fe titanium alloy. *Mater. Sci. Eng. A* **2017**, *694*, 41–47. [CrossRef]
35. Andrade, U.; Meyers, M.A.; Vecchio, K.S.; Chokshi, A.H. Dynamic recrystallization in high-strain, high-strain-rate plastic deformation of copper. *Acta Metall. Mater.* **1994**, *42*, 3183–3195. [CrossRef]
36. Osovski, S.; Rittel, D.; Landau, P.; Venkert, A. Microstructural effects on adiabatic shear band formation. *Scr. Mater.* **2012**, *66*, 9–12. [CrossRef]
37. Wang, B.F.; Yang, Y.; Chen, Z.P.; Zeng, Y. Adiabatic shear bands in alpha-titanium tube under external explosive loading. *J. Mater. Sci.* **2007**, *42*, 8101–8105. [CrossRef]
38. Yan, D.V.P.; Jin, X.L. Characterization of shear band formation and microstructure evolution during orthogonal cutting of Ti-5553: Part I—Shear angle, strain and strain rate. *J. Mater. Eng. Perform.* **2020**, *29*, 4063–4074. [CrossRef]
39. Lieou, C.K.C.; Bronkhorst, C.A. Dynamic recrystallization in adiabatic shear banding: Effective-temperature model and comparison to experiments in ultrafine-grained titanium. *Int. J. Plast.* **2018**, *111*, 107–121. [CrossRef]
40. Yang, Y.; Wang, B.F. Microstructure evolution in adiabatic shear band in α-titanium. *J. Mater. Sci.* **2006**, *41*, 7387–7392. [CrossRef]
41. Cantwell, P.R.; Tang, M.; Dillon, S.J.; Luo, J.; Rohrer, G.S.; Harmer, M.P. Grain boundary complexions. *Acta Mater.* **2014**, *62*, 1–48. [CrossRef]
42. Dillon, S.J.; Harmer, M.P.; Luo, J. Grain boundary complexions in ceramics and metals: An overview. *JOM* **2009**, *61*, 38–44. [CrossRef]
43. Meiners, T.; Frolov, T.; Rudd, R.E.; Dehm, G.; Liebscher, C.H. Observations of grain-boundary phase transformations in an elemental metal. *Nature* **2020**, *579*, 375–378. [CrossRef]
44. Tang, L.; Chen, Z.; Zhan, C.; Yang, X.; Liu, C.; Cai, H. Microstructural evolution in adiabatic shear bands of copper at high strain rates: Electron backscatter diffraction characterization. *Mater. Charact.* **2012**, *64*, 21–26. [CrossRef]
45. Bu, Y.; Li, Z.; Liu, J.; Wang, H.; Raabe, D.; Yang, W. Nonbasal slip systems enable a strong and ductile Hexagonal-Close-Packed high-entropy phase. *Phys. Rev. Lett.* **2019**, *122*, 075502. [CrossRef] [PubMed]
46. Lim, H.; Hale, L.M.; Zimmerman, J.A.; Battaile, C.C.; Weinberger, C.R. A multi-scale model of dislocation plasticity in α-Fe: Incorporating temperature, strain rate and non-Schmid effects. *Int. J. Plast.* **2015**, *73*, 100–118. [CrossRef]

materials

MDPI

Article

Mechanical Behavior of Titanium Based Metal Matrix Composites Reinforced with TiC or TiB Particles under Quasi-Static and High Strain-Rate Compression

Pavlo E. Markovsky [1], Jacek Janiszewski [2], Oleksandr O. Stasyuk [1], Vadim I. Bondarchuk [1], Dmytro G. Savvakin [1], Kamil Cieplak [2], Daniel Goran [3], Purvesh Soni [3] and Sergey V. Prikhodko [4,*]

1 G.V. Kurdyumov Institute for Metal Physics of N.A.S. of Ukraine, 03412 Kyiv, Ukraine; pmark@imp.kiev.ua (P.E.M.); stasiuk@imp.kiev.ua (O.O.S.); vbondar77@gmail.com (V.I.B.); savva@imp.kiev.ua (D.G.S.)
2 Jarosław Dąbrowski Military University of Technology, 00908 Warsaw, Poland; jacek.janiszewski@wat.edu.pl (J.J.); kamil.cieplak@wat.edu.pl (K.C.)
3 Bruker Nano Analytics, Am Studio 2D, 12489 Berlin, Germany; Daniel.Goran@bruker.com (D.G.); Purvesh.Soni@bruker.com (P.S.)
4 Department of Materials Science and Engineering, University of California, Los Angeles, CA 90095, USA
* Correspondence: sergey@seas.ucla.edu; Tel.: +310-825-9735

Abstract: The mechanical behavior of titanium alloys has been mostly studied in quasi-static conditions when the strain rate does not exceed 10 s^{-1}, while the studies performed in dynamic settings specifically for Ti-based composites are limited. Such data are critical to prevent the "strength margin" approach, which is used to assure the part performance under dynamic conditions in the absence of relevant data. The purpose of this study was to obtain data on the mechanical behavior of Ti-based composites under dynamic condition. The Metal Matrix Composites (MMC) on the base of the alloy Ti-6Al-4V (wt.%) were made using Blended Elemental Powder Metallurgy with different amounts of reinforcing particles: 5, 10, and 20% of TiC or 5, 10% (vol.) of TiB. Composites were studied at high strain rate compression ~1–3 × $10^3 \cdot s^{-1}$ using the split Hopkinson pressure bar. Mechanical behavior was analyzed considering strain rate, phase composition, microstructure, and strain energy (SE). It is shown that for the strain rates up to 1920 s^{-1}, the strength and SE of MMC with 5% TiC are substantially higher compared to particles free alloy. The particles TiC localize the plastic deformation at the micro level, and fracturing occurs mainly by crushing particles and their aggregates. TiB MMCs have a finer grain structure and different mechanical behavior. MMC with 5 and 10% TiB do not break down at strain rates up to almost 3000 s^{-1}; and 10% MMC surpasses other materials in the SE at strain rates exceeding 2200 s^{-1}. The deformation mechanism of MMCs was evaluated.

Keywords: titanium matrix composite; titanium carbide; titanium boride; microstructure; mechanical properties; high-strain-rate testing; split Hopkinson pressure bar; quasi-static compression; strain energy

Citation: Markovsky, P.E.; Janiszewski, J.; Stasyuk, O.O.; Bondarchuk, V.I.; Savvakin, D.G.; Cieplak, K.; Goran, D.; Soni, P.; Prikhodko, S.V. Mechanical Behavior of Titanium Based Metal Matrix Composites Reinforced with TiC or TiB Particles under Quasi-Static and High Strain-Rate Compression. *Materials* **2021**, *14*, 6837. https://doi.org/10.3390/ma14226837

Academic Editor: Lijun Zhang

Received: 5 October 2021
Accepted: 8 November 2021
Published: 12 November 2021

1. Introduction

Titanium alloys are an important structural material of modern aerospace, automotive, shipbuilding and military technologies due to the high level of specific strength, fracture toughness, fatigue strength, corrosion resistance, non-magnetization, and some other specific physical mechanical and service properties [1–4]. Conventional titanium alloys alloyed with various elements, which are commonly divided into stabilizers of α- or β-phases, do not always meet the requirements for a set of physical and mechanical properties. For instance, these alloys are characterized by rather low wear resistance, fretting corrosion, and relatively low hardness [1,5–7]. These drawbacks can be overcome by preparing titanium-based Metal Matrix Composites (MMC) containing some fine hard particles, for example, carbides, borides, etc. [8–10]. Another pressing problem is that the choice of materials for new products is usually based on reference data on mechanical properties, which were determined under certain standard, mainly quasi-static, test conditions. In most cases, the standard test settings do not correspond to the actual operating conditions of real equipment exposed to highly dynamic loads. Therefore, to prevent the failure of such structures, designers usually have to apply the so-called "strength margin" approach, which sometimes leads to an increase in the mass of parts and structures at times, and consequently to essential increase in the total weight of products, their price, and operating costs.

The influence of the strain rate on the mechanical behavior of a wide range of titanium alloys has been studied in sufficient details under quasi-static test conditions when the strain rate is somewhere below 10 s^{-1} [11–18], while the number of studies on the same materials in dynamic conditions is more limited [19–22]. Especially studies focusing on factors such as chemical and phase composition, and in particular, the evolution of the microstructure under high strain-rate conditions, have not been widely deliberated. In addition, such studies are practically absent for MMC based on titanium alloys reinforced with hard particles such as TiC or TiB, which have been shown to be a very promising material for various applications, for example, working under conditions of wear-resistant friction parts, or anti-ballistic protection elements [8–10,23,24]. The mechanical behavior of MMC materials at the moment have not been fully investigated with the exception of quasi-static strength conditions. That is why the present study has been devoted to the systematic investigation of the effect of strain rate on mechanical behavior of the widespread two-phase $\alpha + \beta$ titanium alloy, Ti-6Al-4V (wt.%, Ti64 hereafter), produced using Blended Elemental Powder Metallurgy (BEPM) [25–27], as well as MMCs based on this alloy reinforced with particles of TiC or TiB [10]. Dynamic impact compression tests were carried out using a split Hopkinson pressure bar (SHPB) apparatus, and the obtained results were compared with reference data of quasi-static strength experiments.

2. Materials and Methods

The alloy Ti64 was fabricated using BEPM; the details of the used technological protocol are described elsewhere [25–27]. The starting materials were TiH_2 powder (particles size < 40 μm), and standard Al-V (60–40 wt.%) master alloy in powder form (particles size < 63 μm). These powders were mixed, cold pressed in the die under a load 250 MPa, and then sintered at 1250 °C for 4 h in a vacuum of 10^{-3} Pa. The resulting material was designated as Ti64BEPM. Samples of MMC based on the same alloy Ti64 reinforced with 5, 10, and 20 (vol.%) of TiC (particles size below 30 μm), or 5, and 10% of TiB (these quantities of hardening particles TiC and TiB were selected based on the results of works [24,28,29]) particles were prepared via the same BEPM technology. The reinforcing particles were introduced to the blend at the mixing stage [10]. Titanium monoboride, TiB, in the final MMC was obtained by adding into a green mixture of titanium diboride, TiB_2 powder (1 ÷ 30 μm). Diboride is chemically converted to monoboride during sintering according to the reaction: $Ti + TiB_2 = 2TiB$. The resulting MMCs were designated as Ti64BEPM + XTiC, and Ti64BEPM + XTiB, where "X" denotes the quantity of hardening particles in volume %. All sintered samples were in the form of square bars with dimensions 10 × 10 × 60 mm.

For standard quasi-static tensile (QST) tests, the bars were machined into cylindrical specimens with the gage diameter 4 mm and length 25 mm. Flat samples with dimensions 2 × 9 × 60 mm were prepared for the elastic properties and damping capacity measurements. Finally, cylindrical samples with gauge length and diameter of 5 mm for the dynamic SHPB and quasi-static compression (QSC) tests were cut by the electric discharge machining technique.

Tensile properties were determined following the ASTM E8 standard using the INSTRON 3376 machine. Young's, Shear moduli, Poisson's ratio, and Damping capacity of materials were measured with the Resonance-Frequency-Damping Analysis (RFDA) system (IMCE, Genk, Belgium) using Impulse Excitation Technique (IET) following the ASTM E1876-15 protocol. Material microstructures before and after tests as well as the specimens' fractures were studied using scanning electron microscope (SEM), Vega 3 (Tescan, Czech Republic), equipped with energy dispersive X-ray (EDX) spectroscopy allowing the measurement of the chemical composition of materials. Crystal orientation and phase mapping was performed using Electron Back Scatter Diffraction (EBSD) on bulk samples and Transmission Kikuchi Diffraction (TKD) on electron transparent samples. EBSD and TKD measurements were made using the QUANTAX EBSD system (Bruker, Germany) installed on a Merlin FE-SEM (Zeiss, Germany). Electron transparent samples were prepared from specific locations on the bulk samples using the lift-out method on a Lyra 3 Focused Ion Beam (FIB) SEM (Tescan, Czech Republic). Phase composition of the specimens and their crystalline structure measurements including the texture analysis were performed using X-ray diffraction (XRD) Ultima IV (Rigaku, Japan) system. The gas content within the sintered specimens was measured using a gas analyzer ELTRA OH900.

High strain rate compression tests were performed using the SHPB technique [30–32]. The basic components of the bar system are shown in Figure 1. The length of the input and output bar was 1200 mm, the length of the striker bar 250 mm and the diameter of all bars 12 mm. The bars were made of heat-treated maraging steel of grade MS350, providing a yield strength of 2300 MPa and an elastic wave speed of 4960 m/s. The striker bar was driven by a compressed air system with a barrel length and inner diameter equal to 1200 mm and 12.1 mm, respectively. The impact striker bar speed used during the experiments ranged from18 to 29 m/s. The pulse shaping technique was used to minimize the wave dispersion and to facilitate stress equilibrium. A copper pulse shaper with a 3-mm diameter and thicknesses of 0.3 or 0.4 mm were used. More details on the applied SHPB stand are given in [22].

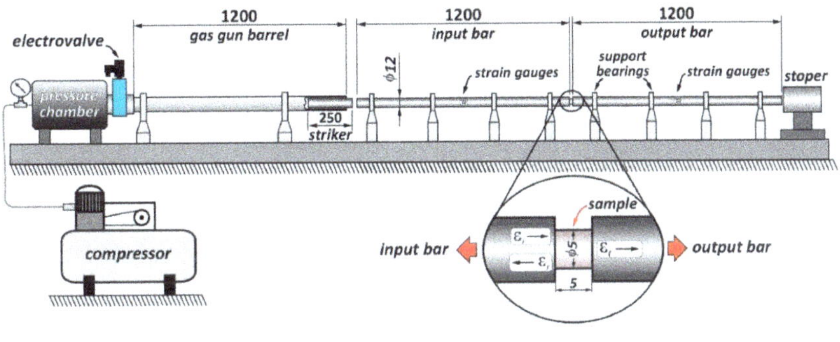

Figure 1. The schematics of the split Hopkinson pressure bar (SHPB) system used in this study [22].

3. Results

3.1. Characteristics of the Original Structures

Typical initial microstructures of the studied materials are presented in Figure 2, and their chemical compositions are listed in Table 1. The measured QST, and 3-point flexure properties, as well as measured elastic characteristics are presented in Table 2. Optimization of sizes of used powders and technological regimes of sintering earlier reported [10] enables to obtain Ti64BEPM alloy homogeneous by chemical composition and having residual porosity below 1.5% (by vol.). As the Figure 2a shows, the prior β-grains' size was relatively small and equaled 100–150 μm. The intragranular α + β structure was sufficiently fine with the α lamellas having average length 20 ÷ 70 μm and thickness 2 ÷ 5 μm. The dimensions of all structural elements, such as β-grains, α + β colonies, and individual α-plates (lamellae) are essentially smaller as compared to the same elements of typical coarse-grained lamellar microstructure of Ti64 produced using conventional cast and wrought technology. More details on such comparison were presented earlier [22,24]. The present study structures provide well-balanced standard mechanical properties (tensile) (p. 1 in Table 2), which are superior to those of the lamellar microstructures alloy obtained using conventional cast and wrought method [1,33].

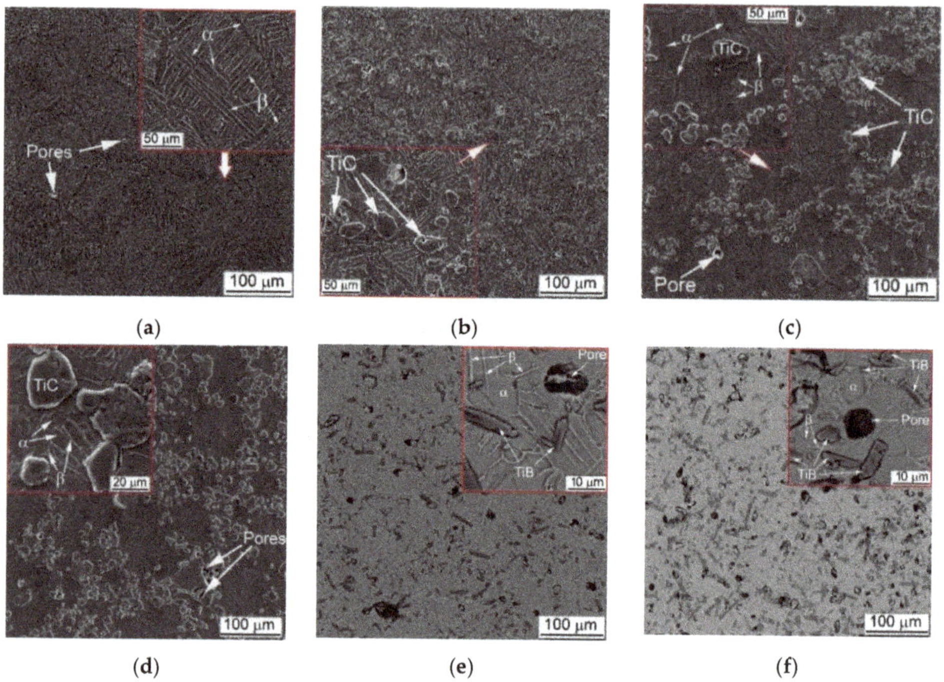

Figure 2. Microstructure of studied materials: (**a**) Ti64BEPM, (**b–f**) MMC on the base of Ti64BEPM containing (**b**) 5 (vol.) % TiC, (**c**) 10% TiC, (**d**) 20% TiC, (**e**) 5% TiB, and (**f**) 10% TiB particles. SEM, (**a–d**) BSE, (**e**,**f**) SE images.

Table 1. Chemical and phase composition of studied materials.

##	Alloying Elements and Impurities, wt.%					TiC, or TiB, vol.%	Ti	Residual Porosity, %	Phase Composition
	Al	V	Fe	O	N				
					Ti64BEPM				
1	5.94	4.06	0.16	0.19	0.007	-	Balance	1.5	α + β
					Ti64BEPM + 5TiC				
2	6.02	3.9	0.14	0.2	0.009	5	÷	2.1	α + β + TiC
					Ti64BEPM + 10TiC				
3	6.0	4.02	0.11	0.21	0.008	10	÷	2.4	α + β + TiC
					Ti64BEPM + 20TiC				
4	6.04	4.07	0.17	0.18	0.007	20	÷	2.8	α + β + TiC
					Ti64BEPM + 5TiB				
5	6.03	4.04	0.12	0.22	0.008	5	÷	3.1	α + β + TiB
					Ti64BEPM + 10TiB				
6	6.02	4.01	0.13	0.21	0.009	10	÷	4.0	α + β + TiB

Table 2. Mechanical Properties of the Obtained Materials.

##	Tensile Properties (QST) (Rate 8 × 10^{-4} s^{-1})				3-Point Flexure		Young Module, GPa	Damping	Sound Frequency, Hz	Vickers Hardness, HV
	YS, MPa	UTS, MPa	El. %	RA, %	Strength, MPa	Strain, %				
					Ti64BEPM					
1	932	1033	7.6	21.2	2088	16.3	123	0.000251	12,022	339
					Ti64BEPM + 5TiC					
2	708	708	≤0.1	-	1680	4.2	135.2	0.000266	13,551	380
					Ti64BEPM + 10TiC					
3	620	620	≤0.1	-	1350	≤0.1	137	0.000309	14,029	403
					Ti64BEPM + 20TiC					
4	567	567	≤0.1	-	180	≤0.1	140.4	0.000219	14,649	425
					Ti64BEPM + 5TiB					
5	844	844	≤0.1	-	980	≤0.1	134.5	0.000284	13,624	362
					Ti64BEPM + 10TiB					
6	522	552	≤0.1	-	645	≤0.1	138.2	0.000322	14,219	388

Introduction of reinforcement particles into Ti64BEPM (Figure 3) significantly affects the alloy structure formation during the sintering process [24,34]. TiC particles are considered relatively stable at the used sintering conditions retaining their close to equiaxed morphology (Figure 2b); however, some possibility of partial diffusion of carbon atoms into the matrix alloy solid solution cannot be completely excluded in the light of recent studies [35,36]. It should also be noted that mixing of the alloy blend with TiC particles often accompanied by agglomeration of the particles forming some aggregates and leaving some singular particles. This phenomenon increases with increasing content TiC (Figure 2c,d). The presence of conglomerates and individual TiC particles provides a pinning effect restricting the growth of β-grains in the Ti64 matrix alloy, which do not exceed 50–70 microns in the sintered state of the composite, against 100–150 microns for material without reinforcing particles (Figure 2b–d vs. Figure 2a). As the fraction of TiC particles increases, the residual porosity also increases slightly as shown in Table 1. The structure of composite was somewhat different in the case of TiB reinforcement. These particles are the result of an in-situ chemical reaction during sintering, which transforms more or less equiaxial TiB_2 particles into needle- or plate-like TiB (Figure 2e,f). As recently shown [37], the TiB particles growth process is accompanied by the Kirkendall effect, which causes a higher total porosity compared to the alloy or its composite with TiC. In addition, the TiB composites had a substantially uniform distribution of reinforcing particles without visible aggregates. All these effects are discussed in more details in previous studies [10,26,27,38].

As previously discussed, [10,26,27] the main reason for the observed microstructure refinement is the presence of residual pores, which pins the β-grain boundaries, thus preventing the grains coarsening during the structure sintering at single β phase field temperatures. A finer β-grain structure in turn affects the size of intra-grain structure that forms in α + β phase field during cooling. The tensile properties of Ti64BEPM alloy compared to the cast and wrought Ti64 with lamellar microstructure [22] were somewhat improved in strength (Table 2, p. 1), most likely due to a higher oxygen content (Table 1, p. 1). In turn, lower ductility can be associated with both: increased oxygen content and residual pores (2% and above). The addition of TiC particles changes the coarseness of both β-grains and intragranular lamellar structures, because these hard particles play the same pinning role as residual pores, and naturally this effect is enhanced as the number of particles increases (compare Figure 2a–d). For instance, in Ti64BEPM + 5TiC alloy in regions relatively free of particles, the grain size is about 100 μm (Figure 2b), while in their presence the grain size varied within the range 10 ÷ 70 μm. The sizes of the colonies and individual lamellas associated with grain size also decreases. A similar trend is generally valid for the higher contents of TiC composites (Figure 2b–d). Except that the clustering of the TiC particles becomes more pronounced at a higher particles content; and it also increases the value of residual porosity from about 1.5% in Ti64BEPM to 2.1%, 2.4%, and 2.8% for Ti64BEPM + 5TiC, Ti64BEPM + 10TiC and Ti64BEPM + 20TiC, respectively (Figure 2a–d). Given the previously established fact that an increase in residual porosity in the range of 1–2% has little effect on the tensile properties of the Ti64 alloy [24–26], the almost absolute brittleness of MMC with 5% and above TiC (compare p. 1 with pp. 2–4 in Table 2) should be related to the presence effect of hard particles. In addition to ductility, the strength of the material also decreases with an increase in the number of particles. The UTS is reduced from 1033 MPa for Ti64BEPM to 708 Mpa, 620 Mpa and 567 Mpa for 5%, 10%, and 20% TiC, respectively.

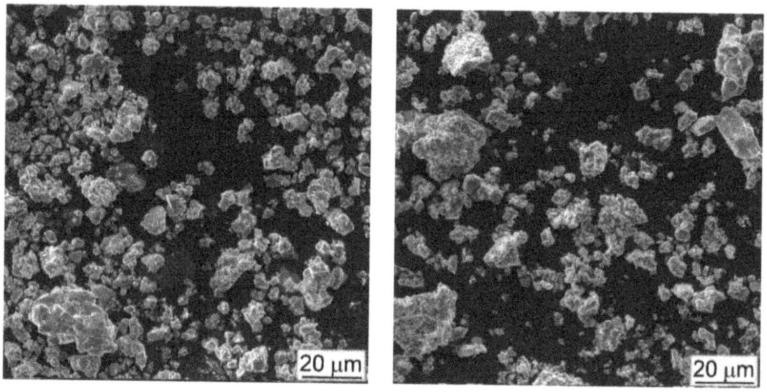

(a) (b)

Figure 3. SEM Images (SE) of (**a**) TiC and (**b**) TiB_2 particles used in preparation of Ti64BEPM + XTiC and Ti64BEPM + XTiB MMCs.

Due to more uniform distribution of TiB particles in the respective composites, the distance between the strengthening particles is shorter than in TiC-containing composites, which causes the β-grains refinement below 50–60 μm. Better refinement of TiB composites causes that in smaller β-grains at 5% TiB the α-phase still has a shape of relatively short lamellae (Figure 2e), while at 10% of TiB particles the α-phase already has a morphology close to globular (Figure 2f). As for the mechanical characteristics of MMCs with TiB under the tension and the 3-point bending test, the strength is markedly lower compared to similar data of TiC composites with alike particles content (Table 2, pp. 5 and 6). This is apparently due to the higher number of residual pores (pp. 5 and 6 in Table 2) and possibly to the effect of the particle morphology, when the sharp edges of TiB plates and needles act as stress concentrators [24].

The XRD results confirm that the matrix of alloy is two-phase α + β base with a small amount of β-phase, and the presence of TiC or TiB phases confirmed if reinforcing particles were added to the blend before the sintering. The typical (102)α pole figures of as-sintered specimens are presented in Figure 4a–c. The Ti64BEPM is characterized by a random crystallographic texture, quite typical of a metal having relatively large (~100 μm) prior β-grains (Figure 4a). The size of these grains determines the size of the α-lamellas packets within the grains that gives corresponding reflections on these pole figures. Compared to a reinforcement-free alloy, composites with 10% of hard particles exhibit a higher density of α-phase spots closer the center of the pole figures than on their periphery (Figure 4b,c vs. Figure 4a). A slight decrease in the size of the spots of the (102) α reflection can also be noted, which can be attributed to already mentioned decrease in the average size of β-grains and α-packets caused by presence of particles TiC or TiB (Figure 2c,f vs. Figure 2a).

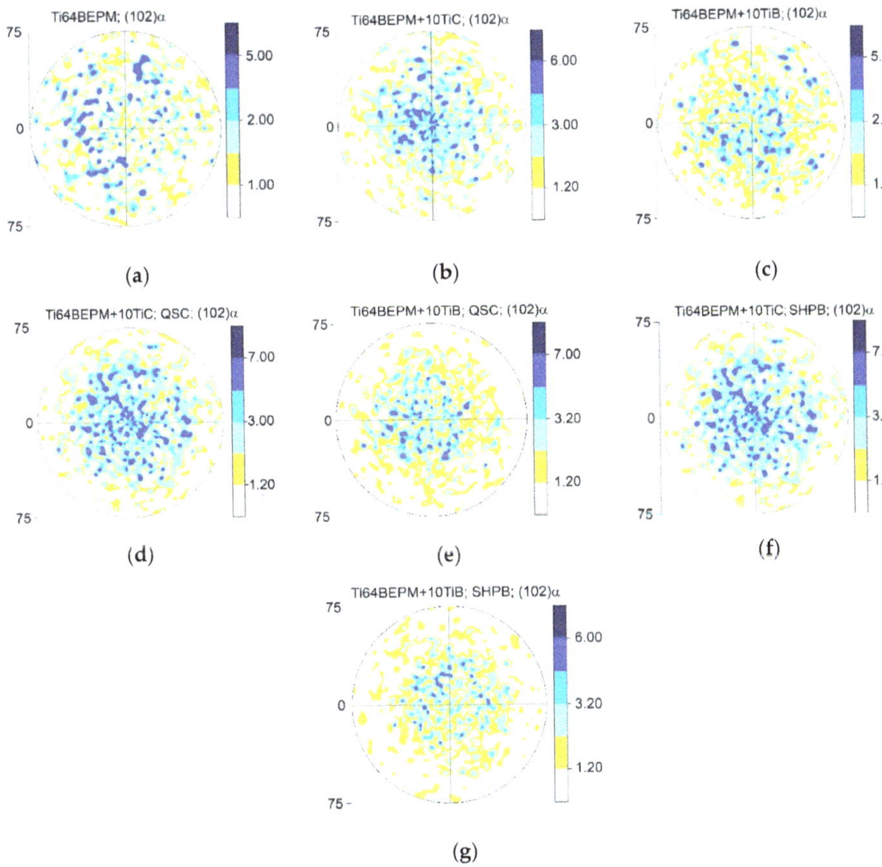

Figure 4. The pole figures (102) α of Ti64BEPM (**a**), Ti64BEPM + 10TiC (**b**,**d**,**f**), and Ti64BEPM + 10TiB (**c**,**e**,**g**); in initial (as-sintered) state (**a**–**c**), and after quasi-static (**d**,**e**), and high-strain rates (**f**) 1550 s^{-1}, and (**g**) 2050 s^{-1} compression.

3.2. Compression Tests

Typical results of compression test with different strain rates of all studied samples BEPM fabricated are shown in Figure 5. It can be emphasized that the general view of the stress-strain curves of BEPM-made materials tested at high- strain-rate is rather the same as for a conventional cast and wrought Ti64 alloy with lamellar microstructure (Ti64LM) as was described in earlier work [22]. A comparison of the two materials manufactured by different technologies shows that for approximately the same strain rate levels, the maximum strength and strain levels were significantly higher for Ti64BEPM material than for conventional ones Ti64LM. As a result, the fracture of Ti64BEPM samples occurs at higher strain rates of more than 2200 s^{-1}, while cast and wrought samples of Ti64LM break at strain rates below 2000 s^{-1} [22]. This behavior was explained by finer microstructure of Ti64BEPM, ensuring its better plasticity. The addition of 5% TiC at first glance causes only minor changes in stress-strain curves compared to non-reinforced with TiC alloy Ti64BEPM (Figure 5b vs. Figure 5a). However, a more detailed analysis of stress-strain curves shows that at similar strain rates the plastic flow stress level increases while the cracking strain values decrease only slightly (compare, for instance curves 1 and 2 in Figure 5a with curves 2 and 3 in Figure 5b). In addition, the introduction of 5% TiC reduces the strain rate at which fracture occurs from 2210 s^{-1} to 2040 s^{-1}. An increase in TiC content of up to 10%

boosts this trend, leading to an increase in plastic flow stress and a decrease in cracking strain (Figure 5c).

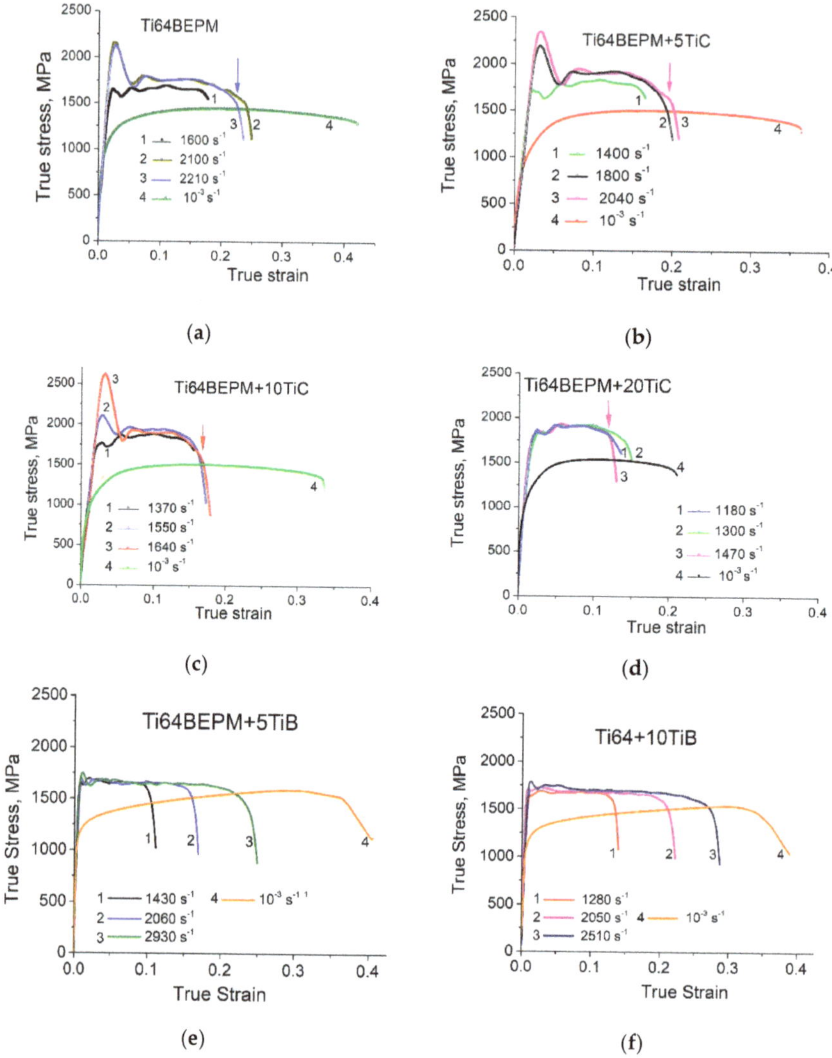

Figure 5. Typical examples of true-stress–true strain curves for SHPB (##1–3), and QSC (# 4) tests obtained with: (**a**) Ti64BEPM, (**b**) Ti64BEPM + 5TiC, (**c**) Ti64BEPM + 10TiC, (**d**) Ti64BEPM + 20TiC, (**e**) Ti64BEPM + 5TiB, and (**f**) Ti64BEPM + 10TiB samples tested at different rates. Arrows (their colors match the colors of the relevant curves) indicate the moments of samples' cracking.

Finally, at 20% of TiC, the plastic flow stress reaches its maximum level of 1800 ÷ 1900 MPa, while the cracking strain is noticeably reduced by about half compared to Ti64BEPM + 5TiC. This results in samples breaking at relatively low strain rate of 1470 s^{-1}, which is the lowest among all the materials studied in the present and previous [22] works.

The high strain-rate curves of the Ti64BEPM hardened with TiB particles have significant differences compared to the MMCs hardened with TiC. First of all, surprisingly, having slightly lower plastic flow stress level (1600–1700 MPa for the Ti64BEPM + TiB versus 1700–1900 MPa for the Ti64BEPM + TiC), these MMCs are characterized by much higher plasticity, and they were not broken even at the strain rates up to 3000 s^{-1} (Figure 5e,f). Such superiority of TiB-hardened MMCs over their counterparts hardened by TiC seems to be quite unexpected. In earlier discussed QST and 3-point bending tests results (Table 2), the relation between characteristics of strength and ductility was the opposite. However, as will be shown below, the compressive load fundamentally changes the nature of the mechanical behavior of these materials when compared to the tensile load.

QSC tests for all MMCs materials hardened with TiC particles show the similar plastic flow stress level of about 1300–1400 MPa, which is slightly higher than for the base alloy Ti64BEPM (curves#4 in Figure 5a–d). The main difference between these four materials is a gradual decrease in the cracking strain: from 0.43–0.5 for Ti64BEPM to 0.22–0.25 for Ti64BEPM + 20TiC. In turn, the QSC stress-strain curves of MMCs with TiB reveal marked differences from previous cases. First, the achievable strength exceeds 1500 MPa in combination with a strain greater than 0.35 (curves 4 in Figure 5e,f). Second, the Ti64BEPM + TiC showed a small strain hardening effect for strain range to 0.1, while Ti64BEPM + TiB revealed a very small negative strain hardening (strain softening effect). Interestingly, the Ti64BEPM + TiB deformed in quasi-static conditions does not show such behavior, whereas in the case of composite Ti64BEPM + TiC, the strain hardening mechanism has almost the same course, both in quasi-static and dynamic deformation. A positive slope of the Ti64BEPM + TiC curves can be associated with either dynamic matrix strengthening or material compaction due to pore collapsing. In turn, negative slope of the high strain-rate curves for the Ti64BEPM + TiB materials can be resulted in cracking the TiB particles under dynamic deformation. It should also be noted that increasing the TiB content from 5 to 10% practically does not change the mechanical characteristics of the material.

4. Discussions

4.1. Strain Energy

In order to carry out more in-depth assessment of the mechanical behavior of the materials tested under dynamic loading, an additional parameter, i.e., elastic–plastic strain energy (*SE*), was used. It is a convenient parameter that allows comparing the mechanical response of materials tested with various methods and strain rates [13–17]. The SE is defined as the internal work performed to deform a material specimen through an action of the externally applied forces. The *SE* was determined by integrating the area under the stress–strain curve. In the case of the cracked specimens, a value of $\varepsilon_{\rm upper}$ corresponded to a value of strain at fracture, whereas for the non-cracked specimens, $\varepsilon_{\rm upper}$ was assumed to be equal to strain at the moment of the specimen unloading (sharp drop in stress–strain curve). The upper integration limit $\varepsilon_{\rm upper}$ for the non-cracked specimens under quasi-static loading was assumed to be 0.5. The *SE* values calculated from the experimentally obtained stress-strain curves and plotted vs. strain rate are shown in Figure 6a. The Ti64BEPM samples demonstrate the highest *SE* values among all previously studied [22] Ti based materials when tested at strain rates up to 2200 s^{-1} (A and C arrows on the curve #1). At higher strain rates (B arrow, ibid.), the Ti64BEPM samples break and the *SE* values drops substantially.

The addition of a reinforcing phase to the alloy significantly alters the mechanical behavior of the sintered material. For the 5% TiC MMC, the *SE*, including its value at maximum, the SE_{max}, is essentially higher compared with the Ti64BEPM alloy not hardened by particles. However, the 5%TiC MMC fractured at a critical strain rate, $\dot{\varepsilon}_{\rm max}$, 1927 s^{-1}, which is significantly lower in comparison to alloy without reinforcement, $\dot{\varepsilon}_{\rm max}$ – 2220 s^{-1} (Figure 6a, curves 1 and 2 respectively). A further increase in the TiC content to 10% and 20% decreased both the SE_{max} and $\dot{\varepsilon}_{\rm max}$ values (Figure 6a, curves 3 and 4).

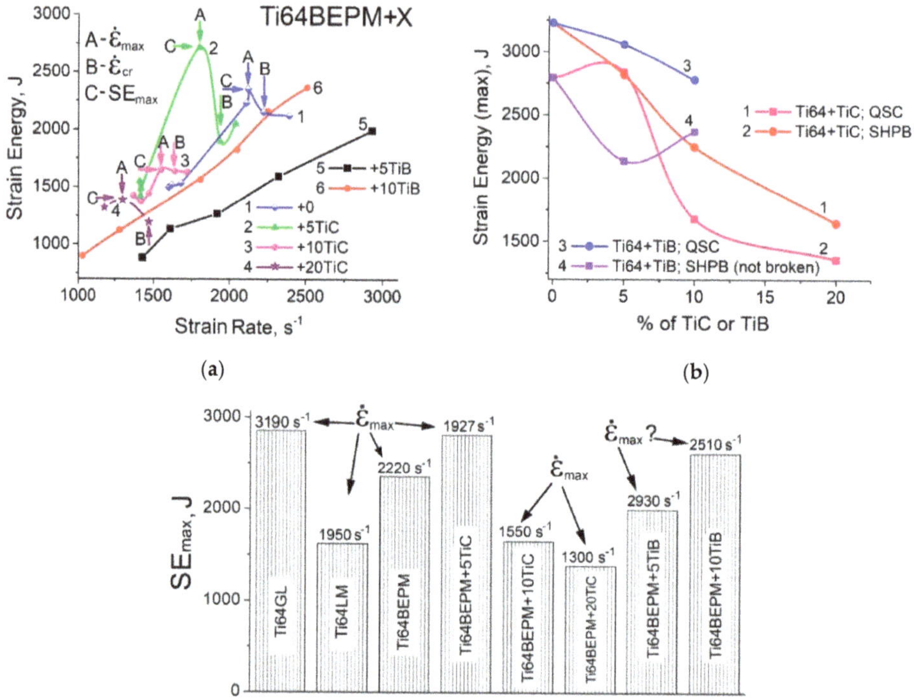

(c)

Figure 6. The strain energy *SE*: (**a**) vs. strain rate dependencies for studied Ti64BEPM based materials without (1) and MMCs with (2–4) TiC, and (5,6) TiB particles; (**b**) The strain energy *SE* for QSC with rate 10^{-3} s^{-1} and high-strain-rates SHPB compressions vs. content of TiC or TiB particles; (**c**) the SE_{max} and $\dot{\varepsilon}_{max}$ values for tested materials compared with the data for Ti64GL and Ti64LM taken from [22]. Vertical arrows in (**a**) indicates the strain rates at which specimens fracture. A, B, and C in (**a**) indicate measured values of $\dot{\varepsilon}_{max}$, $\dot{\varepsilon}_{cr}$, and SE_{max} respectively. The "?" sign in (c) above the Ti64BEPM + 10TiB bar means that it is not known whether this value is the limit, since the sample was not broken and it was not tested at a higher strain rate.

The reinforced TiB MMCs are characterized by lower *SE* values for the same strain rates in comparison with Ti64BEPM containing similar amounts of TiC: curves 5 and 6 vs. curves 2 and 3, correspondingly (Figure 6a). However, the containing TiB MMCs did not fracture in the entire range of tested strain rates, as opposed to titanium carbide reinforced ones. As a result, the lowest *SE* values for 5%TiB at strain rates below 2000 s^{-1} were obtained; however, the *SE* values for 10%TiB MMCs become markedly higher (curves 5 and 6). A further increase in the strain rate of the 10%TiB MMC specimens above about 2250 s^{-1} results in an increasing the *SE* level exceed all other tested materials. If we compare these materials in terms of the SE_{max}, then the Ti64BEPM + 10TiB MMC has superiority over Ti64BEPM + 10TiC (curves 4 and 2 in Figure 6b). In turn, for the QSC testing condition, the *SE* values for both Ti64BEPM + TiB MMCs are higher than for Ti64BEPM + TiC with the same amount of TiB and TiC used. The difference in the SE_{max} values increase with increasing content of reinforcing particles (Figure 6b, curve 3 vs. curve 1). It can also be noted that the SE_{max} values for the SHPB and QSC tests were approximately the same only for Ti64BEPM + 5TiC material (Figure 6b, curves 1 and 2). For all others materials, the SE_{max} values were much higher for QSC compared to the obtained in the SHPB tests (curves 1 and 3 vs. 2 and 4, ibid.).

The comparative overview of the SE_{max} values at the corresponding maximum strain rates, $\dot{\varepsilon}_{\max}$, for different materials on the base of Ti64 alloy studied here and also earlier reported data are shown in Figure 6c. It can be seen that the best combination of parameter SE_{max} and $\dot{\varepsilon}_{\max}$ corresponds to cast and wrought Ti64 with an "ideal" globular microstructure (2795 J at 3190 s^{-1}), and Ti64BEPM + 5TiC (2840 J at 1927 s^{-1}). The next materials in decreasing order of SE_{max} and $\dot{\varepsilon}_{\max}$ parameters are Ti64BEPM + 10TiB (2370 J at 2510 s^{-1}) and Ti64BEPM + 5TiB (2140 J at 2930 s^{-1}), and finally Ti64BEPM + 10TiC and Ti64BEPM + 20TiC, for which the lowest values were obtained as a result of premature fracture of the samples (Figure 6c).

4.2. Changes in Phase Composition, Microstructure and Micro-Texture

4.2.1. MMC Reinforced with TiC

The X-ray diffraction patterns of TiC MMCs in the initial state, after the QSC and the high-strain-rate SHPB tests, are compared with the base Ti64BEPM material in Figure 7.

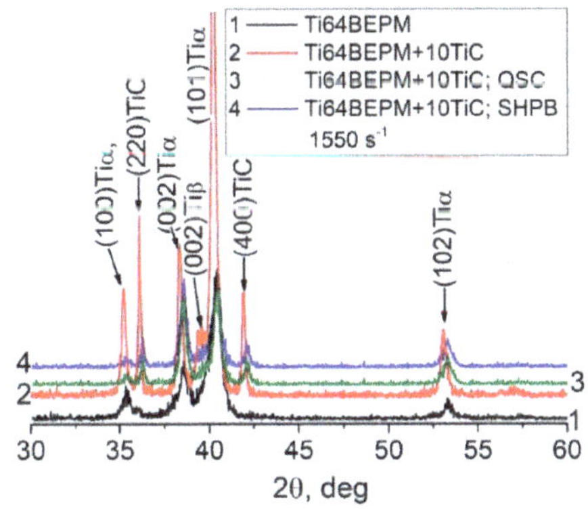

Figure 7. The X-ray diffraction patterns of (curve 1) Ti64BEPM, and Ti64BEPM + 10TiC in (curve 2) as-sintered state, and after (curve 3) QSC ($\dot{\varepsilon}$ = 10^{-3} s^{-1}) and (curve 4) high-strain-rate SHPB ($\dot{\varepsilon}$ = 1550 s^{-1}) tests.

The as-sintered Ti64BEPM material is predominantly characterized by the presence of α-phase, and a very small amount of β-Ti (less than 5% by vol.), which is difficult to see and can be traced by the presence of {002}Tiβ peak (Figure 7, curve 1). The addition of TiC particles causes the appearance of the TiC phase diffraction, and a significant increase in peaks intensity of the α-Ti phase (Figure 7, curve 2). There is a slight shift in the position of the α-Ti peaks compared to the peaks of the base material without TiC, which is likely associated with some changes in the chemical composition of the titanium-based matrix. (As will be shown below (see Section 4.2, Figure 21, and Table 4), an interaction of titanium with TiC particles took place that leads to some depletion of the matrix with titanium.) The relatively weak peak {002} of β-phase becomes distinguishable. After both types of compression tests, the magnitude of all peaks decreased to approximately the level that existed for the initial sintered state (curves 3 and 4). There was no broadening of the peaks observed after both, the quasi-static and dynamic compression tests. Such observation suggests that [39] there has been no accumulation of residual elastic stresses in either the α phase or the TiC particles. The whole deformation energy dissipates probably by crushing the hard particles and plastic flow of matrix phases and formation of a dislocation substructure in therein. There are no significant changes in the crystallographic texture of the samples after both types of compression tests used, as shown by the (102) pole figures of α-Ti (compare Figure 4d,f with Figure 4c).

Certain zones outstanding in the direction of compressive force in the longitudinal section of the tested cylindrical samples during the SHPB and the QSC compression tests were determined in a previous study performed on an alloy Ti64 with different microstructures [22]. The resulting microstructures were quite explicit for the identified zones where the stresses act differently on parts of tested sample (Figure 8). It has been shown that the deformed structure results from the action of the high-strain rate compression depending on the type and dispersion of the original microstructure [22].

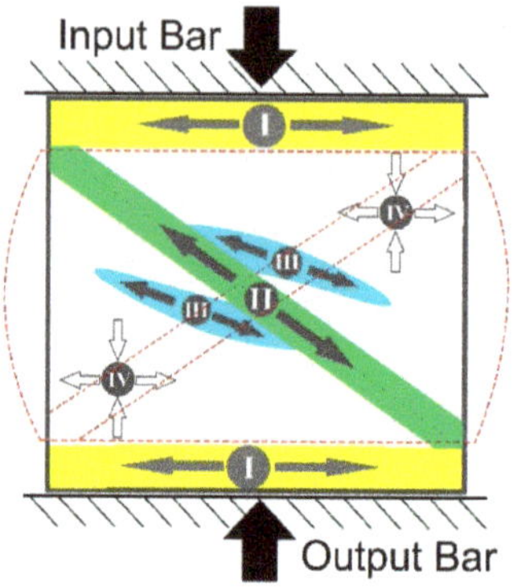

Figure 8. Schematic map of the location of the different zones in the longitudinal section of the test cylindrical samples. Roman numbers denote a certain zone. Black arrows indicate the direction of compression force.

Suggested implication is entirely true for the present evaluation of the metal-matrix composites obtained using BEPM. The Ti64BEPM alloy has been studied in details in this earlier work and it would be useful to assess the structures obtained in present study similarly, to specify the following important arguments. The samples Ti64BEPM showed signs of plastic deformation mainly in Zone I, i.e., in the areas of contact with the surfaces of input and output bars (Figure 1). This deformation zone was most evident on non-fractured samples and was obviously near the specimen contact surface with the front face of the input bar (Figure 9a). Collapsed residual pores were also observed at various locations across the sample (Figure 9a).

The compression applied at higher strain rate (with 2390 s^{-1}) causes samples to fracture. The fracture surface is characterized by a typical rectilinear zone, which indicates the main crack propagation (A arrowed in Figure 9b) and adjacent to it zones of the secondary cracks spread (B, ibid). A higher magnification images reveal additional features, namely, residual pores (Figure 9c), generally absent in the cast and wrought alloy [22,39,40]. The microstructure of such Ti64BEPM specimens was quite similar to the alloy Ti64 with a lamellar structure, in which a plastically deformed Zone I has approximately the same depth close to 20–30 μm (Figure 9d). The main crack initiates and propagates within the Zone II running through the entire cylinder at an angle of about 45° to its vertical axis. This zone is a very narrow layer of adiabatic shear band (ASB), which has a thickness of about 5–6 μm near the edge of the cylinder (Figure 9e). The smaller secondary cracks associated with β-grain and α-colony boundaries are observed in the Zone III (Figure 9f). Specific evidence of plastic deformation was not found within Zone IV, while fine α-needles were observed inside individual α-lamellae and identified as α′-martensite (details are discussed elsewhere [22]).

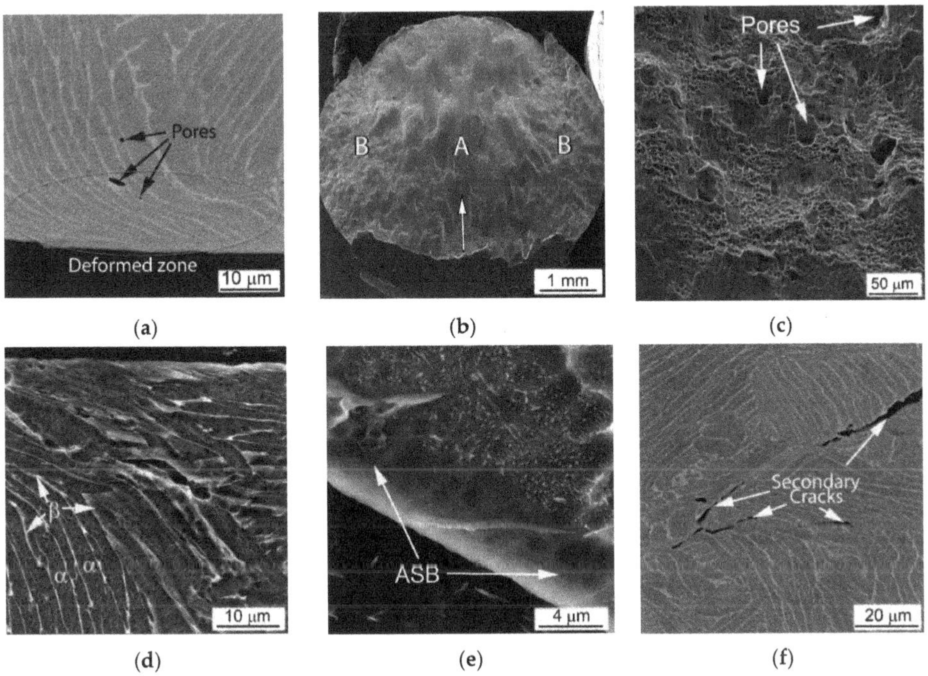

Figure 9. Microstructure (**a**,**d**–**f**), and fracture surface (**b**,**c**) of Ti64BEPM specimens tested with the strain rates: (**a**) 2220 s^{-1} (not broken), and (**b**–**e**) 2390 s^{-1} (broken). (**a**,**d**) Zone I, (**e**) Zone II, (**f**) Zone III. SEM, (**a**,**f**) SE, (**b**–**e**) BSE. The specimens were compressed along the vertical axis as shown in Figure 8, and images in (**a**,**d**–**f**) are aligned with their vertical axis along with the acting external load like in Figure 8.

The MMCs containing TiC particles were fractured at relatively lower strain rates, and their fracture morphology and microstructure differed significantly compared to the base Ti64BEPM material. The representative microstructures of Ti64BEPM + 5TiC are shown in Figure 10. During the fracture, the growth direction of the main (A in Figure 10a) and peripheral (B, ibid.) cracks differs in this composite from the base BEPM alloy (without TiC). As can be seen the main crack can actually change the direction of its further propagation when meets the clusters of particles TiC (Figure 10b). Similarly, the change in the initial direction of the crack growth is observed for smaller secondary cracks (Figure 10c).

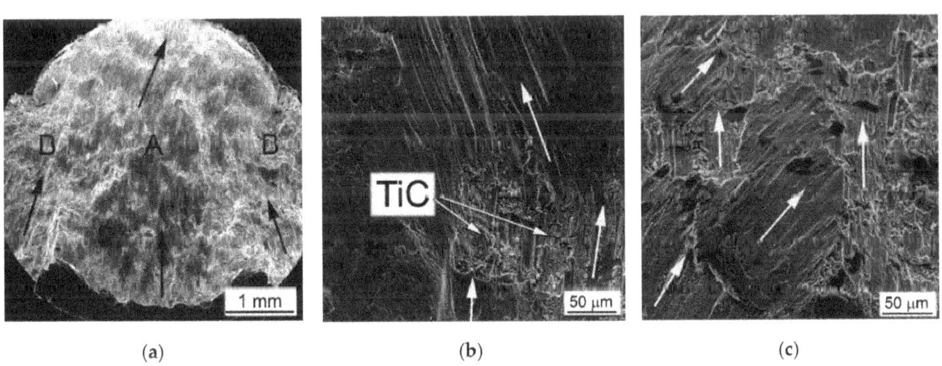

Figure 10. *Cont.*

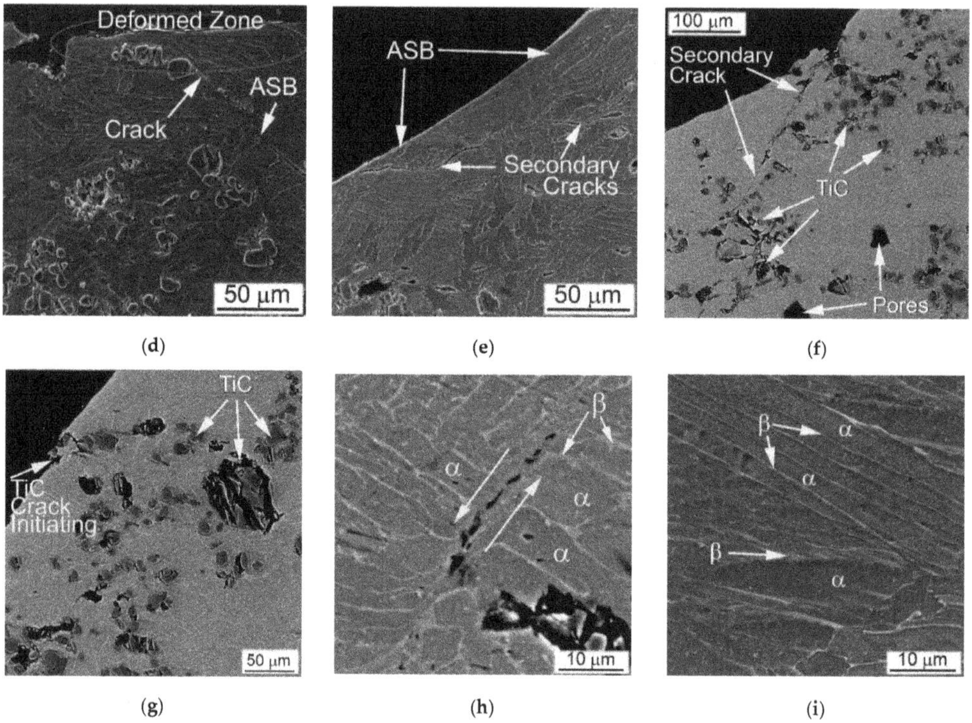

Figure 10. Typical fracture surface (**a–c**) and microstructure (**d–i**) of Ti64BEPM + 5TiC MMC SHPB tested at 2040 s^{-1} strain rate. Arrows in (**a**,**c**) indicates directions of cracks growth. (**d**) Zone I, (**e–g**) Zone II; (**h**) Zone III (**i**) Zone IV. SEM, (**a–d**,**g**,**i**) SE, (**e**,**f**,**h**) BSE. (**f**,**g**,**h**) polished; (**d**,**e**,**i**) polished and etched. Images (**d–i**) are aligned with their vertical axis along with the acting external load like in Figure 8.

The deformed Zone I was less pronounced in MMC than the Ti64BEPM TiC-free (Figure 10d vs. Figure 9a). Rather thin, not more than 5–7 μm, ASB was observed in Zone II in some places free of TiC particles close to the main crack (Figure 10e). In most places where TiC particles lie on the fracture surface, the ASBs were not observed. Moreover, these particles not only changed the direction of the crack propagation, but also served as initiators of secondary cracks (Figure 10e,f). The majority of the residual pores appear unchanged (Figure 10f), but particles TiC, especially larger particles or aggregates of smaller ones, break down (Figure 10g). The microstructure of the matrix alloy in the Zone III shows signs of quite strong plastic deformation, including shear displacement inside the $\alpha + \beta$ packets within some grains, probably with some involvement of residual pores in this process (Figure 10h). Finally, no evidence of deformation was found in Zone IV (Figure 10i).

The MMC with a higher content of TiC particles after the high-strain-rate test demonstrates a similar fracture surface (Figure 11a–d) and microstructure (Figure 11e–i), as discussed above. As in the previous case, the propagation directions of the main and peripheral cracks passing through the coarse TiC particles and their agglomerates have changed (Figure 11a), but a more distinct secondary cracking has occurred (Figure 11b). Several cases of melting of small areas on the fracture surface were found (Figure 11c). Their presence obviously implies significant local heating, which can be observed even for specimens deformed at this relatively slow strain rate (1640 s^{-1}). In Ti64BEPM + 10TiC, similar to 5% composite, carbides strongly affect the propagation of cracks, although the fracture character within the matrix alloy stays ductile (Figure 11d). Alike the former case, carbide particles, especially larger ones, which are located near the main crack, were broken up (Figure 11e), and thin ASB was observed near the crack surface free of TiC particles (Figure 11f).

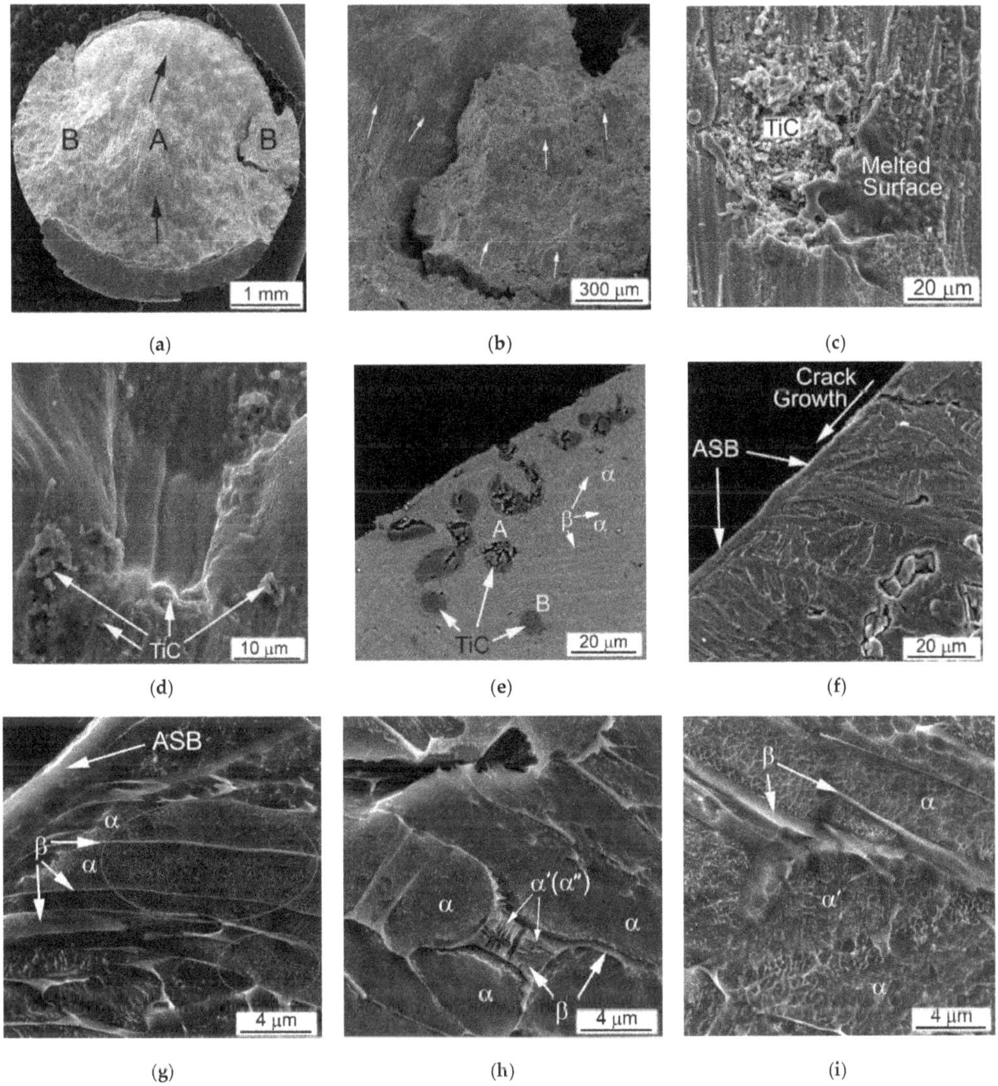

Figure 11. Fracture surface (**a**–**d**) and microstructure (**e**–**i**) of Ti64BEPM + 10TiC, SHPB tested with rate 1640 s^{-1}. Arrows in (**a**,**d**) indicate the direction of cracks growth. (**e**–**g**) Zone II; (**h**,**i**) Zone IV. Ovals in (**g**) indicate the areas within α- lamellas containing martensitic needles. The broken TiC particles near the main crack are A labeled in (**e**), while not broken particles a bit away from the main crack are B labeled. SEM, (**a**–**d**) (**f**–**i**) SE, ((**e**), polished) BSE; (**f**–**i**) polished and etched. Images (**e**–**i**) are aligned with their vertical axis along with the acting external load like in Figure 8.

Remarkable structural features have been found in this sample at various locations within Zones II, III, and IV shown in Figure 11g–i, respectively. They resemble a highly dispersed needles, mutually oriented to each other at an angle of about 60°. Previously, such structures were reported for Ti64BEPM, as well as for the cast and wrought Ti64GL after SHPB tested at much higher strain rates, namely, 2390 s^{-1} and 3330 s^{-1}, respectively [22].

The observed structures were explained as α′-martensite formation inside untransformed broad α-lamellae. The occurrence probability of such a transformation was explained due to the dynamic loading conditions, which cause local rapid heating of relatively small samples, and then their rapid cooling, resulting from heat dissipation into SHPB metal bars with relatively good thermal conductivity. In this case, two types of martensitic needles with slightly different dispersion have been found, i.e., more dispersed as shown in Figure 11g and somewhat coarser as can be seen in Figure 11h. The chemical microanalysis data presented in Table 3 show that the coarser martensitic needles had a higher content of the β- stabilizing elements of vanadium and iron, while, in contrast, α- stabilizing aluminum was present in a smaller amount (p. #2 vs. p#1 in Table 3). The observed increased content of β-stabilizers should lead to the formation of orthorhombic α″- martensite rather than hexagonal α′- martensite [1,41–44]. Most likely, this occurs due to some cooling delay within the local micro-volumes, which leads to the redistribution of alloying elements between the β- and α-phases, enriching the last phase with vanadium. There is also the possibility of some micro-inhomogeneity of the matrix alloy, which was formed during the synthesis of MMC. The second explanation can also be supported by the earlier discovered ability of partial carbon diffusion from titanium carbide particles into the matrix alloy during sintering [32,35,36]. This mechanism undoubtedly leads to a certain diffusion redistribution of titanium and alloying elements between phase constituents. In a certain sense, it may seem surprising that this phenomenon (formation of martensite) has been observed in α-lamellae of Ti64 alloys free of hard particles at least twice the strain rates than in the present Ti64BEPM + 10TiC MMC. It is also obvious that the presence of solid particles TiC in the structure during intense plastic deformation can cause additional localization of the plastic flow even at lower rates, since these particles play the role of additional local stress concentrators, which in turn can cause more intense local heating.

Table 3. Microanalysis EDS data of the areas within α-lamellae containing martensitic needles of two different dispersions.

Location (in Figure 11)	Chemical Composition, wt.%				
	Al	V	Fe	Ti	Sum of Impurities
1 (11g)	6.2	4.2	1.2	Balance	<0.4
2 (11h)	4.8	11.8	1.8	Balance	<0.4

Important conclusions on the deformation mechanism can be drawn from EBSD results revealing the local crystal structure and some of its defects. The images presented in Figure 12 show the local microstructure of the cylindrical sample Ti64BEPM+10TiC fractured along Zone II (Figure 8) during the SHPB test with the strain rate 1640 s^{-1}. The fractured surface (cross-section) can be seen in the upper left corner of Figure 12a–c. In an earlier study of the SHPB tested Ti64 alloy, acceptable EBSD quality patterns/results were obtained for lamellar microstructure samples deformed only at relatively low strain rates [22]. The pattern quality map (a) shows regions of darker gray levels in the vicinity of the TiC particles indicating that these regions contain significantly larger dislocation densities than the rest of the sample covered in this EBSD map. The high dislocation density decreases the local crystallinity state, therefore these regions appear darker in the pattern quality map and non-indexed in the phase and orientation maps. These results are very consistent with the expected increased strain rate seen by the matrix in the vicinity of the TiC particles. The structure adjacent to the fractured surface depicted in Figure 12d shows that some of the α-phase lamellae become highly fragmented (A-labeled) with the submicron blocks′ size, while some other lamellae (B-labeled) just next to them practically remain unchanged. The structure at a distance within of several hundred microns from Zone II is also very heterogeneous, especially near particles TiC or the α-lamellae colonies boundaries or β-grains boundaries. For example, lamellae are finer and more deformed close to particles TiC, multiple twinning in the α-lamellae can be seen close to TiC particles and boundaries of β-grains as shown in Figure 12e. Areas of larger plastic deformation in some lamellae can be seen away from Zone II (about 300 μm), as shown by a large green lamella at the center of Figure 12f. The change of orientation of this lamella is about 20 deg. within 10 μm space and its tip close to particle TiC is highly deformed, whereas the rest of lamella is deformed only slightly. The reasons of observed high heterogeneity of plastic deformation are obviously related to the existing structural defects, such as: particles TiC, β-grains boundaries, and α-lamellae colonies boundaries. However, high deformation heterogeneity can be also resulted from effect of orientation of lamellae toward the acting load, favoring dislocation motion mechanism, subblocks building, twinning, etc.

Figure 12. SEM EBSD images of the fractured composite Ti64BEPM + 10TiC after SHPB test with the rate 1640 s^{-1}. Pattern Quality Map (**a**), phase map (**b**), and orientation map (**c**–**f**) images. The α-phase is shown in (**b**) using green, β-phase blue, and TiC red colors. All images are taken in vicinity of the Zone II (Figure 8); the images (**a**–**c**) centers are within 200 μm away from the place of fracture (main crack) visible in the left top corner of the images (**a**–**c**). The numbered areas in (**c**) are shown at higher magnification images: (1) in (**d**), (2) in (**e**) and (3) in (**f**).

In general, the fracture surface and internal microstructure of the SHPB tested Ti64BEPM + 20TiC samples (Figure 13) were very similar to above discussed results for the composites containing smaller amounts of titanium carbide particles. The growth directions of the main and peripheral cracks changed during the high strain-rate compression (Figure 13a). Carbide particles affect the direction of cracks propagation not only in the plane propagation of the main crack, but also at an angle to it (Figure 13b). A plurality of dimples can be seen inside the titanium-based matrix, which are typical of the ductile fracture, but a large number of carbide particles splinters are also observed on the fracture surface, which are obviously result of the crack spread (Figure 13c). The internal microstructure of the sample had quite distinctive features. In the immediate vicinity of the surface of the main crack, as well as away from it, only large TiC agglomerates were destroyed (some of them felt out of during samples preparation because of relatively poor bond with matrix), while several smaller individual particles had some cracks (Figure 13d,e).

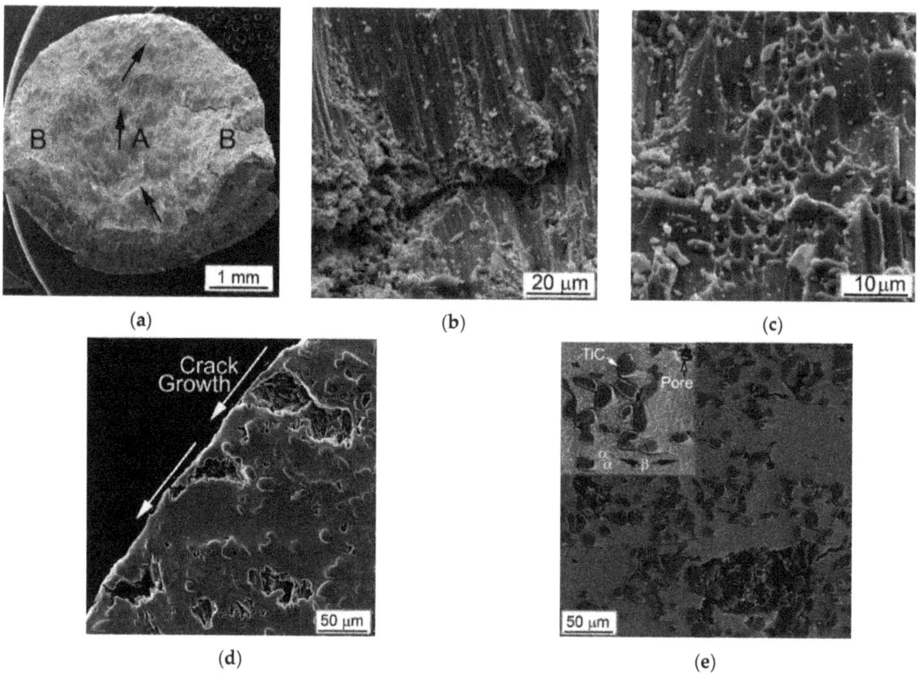

Figure 13. Fracture surface (**a–c**) and microstructure (**d**,**e**) of Ti64BEPM + 20TiC sample, SHPB tested with rate 1470 s^{-1}. Arrows in (**a**,**d**) indicates directions of cracks growth. (**d**) Zone II; (**e**) Zone IV. SEM, (**a–c**) SE, (**d**,**e**) BSE images.

The above results suggest that the presence of hardening particles TiC, even in such small amounts as 5%, plays a critical role in the deformation and fracture of the MMC. The hard (but brittle) particles are broken under the applied impact load and predetermine the deformation and damage of the Ti-based alloy matrix. From a comparison of the results obtained for the Ti64BEPM alloy and MMCs containing 5% and 10% TiC, it can be concluded that the presence of the particles and an increase in their content leads to the switching of the "weakest" microstructural element responsible for the bulk of the *SE* dissipation: from purely plastic flow (deformation) of the matrix to brittle fracture of the TiC particles that initiate the onset of cracks and promote their propagation. In the Ti64BEPM alloy, in the absence of TiC, the role of the "weakest link" is played by the boundaries of β- grains covered by coarse α-lamellae and the boundaries of α + β colonies (Figure 6f) already discussed in more details elsewhere [22]. The introduction to Ti64BEPM alloy of particles of TiC can change the nature of the mechanism of deformation and fracture. At 5% of the carbide particles TiC, which mainly caused an increase in the strength of the material (compare Figure 5a with Figure 5b, and Curves ## 1 and 2 in Figure 6a), and only when the strain rate reached $\dot{\varepsilon}_{cr}$ value, most of the particles begin to break, thereby causing the formation of cracks fracturing the specimen. The structure embrittlement effect grows with further increase of TiC portion within the MMC resulting in decrease in V_{cr} at which material cracks. For example, at 20%TiC the reserve of plasticity of Ti64 matrix is insufficient for prevention of destruction of material even at quite low strain rate such as 1470 s^{-1}, not talking about higher rates (Figure 5d, and curve #4 in Figure 6).

The particles of TiC have a very similar effect on the mechanical behavior in QSC condition. The introduction of 5%TiC in Ti64BEPM resulted in a slight increase in strength as well as a slight decrease in plasticity, as seen from comparison of curves #4 in Figure 5a,b. The strength level remains rather the same at 10% and 20% of TiC, while the plasticity reduction becomes more evident (Figure 5c,d). Such differences in mechanical behavior are directly related to the specifics of the deformation mechanism, which is affected by the microstructure of the specimens (Figure 14). The Ti64BEPM structure after QSC tests was characterized by clear signs of uniform plastic deformation throughout the entire sample volume (Figure 14a). At the same time, small cracks were also observed at various places of interfacial α/β- boundaries, or at the boundaries of α-Ti colonies. In turn, the residual pores, they mainly collapsed (at least partially) without cracks initiation (Figure 14b).

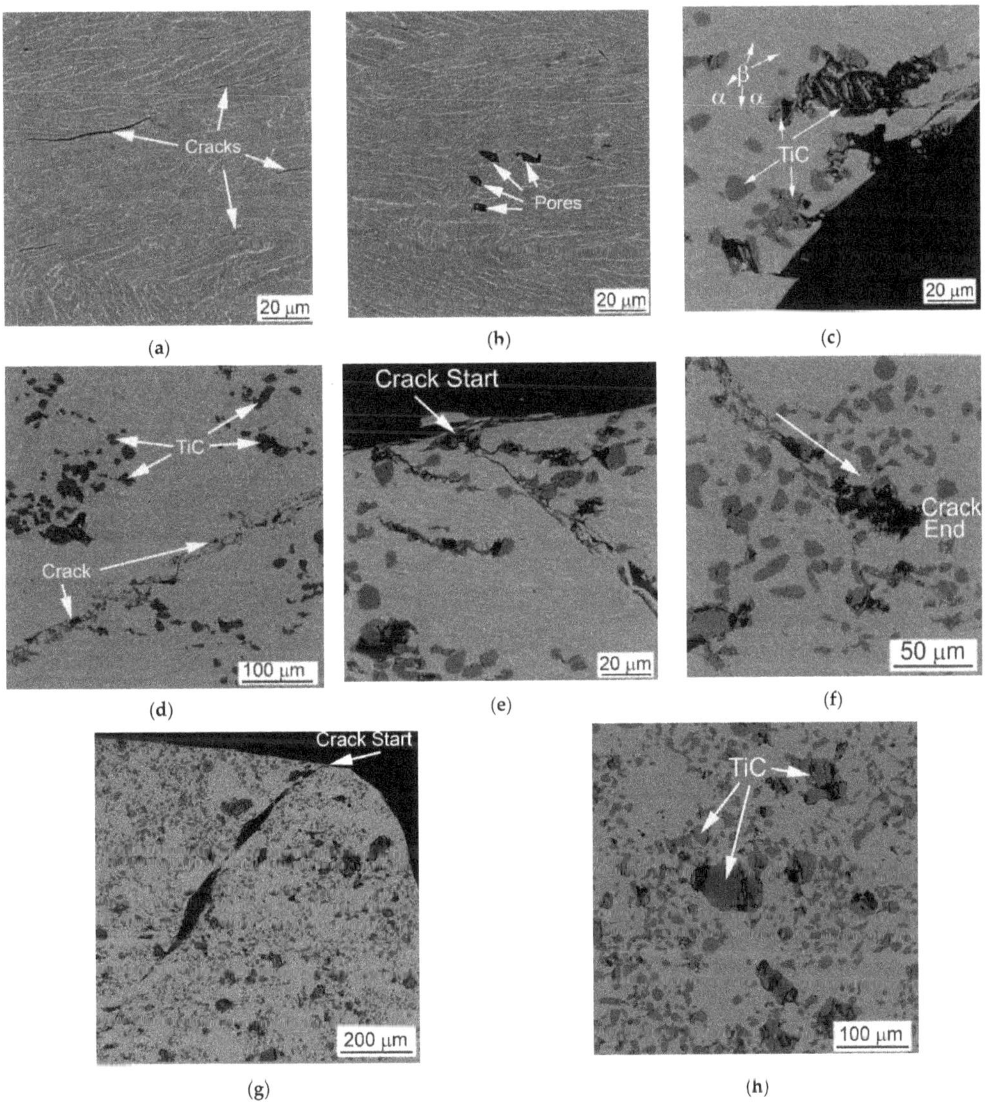

Figure 14. SEM images of samples after QSC tests with rate 10^{-3} s^{-1}: (**a**,**b**) Ti64BEPM, (**c**,**d**) Ti64BEPM + 5TiC, (**e**,**f**) Ti64BEPM + 10TiC, and (**g**,**h**) Ti64BEPM + 20TiC. (**a**,**b**,**d**,**h**) Zone IV, (**c**) Zone II, (**e**,**g**) Zone I, and (**f**) Zone III. BSE.

The situation changes drastically in the presence of TiC particles. First of all, all samples, as a rule, are destroyed by the main crack propagating at an angle of about 45° to the applied load (Figure 14c). There are numerous secondary cracks appears in addition to the main crack. (Figure 14d–g). The carbide particles located on the surface of the specimens can initiate these cracks (Figure 14e,g), while the particles inside the specimens can stops the cracks propagation (Figure 14f). At the same time, predominantly large TiC particles and their conglomerates breaks (Figure 14c,f,h), while the small particles remain almost unchanged. The signs of plastic flow are observed in the matrix of composite only at 5% and 10% of TiC particles (Figure 14d,e).

4.2.2. MMC Reinforced with TiB

A comparison of the X-ray diffraction patterns of the base Ti64BEPM alloy and its composite reinforced with 10% particles of TiB in the sintered state, as well as after deformation at different strain rates is shown in Figure 15. The formation of TiB phase during sintering is reflected by the presence of corresponding X-ray peaks in the diffraction pattern, which are particularly well resolved on untested specimens (compare curves 1 and 2 in Figure 15). These peaks become difficult to distinguish for specimens deformed under quasi-static and high strain-rate compression (curves 3 and 4, ibid.).

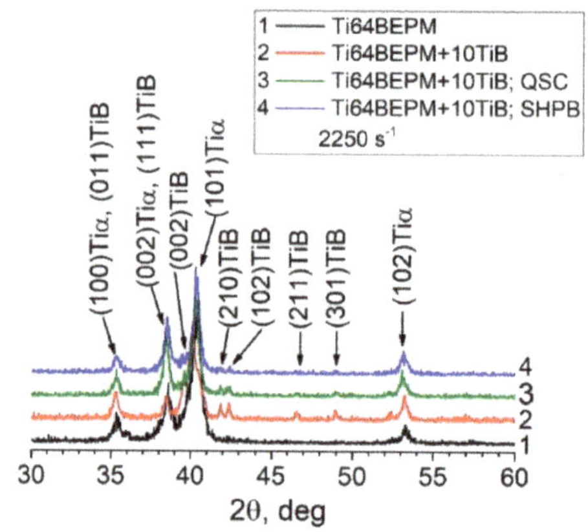

Figure 15. The X-ray diffraction patterns of (curve 1) Ti64BEPM, and Ti64BEPM + 10TiB in (curve 2) as-sintered state, and after QSC ($\dot{\varepsilon} = 10^{-3}\ s^{-1}$) (curves 3 and 4), and SHPB ($\dot{\varepsilon} = 2250\ s^{-1}$) tests.

The textures of the α-phase crystallites of these composites after both types of compression are more distinct than the initial states after sintering (including Ti64BEPM + XTiC), showing a relatively higher (102)α pole density at the center of the pole figures (Figure 4e,g vs. Figure 4c).

As noted earlier, MMC strengthened by TiB particles did not break down during high strain-rates compression tests. This behavior was adequately reflected in the assessment of microstructure after compression tests. There was no evidence of significant plastic deformation within α + β- matrix found by SEM for 5% of TiB composites, even in areas within cylinder samples where deformations were expected to be greatest, such as Zones I and II; as was no observed either ASB or intensive cracking (Figure 16). The only signs of deformation were cracks in TiB particles (Figure 16a,c), which at higher strain rates also spread to the α + β- matrix (Figure 16d). In fairness, some elongation of the α + β packets as a result of the plastic flow can be noted at higher strain rates, as shown in Figure 16d.

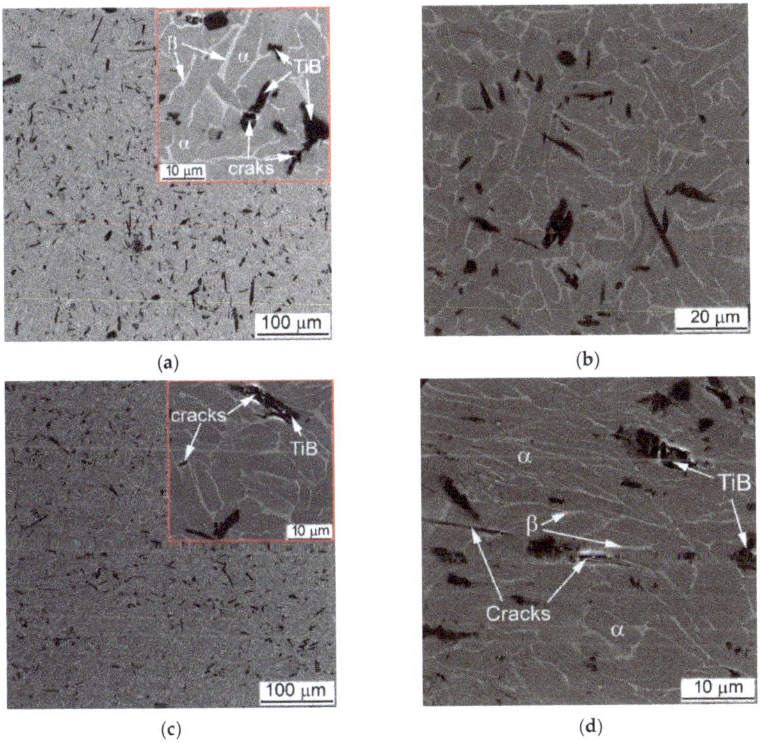

Figure 16. SEM BSE images of Ti64 + 5TiB after the high-strain-rate SHPB test with rates (**a**,**b**)—1430 s^{-1}, (**c**,**d**)—2930 s^{-1}. (**a**,**c**)—Zone II; (**b**,**d**)—Zone I.

The structure of 10% of TiB MMC after the high strain-rates test also did not show noticeable signs of plastic deformation in α + β-matrix (Figure 17). As for Ti64BEPM + 10TiB, SEM finds mainly cracking of TiB lamellae (Figure 17a,c). There was also no principal difference observed between the structures of Zones I and II, and no differences after a twofold strain rate increase as shown in Figure 17a,b vs. Figure 17c,d. Such "microstructural resistance" of MMCs reinforced by particles of TiB to high strain-rate compression—in contrast to TiC MMCs—explains the stability of the strain energy characteristics with increasing the strain rate (curves 5 and 6 vs. curves 2 and 3 in Figure 6a).

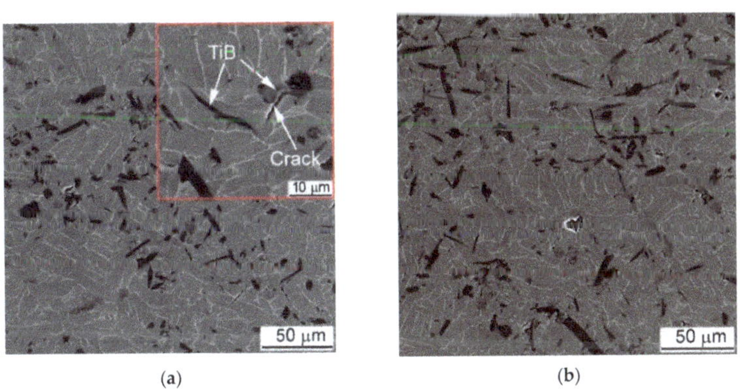

Figure 17. *Cont.*

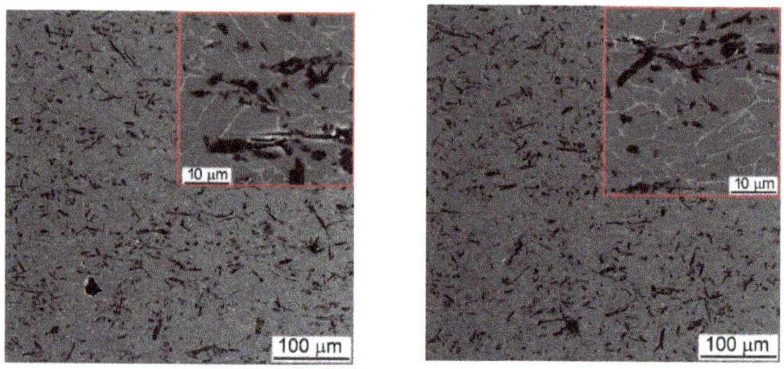

(**c**) (**d**)

Figure 17. SEM BSE images of Ti64 + 10TiB after the SHPB tests with strain rates (**a**,**b**)—1280 s^{-1}, (**c**,**d**)—2510 s^{-1}. (**a**,**c**)—Zone II; (**b**,**d**)—Zone I.

The EBSD data of Ti64BEPM + 10TiB sample after SHPB test are shown in Figure 18. While this sample maintained macroscopic integrity after the test, the EBSD results indicate that the sample experience localized shear deformation at the microscopic level. The dark band labeled A in Figure 18c shows a major shear band where a much higher dislocation density has lowered the local crystallinity of the matrix resulting in a high ratio of non-indexed points. Most of the few points indexed inside this shear band are from the TiB particles which were less affected by the severe plastic deformation due to their much higher hardness. Additional, minor shear bands, labeled B in Figure 18c, occur similarly in the vicinity of particles TiB. EBSD observation also reveals the fragmentation of the large TiB particles. However, conventional EBSD technique has difficulty to fully resolve the β-Ti phase expected in this alloy and that is obviously due to the sub-micron dimension of this phase regions. Nevertheless, some larger β-Ti lamellae can be seen, for instance, the one labelled C in Figure 18f. Resolving the small or highly deformed crystals was even harder after the SHPB test when the structure refined to domains smaller than the spatial resolution of conventional EBSD technique. This was one of the reasons to use the Transmission Kikuchi Diffraction (TKD) technique to investigate these microstructures at the nanometer scale to confirm the crystallinity of shear bands and very fine α- and β-Ti grains that have been modified beyond the resolution reach of conventional EBSD.

As can be seen from Figure 18, despite the absence of the cracks, it was possible to identify a band of severe plastic deformation, A and B labeled in Figure 18a–c. It can be noted that all these bands of localized plastic deformation seem to evade microvolumes with large particles of TiB. At higher magnification, it becomes clear that the TiB particles have no internal defects. The largest particles break up into individual fragments, while the many smaller ones seem to retain their original shape and size (Figure 18d–f). Dislocations density (affecting pattern quality in EBSD and TKD maps) inside the α-phase lamellae seems increased (Figure 18g). Intense plastic deformation is localized in microvolumes surrounding TiB particles, and these areas look like dark regions with poorly resolved Kikuchi patterns obscuring a clear orientation map (Figure 18g–i).

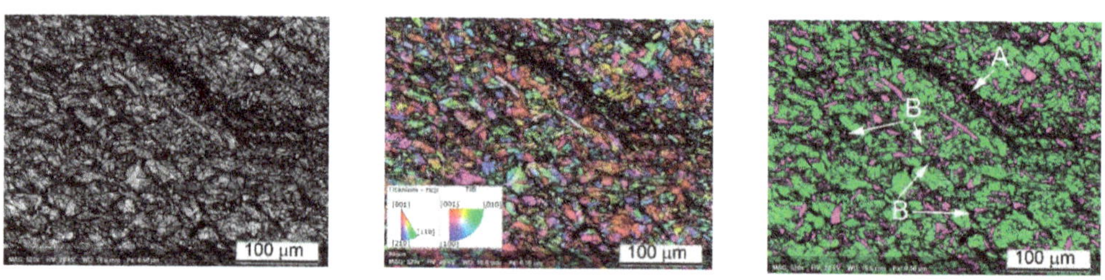

(**a**) (**b**) (**c**)

Figure 18. *Cont.*

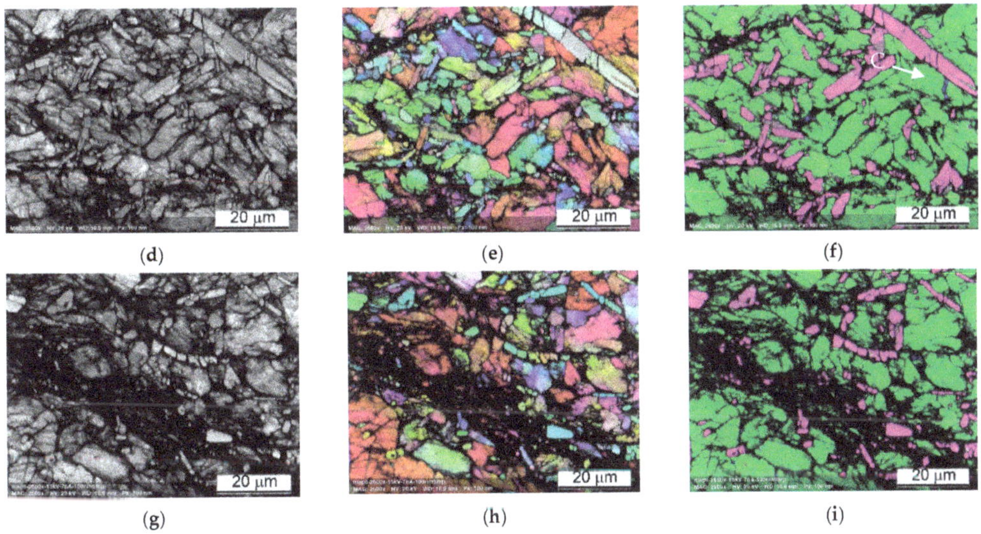

Figure 18. SEM EBSD images of Ti64BEPM + 10TiB sample after SHPB test with rate 2510 s^{-1}. The sample stays unbroken after the test, but forms the ASB area in the Zone II (Figure 8). Pattern Quality Maps (**a**,**d**,**g**); orientation maps (**b**,**e**,**h**) phase maps (**c**,**f**,**i**) images. The green color in phase map images represents α-phase, blue β-phase, and magenta TiB particles.

TiB MMC specimens deformed with a quasi-static strain rate show a macroscopic cracking, in contrast to dynamically loaded samples (Figure 19). That was the obvious result of essentially higher general strain achieved (curves 4 vs. curves 1–3 in Figure 5e,f). After quasi-static loading, both composites strengthened with TiB formed a main crack crossing samples at an angle 45° to the main axis of the cylinder (Figure 19a). In this case, the main plastic deformation was located exactly near the main crack (Figure 19b,c). All other regions remote from the crack were characterized with practically uniformly deformed structure. The specimen deformation was mainly caused by crushing the TiB particles (Figure 19d). Comparing the samples with 5% and 10% of TiB (Figure 19a,d), it can be concluded that such a wide (up to 100 µm) band of cracks crossing the samples was formed by merging two separate cracks growing from opposite ends of the cylindrical samples. The QSC test showed more intense cracking of the TiB particles at a higher their content, which in turn caused more intense secondary cracking of the matrix phase (Figure 19f). Comparison of the matrix structure of the alloy Ti64BEPM + TiB in the initial state (Figure 2e,f), with its structure after the QSC test (Figure 19d,f) reveals noticeable differences resulting from plastic deformation, while the similar comparison of the structures after SHPB tests (Figure 17b) does not show significant changes in most zones.

EBSD results shown in Figure 20a,b indicates the presence of two large cracks (arrowed in Figure 20a) surrounded by regions containing very fine grains and a high fraction of non-indexed points which are the key indicators of severe plastic deformation associated with localized shear. Furthermore, the two cracks seem to have been in the process of merging when the quasi-static deformation process stopped. It is likely that the strain rate experienced by this sample was just below the threshold value that would have resulted in a catastrophic break-up. As in the case of a high strain-rate deformation, the TiB particles loaded quasi-statically were also fragmented. The TiB fragments were additionally bent by the plastically deformed α-crystals with well-defined orientation, and showing developed dislocation substructure (Figure 20d,e).

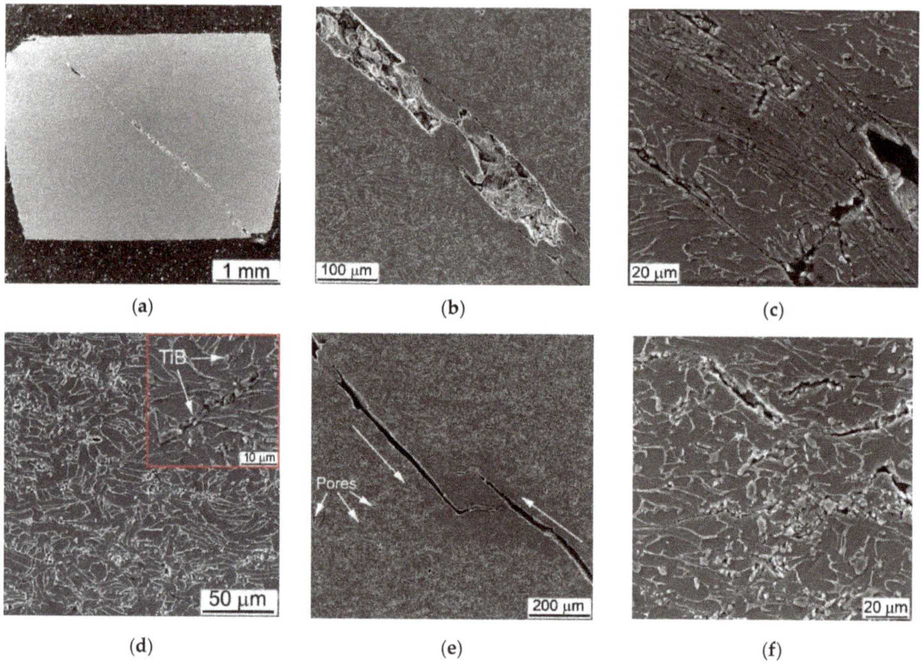

Figure 19. SEM images of (**a–d**) Ti64BEPM + 5TiB, and (**e**,**f**) Ti64BEPM + 10TiB after QSC tested with the rate 10^{-3} s^{-1}. (**a**) General view of the specimen, (**b**,**c**,**e**) Zone II, (**d**,**f**) Zone III. SEM, (**a**) SE, (**b–f**) BSE. Large arrows in (**e**) indicate the shift direction of the sample's halves.

Figure 20 demonstrates very good correlation between the predicted deformation pattern and that experimentally observed. It shows the crack formation exactly along Zone II (Figure 8), of predicted maximum stress zone. It also can be seen that the experimentally observed load distribution is slightly more complicated than predicted, and EBSD results prove this. We see an area, a green stripe in (highlighted by red dashed lines in Figure 20a), where the plastic deformation was lowest. This stripe region is denoted by a much higher indexing rate than the rest of the map, the grey regions being not indexed due to a weak diffraction signal or too fine structure. It propagates in a direction close to the orthogonal direction to Zone II; however, for confirming this statement, a further sectioning of the sample should be also considered. Nevertheless, it is most likely that the entire strain localization feature runs diagonally across the sample, and it is probably more correctly show its propagation in the sample from corner to corner, as it is sketched with red dashed lines in Figure 8; Figure 20a. It becomes evident that the main crack is the result of two cracks formed simultaneously: one is at the contact with the Input Bar of the sample top and the second is at the Output Bar contact in the lower part of the cylinder sample. Since two cracks meet right in the middle of the cylinder, it suggests that they form approximately simultaneously. It also suggests that Zone II should have been formed at a very specific angle. This angle corresponds to the typical 45-degree deformation localization in compression test, which is valid to most materials, such as metals, plastics, etc. toward the main axis of the cylinder. Thus, we can only expect to see one crack if a certain ratio between the height and diameter of the cylinder is held. This sample was a successful combination of test parameters as it shows the onset of cracking at both ends and their encounter in the center. If the strain rate were slightly higher, the sample would break in two, and it would be less convincing or easy to reach this conclusion. The fact that the two cracks are misaligned by ~100 µm may be due to the shape ratio of the initial undeformed sample (not having an ideal square cross section) or to the presence of TiB particles which may divert the crack propagation slightly from the ideal 45 degrees tilt from the cylinder axis. The higher magnification orientation maps explain what the darker regions in (a) and (b) represent. There are areas of highly fragmented and plastically deformed matrix alloy structure. One of such area is marked by an arrow in (d). Crystals are fragmented into sub-100 nm domains and are very difficult to resolve by conventional EBSD. These questions can be answered by orientation mapping in transmission mode (TKD in SEM). The image (d) also shows that some grains have experienced a relatively high deformation. Despite the grains' size seeming to not have changed significantly from their initial state, the EBSD results reveal lattice rotations of 15–20 degrees within a few micrometers across the same grain, which in the context of QSC process, can be considered as relatively high values. The grains of the Ti64 matrix alloy demonstrate a developed dislocation system. The smaller particles of TiB were unaffected by the deformation process (two magenta needles in the center of (e)). The TiB particles ~5 µm and larger become fragmented with no signs of plastic deformation. It clearly defines two different ways of dissipating the applied load energy: plastic deformation inside the ductile matrix and structure fragmentation inside the brittle reinforcement particles.

The lift-off sample for TKD was cut with focused ion beam system from the ASB region of the Ti64BEPM + 10TiB sample after the SHPB test with the strain rate 2510 s^{-1} as shown in Figure 21a with a red AB section. The region of interest was first identified using the conventional EBSD as shown in Figure 21a, allowing the detection of physically distinct structures within the ASB. The section had a width of about 10 µm and a depth of about 5 µm and from point A to point B it covered four distinct structures:

(1) TiB particle (magenta);
(2) Relatively preserved, but most likely plastically deformed α-grains (green);
(3) ASB I is a region that is not resolved by conventional EBSD due to extremely high fragmentation of the initial α-grains and/or high dislocation density reducing pattern quality, as well as a region of possible recrystallization;
(4) ASB II is a region of a less deformed initial α-grains structure or possibly a region with a relatively larger recrystallized α-grains.

It is important to note that based on the resolution limit of the field-emission gun SEM-EBSD, the dimension of the crystallites not resolved within the ASB I area is 20–30 nm and smaller [45]. For the same reason, most β phase regions are not resolved by conventional EBSD, except for very few larger β-lamellae. In contrast, the β-phase lamellae are well resolved in the TKD images performed with 3 nm steps and shown in Figure 21b. Orientation maps shows highly heterogeneous plastic deformation in the ASB region. The α-grains experienced the highest deformation level in the vicinity of the TiB particles.

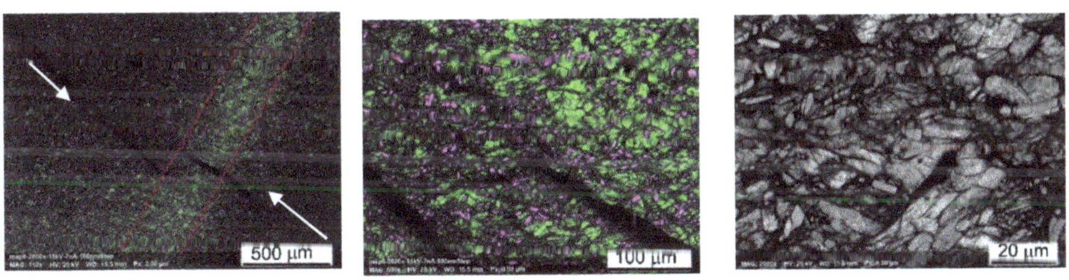

(a) (b) (c)

Figure 20. *Cont.*

(**d**) (**e**)

Figure 20. SEM EBSD results of Ti64BEPM + 10TiB sample after QSC tested with rate 10^{-3} s^{-1}. The shown area is the same as in Figure 19e. Phase map images (**a**,**b**,**e**), where α-phase is shown in green, β-phase in blue and TiB in magenta. (**c**)—Pattern Quality Map. (**d**)—Grain Misorientation Map.

The Grain Average Misorientation (GAM) map shows misorientation of some grains around the TiB particle up to 20+ degrees in the 0.5 µm scale range (color changes from blue to orange), while at a distance of few microns from the particle the internal grains' misorientation is usually twice as small (color changes from blue to green). Most importantly, recrystallization also begins in this area, due to extreme condition of the plastic deformation. In highly deformed sites surrounding the TiB particle, very small (nm) grains of α-phase are formed. Interestingly, the β-phase are also recrystallized. A small β-region on the upper left of the image (Figure 21b) clearly shows recrystallized β-grains. Very large recrystallized grains are not observed; therefore, it can be argued that a very early stage of recrystallization was found here. It can be noted that TKD results clearly show the different deformation mechanism of the structures involved. Brittle particles of TiB dissipate impact energy through fragmentation without signs of their plastic deformation, while crystallites of α and β-phase are plastically deformed.

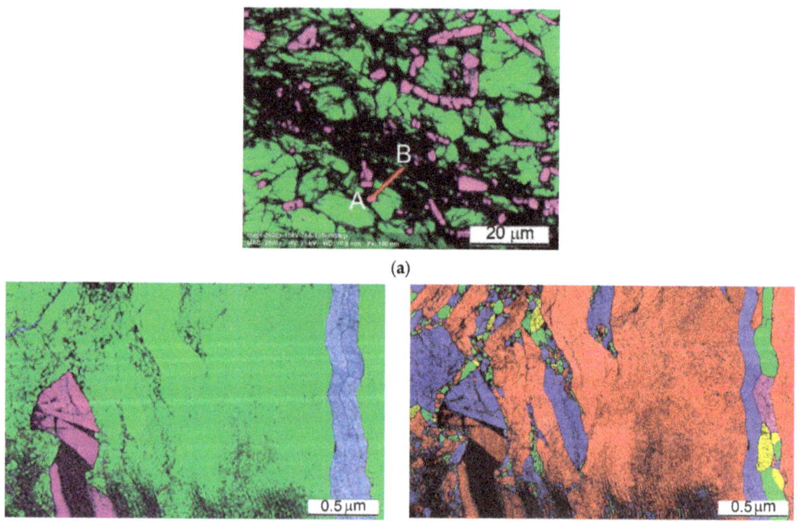

(**a**)

(**b**) (**c**)

Figure 21. *Cont.*

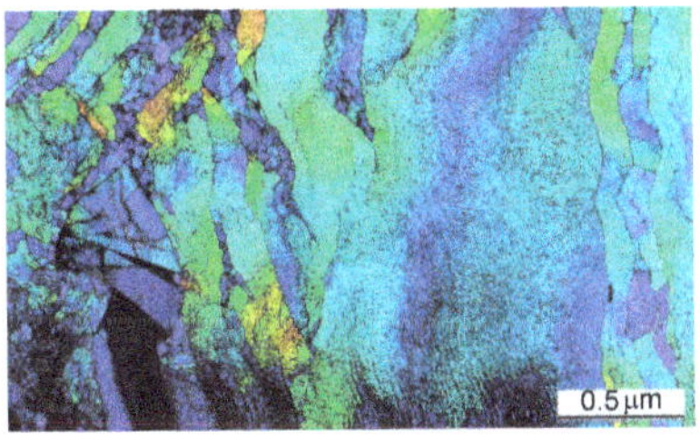

(**d**)

(**e**)

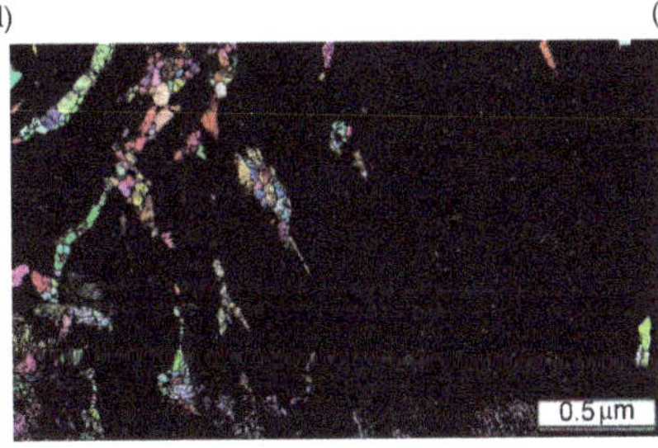

(**f**)

Figure 21. TKD results of the heavily deformed region (ASB area) on the sample Ti64BEPM + 10TiB after SHPB test with the strain rate 2510 s^{-1}. EBSD (conventional) phase map (**a**) showing the area from where the lift-off sample (AB section) for TKD was taken; within α-phase is shown in green, β-phase in blue and TiB is in magenta colors. The rest of the images (**b–f**) show TKD results: phase map (**b**), similarly color coded as in (**a**); grains map, in random color (**c**); GAM map (**d**); IPFz (**e**); IPFz map showing the grains <100 nm (**f**).

4.3. The Effect of Type of Reinforcing Particles and Their Volume Fraction on Deformation Mechanism and Deformation Energy

As shown above, the presences of the hard particles and pores has a significant effect on the deformation processes and mechanical properties of MMCs. It should also be emphasized that the presence of these structural elements has very different effect under tension test compared to compression (Figure 22). Both structural elements play the role of stress concentrators, primarily causing a drop in tensile ductility, which has been discussed previously [10,25,26,46]. This effect explains the decrease in the observed plasticity characteristics as shown in Table 2 and can be schematically illustrated as shown in Figure 22a,b. In case of the simultaneous presence of pores and particles TiC within the MMCs, the decrease in plasticity is more pronounced, due to their cumulative effect, resulting in fracture that obviously taking place in the region of elastic deformation rather far from reaching the values of Yield Stress (compare p. 1 with pp. 2–4 in Table 2). At the same time, these two structural elements have different effects of the stress concentration under compression. Residual pores collapse in the same way as in foam metals but with less pronounced effects on properties due to their relatively small fraction. While carbide particles break down, they initiate stress concentration at locations where cracks form in the metal matrix (compare Figure 22b,e). At the same time, TiB particles under tension cause a concentration of stress at the tips of their plates and/or needles, catastrophically embrittling MMCs (Figure 22c). However, with compression, such an effect is unlikely (Figure 22f). The suggested mechanisms allow to rationalize observed pronounced differences in mechanical properties: compare Figure 5b–d with Figure 5a, and p. 2–4 with p. 1 in Table 2.

The difference between the effect of TiC and TiB particles on the mechanical properties under tension and 3-point bending was discussed elsewhere [10,24]. Due to the needle or plate morphology TiB particles, it has been pointed out that, their sharp edges play more effective role as stress concentrators compared to the rounded edges of the TiC particles, especially when the plates or needles are specifically oriented relative to the applied external load, favoring an easy cracks nucleation. The presence of TiB particles has a less pronounced fracture embrittlement effect than TiC in the case of compression loading, especially at a high-rate strain. Obviously, this can be explained by the fact that titanium boride particles are formed in-situ by a chemical reaction forming particles of TiB with a shape far from equiaxial and a size varying over a wide range (Figure 3b). At the same time, the particles of TiC (Figure 3a) do not change their morphology and size within the matrix alloy during sintering or, as noted above, they tend to form conglomerates during mixing with the titanium hydride and the master alloy.

As a result, the particles size of TiB in both MMC varies in a rather narrow range: the thickness is 2–4 μm, and the length is 10–80 μm, while the particles size of TiC varies from several to hundreds of microns. In addition, the TiB particles are grown in-situ reacting with the surrounding Ti, and as a result have a stronger bond with the surrounding metal matrix. At the same time, globules of TiC partially joint into conglomerates, during mixing with metal components before pressing, and then during subsequent sintering, their interaction with the matrix metal is reduced only to a certain redistribution of carbon and titanium atoms between the matrix alloy and the particle without changing the shape and size of carbide inclusions [35,36]. Figure 23 shows examples of microstructures of MMCs with particles of TiC (Figure 23a) and TiB (Figure 23b) on which the chemical composition inside the particles and inside the matrix was measured. The results of these measurements are collected in Table 4.

As can be seen from the presented EDS data measured at the center of the coarse particle of TiC (or possibly the conglomerate of smaller ones at spots 1 and 4 in Figure 23a), the ratio between Ti and C is approximately 64/36 at. % (pp. 1 and 4 in Table 4), which does not correspond to the stoichiometric composition of titanium monocarbide. The peripheral portion of this particle is even more depleted with carbon (spots 2 and 3 in Figure 23a, and pp. 2 and 3 in Table 4). The smaller carbides have approximately the same carbon content as the periphery of the larger carbide: for example, spot 5 in Figure 23a and p. 5 in Table 4. As for the $\alpha + \beta$ matrix; its chemical composition is almost the same both in the areas adjacent to the carbides, and far from them (spots 6 and 7 in Figure 23a, and items 6 and 7 in Table 4, respectively). In addition, a very small amount of aluminum was found inside the carbide particles, while no vanadium was observed (items 1–5 in Table 4).

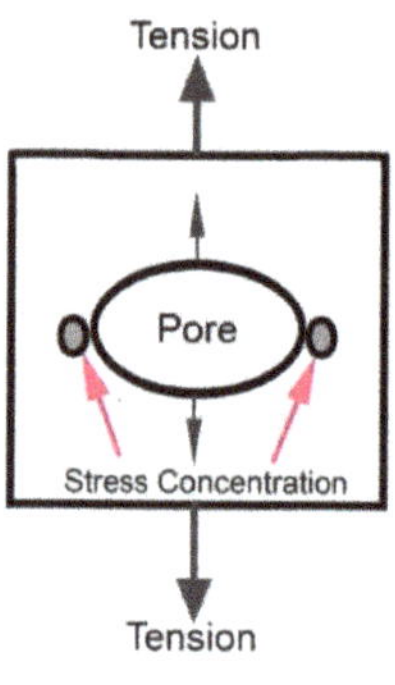

(a)

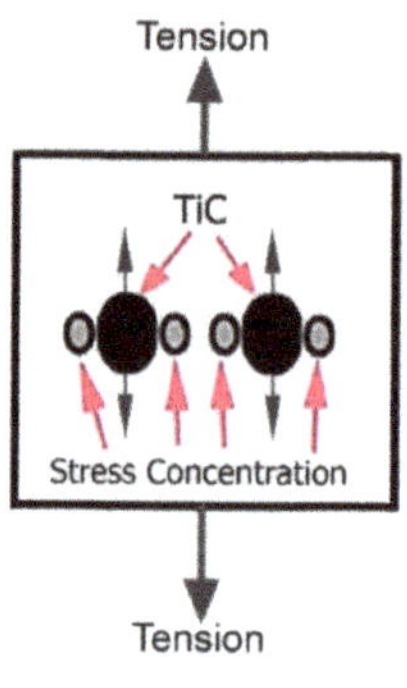

(b)

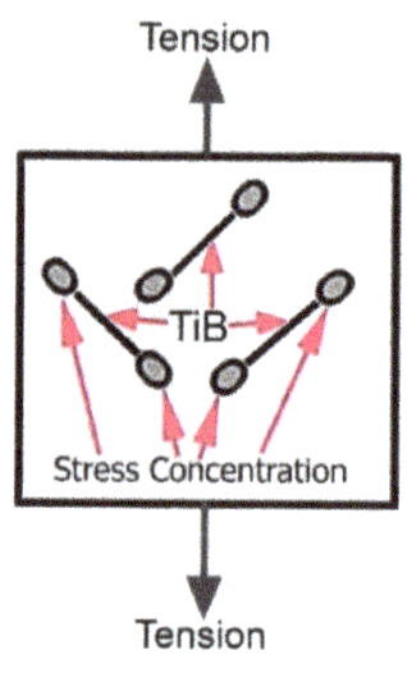

(c)

Figure 22. *Cont.*

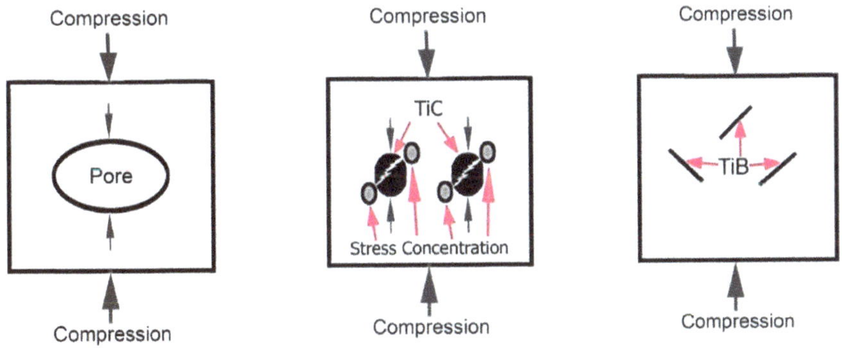

(**d**) (**e**) (**f**)

Figure 22. Schematic illustration of the stress concentration mechanisms involving pores (**a**,**d**), carbides (**b**,**e**), and borides (**c**,**f**) under tensile (**a**–**c**) and compression (**d**–**f**) test.

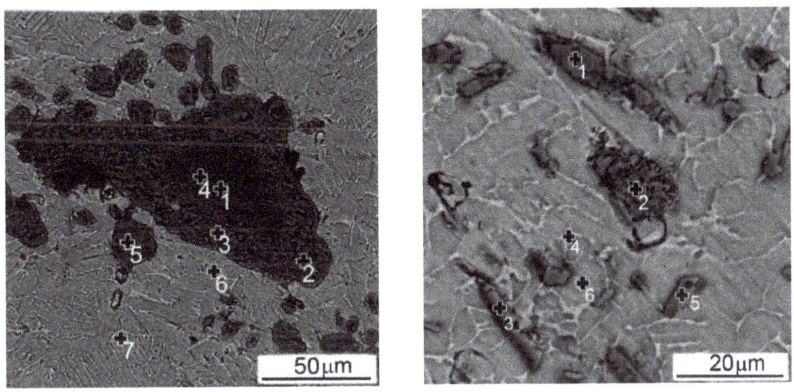

(**a**) (**b**)

Figure 23. SEM images (BSE) of (**a**) Ti64BEPM + 10TiC and (**b**) Ti64BEPM + 10TiB with the marked locations where the EDS chemical composition point analysis was taken (see results in Table 4).

The chemical analysis carried out on particles of TiB within the composite showed that all particles of TiB, despite their size (spots 1–3 in Figure 23b), have a composition that does not correspond to the stoichiometric ratio and is rather enriched in titanium (pp. 8–10 in Table 4). It should be noted that unlike the above-mentioned case of titanium carbide, these borides have, in addition to a small amount of aluminum, also a minor vanadium presence. Another difference compared to MMC with 10% TiC is that in Ti64BEPM + 10TiB, small titanium-enriched particles can have relatively high aluminum content (spot 5 in Figure 23b and item 12 in Table 4). However, it is not high enough to claim the formation of titanium aluminide. It should be noted that $\alpha + \beta$- matrix is boron free, supporting its good bonding with titanium boride (spot 6 in Figure 23b, and p. 13 in Table 4).

In other words, in the both cases of reinforcement with particles of carbides or borides, the particles are enriched with titanium, but depleted (relative to stoichiometric composition) with carbon and boron, respectively. At the same time, almost all boron goes to the formation of TiB plates or needles and is not found in the $\alpha + \beta$ matrix, while carbon quite actively diffuses into the titanium matrix.

Table 4. Local chemical analysis of MMCs' in spots shown in Figure 22.

##	Spot #	Phase	Composition, %							
			Ti		Al		V		C/B	
			Atomic	Weight	Atomic	Weight	Atomic	Weight	Atomic	Weight
Ti64BEPM + 10TiC (Figure 22a)										
MMC General Composition			66.18	80.63	10.68	7.43	4.62	5.98	17.51	5.35
1	1	TiC	63.89	87.02	0.07	0.05	-	-	35.91	12.27
2	2	TiC	75.63	91.97	0.68	0.47	-	-	23.61	7.20
3	3	TiC	71.80	90.56	0.36	0.26	-	-	27.76	8.78
4	4	TiC	65.42	87.43	0.11	0.09	-	-	34.27	11.49
5	5	TiC	76.81	92.84	0.16	0.11	-	-	23.01	6.98
6	6	$\alpha + \beta$	82.17	88.94	9.73	5.94	3.32	3.82	4.79	1.30
7	7	$\alpha + \beta$	80.89	87.15	9.08	5.53	5.24	6.02	4.79	1.30
Ti64BEPM + 10TiB (Figure 22b)										
MMC General Composition			63.21	82.32	11.68	8.03	5.15	6.24	19.17	6.01
8	1	TiB	63.01	87.38	0.10	0.08	0.85	1.26	36.04	11.28
9	2	TiB	63.80	87.22	0.78	0.60	0.09	1.59	34.32	10.59
10	3	TiB	58.50	83.99	0.80	0.65	1.80	2.75	38.89	12.61
11	4	$\alpha + \beta$	76.65	83.14	5.44	3.33	8.79	10.15	-	-
12	5	TiAl (?)	82.58	92.12	9.14	5.74	0.02	0.03	-	-
13	6	$\alpha + \beta$	77.66	85.88	6.09	7.16	5.67	3.53	-	-

The following viewpoints can be emphasized in the summary. Due to the processes involved in forming the structure of the composites during their sintering, the metal matrix composite strengthened with TiB particles is characterized by a relatively stronger bond between the metal matrix and the particles. In addition, TiB composites have relatively smaller particle sizes that do not exceed a few microns. Compared to the particles size in TiC MMC, when they sometimes reach several tens of micrometers, this is more than an order of magnitude greater than for borides. On a top of it, TiC particles tend to form relatively big aggregates of particles and form structural heterogeneity in the distribution of particles within the matrix. As a result, the Ti64BEPM + (X)TiB material is more resistant to high-strain-rate (dynamic) compression as opposed to TiC-containing analogs in a wide range of strain rates and is less prone to the formation of localized plastic deformation zones with the formation of adiabatic shear bands, followed by cracking. In a certain way, this is confirmed by data on sound frequency measurement and vibration damping (Table 2). According to these data, when the particles TiC are added into the Ti64BEPM matrix their fraction increase causes the Sound Frequency gradually increases, and the Damping increases with the particles TiC content raises up to 10%, and then falls (pp. 2–4 vs. p. 1, Table 2). If we compare MMCs containing TiB with those containing TiC, then for equal volume fractions of particles these characteristics in both cases will be higher for Ti64BEPM with TiB, namely: 13,624 Hz and 0.000284 versus 13,551 Hz and 0.000266 for 5% (p. 5 vs. p. 2, Table 2), and 14,219 Hz and 0.000322 versus 14,029 Hz and 0.000309 for 10% (p. 6 vs. p. 3, ibid); and, since the values of Young's moduli remain almost equal for MMCs with an equal fraction of strengthening particles (compare pp. 2 and 3 with 5 and 6, respectively, Table 2), then according to [47,48], the above differences in response to the impulse excited vibrations will be more likely of all is due to the peculiarities of both the microstructure of the composites as whole, and the peculiarities of the interphase boundaries between the Ti64 matrix and solid particles inside it. The action of the latter interphase boundaries, which have a different nature than the α/β- boundaries present in the matrix, complements and enhances the action of the boundaries between the α- and β-phases, which play a decisive role in the formation of pores and cracks during deformation of the free of solid particles Ti64 alloy [22].

The above-mentioned features of MMCs hardened with different types of solid particles have a fundamentally different effect on the localization of deformation under high strain-rate loading. The TiB particles having much better bound to the matrix, and due to the significant difference with latter in elastic moduli, strength, and plasticity [10,49], as well as differences in their tribological properties [50], during deformation undergo intense friction with the flowing and enveloping the broken fragments of TiB particles matrix [23]. This friction causes generation of additional intense heating at the TiB/α + β interfaces. Moreover, an interface between dissimilar materials, even without defects on the contact surfaces, creates an impedance to thermal transport that depends upon the differences in the densities and phonon propagation speeds for the two materials [51], which additionally contributes to local temperature raise. The consequence of these processes is the strong heating in small regions near individual TiB particles oriented in a certain way with respect to the applied load, which leads to local recrystallization of the matrix phases triggered adjacent to TiB particles. The situation intensifies under dynamic conditions, when the load takes place in a very short time (milliseconds), leaving no stage for the structure to recover sufficiently, all this leads to the recrystallization processes that began, which leads to the formation of nano size grains observed in Figure 21.

5. Conclusions

Based on the experimental data presented and their analysis, the following conclusions could be drawn:

1. The mechanical behavior of the alloy Ti64 based composites made using BEPM with different amounts of reinforcing particles: 5, 10, and 20% of TiC or 5, 10% (vol.) of TiB was analyzed taking into account the Strain Energy, phase composition, microstructure, and textures formed under dynamic and quasi static compression using strain rates in the range 10^{-3}–10^{3}·s^{-1}.
2. All materials tested under dynamic conditions using SPB show high localization of plastic deformation in matrix alloy, mainly near reinforcement particles. Such localization was absent in materials tested under QSC.
3. The measured Strain Energy and Maximal Strain Rates demonstrate the highest values for the cast and wrought alloy Ti64, followed by Ti64BEPM + 5TiC and Ti64BEPM + 10TiB, and then Ti64BEPM free of reinforcing particles.
4. At high-strain SHPB tests, Ti64BEPM + XTiC composites have higher strength then a conventional cast and wrought Ti64 alloy if the content of TiC particles does not exceed 5% (vol.) and the strain rate does not exceed 1800 s^{-1}. SE for Ti64BEPM + 5TiC and Ti64BEPM + 10TiC gradually increases with the strain rates reaching a maximum at 1800 s^{-1} and 1640 s^{-1}, respectively, and then drops sharply as a result of brittle fracture. Ti64BEPM + 20TiC fractures at progressively lower strain rates, less than 1470 s^{-1}, and due to a significant decrease in strength and ductility, its SE level is lower than that of Ti64BEPM + 5TiC. At lower strain rates, SE of Ti64BEPM + XTiB composites is lower compared to Ti64BEPM + XTiC equally reinforced, but due to higher plasticity of the composites Ti64BEPM + XTiB they exceed Ti64BEPM + XTiC in SE at strain rates above 2200 s^{-1}.
5. For Ti64BEPM alloy that does not contain reinforcing particles, the deformation localization mechanism changes from macro to micro level when the strain rate increases above 1800–2000 s^{-1}. The localization of the macro level is characterized by plastic flow affects multiple grains forming the adiabatic shear bands. The plastic flow at the micro level is limited to separate α-phase lamellae. In Ti64BEPM +XTiC,

the presence of particles gives emphasis to the localization of plastic deformation at the micro level. This is due to appearance of interfacial boundaries TiC/α and TiC/β, in addition to the interfacial boundaries α/β available in Ti64BEPM and the cast and wrought Ti64. With a higher content of TiC (20 vol.%), carbide particles became the weakest microstructure elements, which led to the brittleness of MMC and its fracture at relatively low rates (less than 1500 s^{-1}) due to the simultaneous crushing of carbide particles and the formation of cracks inside the matrix.

6. The negative effect of TiC is related to the high brittleness of particles or their conglomerates, which can cause the MMC to fracture under compression with relatively low strain rates. Stress concentration is formed at the interface between the carbide fragments and the matrix. Stress concentration effect under compression is less pronounced for MMC with TiB particles, which leads to an improvement in mechanical behavior of MMC composites at a higher deformation rate.
7. Localization of plastic deformation at a high strain-rates compression of MMCs with TiB, causes intense local overheating at the interface between the alloy matrix and the particles of TiB. As a result, rapid dynamic recrystallization of matrix phases can occur locally to form grains with a size of not more than 10 nm.
8. The damping ability of materials, which is analogous to the absorption of strain energy, was measured by the analysis of resonance-frequency-damping capacity. It shows that MMCs containing TiC or TiB particles absorb sound waves better in comparison with Ti64BEPM alloy free from particles. It was explained by the presence of interphases of TiC(TiB)/α, and TiC(TiB)/β in MMCs in addition to α/β interphases in the matrix alloy. The extra interfaces additionally inhibit the dislocation motion.

Author Contributions: Conceptualization, validation, and project administration P.E.M., J.J., and S.V.P.; methodology, P.E.M., J.J., S.V.P., O.O.S., V.I.B., D.G., D.G.S.; formal analysis, V.I.B., P.S., K.C.; investigation, P.E.M., J.J., K.C., O.O.S., V.I.B., S.V.P., D.G., P.S.; writing—original draft preparation, P.E.M.; writing—review and editing, J.J., S.V.P., D.G., V.I.B., D.G.S.; visualization, O.O.S., S.V.P., D.G. All authors have read and agreed to the published version of the manuscript.

Funding: The following authors, P.E.M., O.O.S., D.G.S., S.V.P., acknowledge funding from the NATO Agency Science for Peace and Security (#G5787). Investigations of compression tests influence on phase composition and microstructure of materials were performed within the frames of the Agreement of Cooperation between G.V. Kurdyumov Institute for Metal Physics of N.A.S. of Ukraine and Jarosław Dąbrowski Military University of Technology, Poland without special funding. Separate works made in G.V. Kurdyumov Institute for Metal Physics of N.A.S. of Ukraine were partially financially supported by N.A.S. of Ukraine within the frames of project #III-09-18.

Institutional Review Board Statement: Not available.

Informed Consent Statement: Not available.

Data Availability Statement: The data presented in this study are available on request from the corresponding author. The data are not publicly available because of the potential for the future invention disclosure.

Acknowledgments: The authors acknowledge assistance from the administrations of their Institutions as well as the Bruker Nano Analytics in carrying some experimental work. Special thanks to the Presidium of the National Academy of Sciences of Ukraine (project #III-09-18) and the program "Science for Peace and Security" (project NATO G5787) for funding.

Conflicts of Interest: The authors declare no conflict of interest.

List of Acronyms

Acronym	Term
Ti64	Commercial titanium alloy containing 6 (wt.)% Al and 4% V
Ti64GL	Ti-6Al-4V alloy with globular microstructure
BEPM	Blended Elemental Powder Metallurgy
Ti64BEPM	Titanium based alloy Ti-6(wt.)%Al-4V produced using BEPM
MMC	Metal Matrix Composite
Ti64BEPM + (X)TiC	MMC on the base of Ti-6Al-4V alloy containing X(vol.%) of Titanium Carbide (TiC) produced using BEPM
Ti64BEPM + (X)TiB	MMC on the base of Ti-6Al-4V alloy containing X(vol.%) of Titanium Boride (TiB) produced using BEPM
SE	Strain Energy
QSC	Quasi-Static compression
QST	Quasi-Static tension
SHPB	Split Hopkinson Pressure Bar
SEM	Scanning Electron Microscopy
SE	Secondary Electrons
BSE	Back Scattered Electrons
EBSD	Electron Backscatter Diffraction
ASB	Adiabatic Shear Band
GAM	Grain Average Misorientation
IPFz	Inverted Pole Figure in "z" axis
TKD	Transmission Kikuchi Diffraction
$\dot{\varepsilon}_{max}$	Maximal Strain Rate for which specimens were not fractured during the SHPB tests
$\dot{\varepsilon}_{Cr}$	Critical Strain Rate for which specimens became fractured during SHPB tests first time
ε_{upper}	Maximum strain at fracture, or for the non-cracked specimens, equal to strain at the moment of the specimen unloading
SE_{max}	Maximal Strain Energy value obtained on SHPB impact tests at Maximal Strain Rate
YS	Yield Strength (on Tension)
UTS	Ultimate Tensile Strength (on Tension)
El	Relative elongation (on Tension)
RA	Reduction in Area (on Tension)

References

1. Lutjering, G.; Williams, J.C. *Titanium*, 2nd ed.; Springer: Berlin/Heidelberg, Germany; New York, NY, USA, 2007. [CrossRef]
2. Boyer, R.R.; Briggs, R.D. The Use of β Titanium Alloys in the Aerospace Industry. *J. Mater. Eng. Perform.* **2005**, *14*, 681. [CrossRef]
3. Niinomi, M. Recent metallic materials for biomedical applications. *Metall. Mater. Trans.* **2002**, *33A*, 477. [CrossRef]
4. Fanning, J. Military Application for b Titanium Alloys. *J. Mater. Eng. Perform.* **2005**, *14*, 686–690. [CrossRef]
5. Budinski, K.G. Tribological properties of titanium alloys. *Wear* **1991**, *151*, 203–217. [CrossRef]
6. Sivakumar, B.; Kumar, S.; Sankara Narayanan, T.S.N. Comparison of fretting corrosion behaviour of Ti–6Al–4V alloy and CP-Ti in Ringer's solution. *Tribol. Mater. Surf. Interfaces* **2011**, *5*, 158–164. [CrossRef]
7. Poondla, N.; Srivatsan, T.S.; Patnaik, A.; Petraroli, M. A study of the microstructure and hardness of two titanium alloys: Commercially pure and Ti–6Al–4V. *J. Alloy. Compd.* **2009**, *486*, 162–167. [CrossRef]
8. Eriksson, M.; Salamon, D.; Nygren, M.; Shen, Z. Spark plasma sintering and deformation of Ti–TiB_2 composites. *Mater. Sci. Eng. A* **2007**, *475*, 101–104. [CrossRef]
9. Fan, Z.; Guo, Z.X.; Cantor, B. The kinetics and mechanism of interfacial reaction in sigma fibre-reinforced Ti MMCs. *Compos. Part A Appl. Sci. Manuf.* **1997**, *28*, 131–140. [CrossRef]
10. Ivasishin, O.M.; Markovsky, P.E.; Savvakin, D.G.; Stasiuk, O.O.; Norouzi Rad, M.; Prikhodko, S.V. Multi-layered structures of Ti-6Al-4V alloy and TiC and TiB composites on its base fabricated using blended elemental powder metallurgy. *J. Mater. Process. Technol.* **2019**, *269*, 172–181. [CrossRef]
11. Stefansson, N.; Weiss, I.; Hutt, A.J. *Titanium'95: Science and Technology*; Blenkinsop, P.A., Evans, W.J., Flower, H.M., Eds.; The University Press: Cambridge, UK, 1996; Volume 2, pp. 980–987.
12. Bhattacharjee, A.; Ghosal, P.; Gogia, A.K.; Bhargava, S.; Kamat, S.V. Room temperature plastic flow behaviour of Ti–6.8Mo–4.5Fe–1.5Al and Ti–10V–4.5Fe–1.5Al: Effect of grain size and strain rate. *Mat. Sci. Eng. A* **2007**, *452–453*, 219–227. [CrossRef]

13. Markovsky, P.E.; Matviychuk, Y.U.V.; Bondarchuk, V.I. Influence of grain size and crystallographic texture on mechanical behavior of TIMETAL-LCB in metastable β-condition. *Mater. Sci. Eng. A* **2013**, *559*, 782–789. [CrossRef]
14. Markovsky, P.E.; Bondarchuk, V.I.; Herasymchuk, O.M. Influence of Grain Size, Aging Conditions and Tension Rate on Mechanical Behavior of Titanium Low-Cost Metastable Beta-Alloy in Thermally Hardened Condition. *Mater. Sci. Eng.* **2015**, *A645*, 150–162. [CrossRef]
15. Markovsky, P.E.; Bondarchuk, V.I.; Shepotinnyk, O.V.; Gavrysh, I.M. Influence of phase composition and microstructure on mechanical behavior of Ti-3Al-4.5Fe-7.2Cr and VT22 titanium metastable β alloys on tension with different rates. *Metallofiz. Noveyshye Tekhnologii* **2016**, *38*, 935–952. [CrossRef]
16. Markovsky, P.E.; Bondarchuk, V.I. Influence of strain rate, microstructure, chemical and phase composition on mechanical behavior of different titanium alloys. *J. Mater. Eng. Perform.* **2017**, *26*, 3431–3449. [CrossRef]
17. Markovsky, P.E. Mechanical Behavior of Titanium Alloys under Different Conditions of Loading. *Mater. Sci. Forum THERMEC-2018* **2018**, *941*, 839–844. [CrossRef]
18. Peirs, J.; Verleysen, P.; Degrieck, J.; Coghe, F. The use of hat-shaped specimens to study the high strain rate shear behavior of Ti–6Al–4V. *Int. J. Impact Eng.* **2010**, *37*, 703–714. [CrossRef]
19. Meyer, H.W., Jr.; Kleponis, D.S. Modeling the high strain rate behavior of titanium undergoing ballistic impact and penetration. *Int. J. Impact Eng.* **2001**, *26*, 509–521. [CrossRef]
20. Zheng, C.; Wang, F.; Cheng, X.; Liu, J.; Fu, K.; Liu, T.; Zhu, Z.; Yang, K.; Peng, M.; Jin, D. Failure mechanisms in ballistic performance of Ti6Al4V targets having equiaxed and lamellar microstructures. *Int. J. Impact Eng.* **2015**, *85*, 161–169. [CrossRef]
21. Tianfeng, Z.; Junjie, W.; Jiangtao, C.; Ying, W.; Xibin, W. Dynamic shear characteristics of titanium alloy Ti-6Al-4V at large strain rates by the split Hopkinson pressure bar test. *Int. J. Impact Eng.* **2017**, *109*, 167–177. [CrossRef]
22. Markovsky, P.E.; Janiszewski, J.; Bondarchuk, V.I.; Stasyuk, O.O.; Skoryk, M.A.; Cieplak, K.; Dziewit, P.; Prikhodko, S. Effect of Strain Rate on Microstructure Evolution and Mechanical Behavior of Titanium-Based Materials. *Metals* **2020**, *10*, 1404. [CrossRef]
23. Ivasishin, O.M.; Teliovych, R.V.; Ivanchenko, V.G.; Tamirisakandala, S.; Miracle, D.B. Processing, microstructure, texture, and tensile properties of the Ti-6Al-4V-1.55 B eutectic alloy. *Metall. Mater. Trans. A* **2008**, *39*, 402–426. [CrossRef]
24. Markovsky, P.E.; Savvakin, D.G.; Ivasishin, O.M.; Bondarchuk, V.I.; Prikhodko, S.V. Mechanical Behavior of Titanium-Based Layered Structures Fabricated Using Blended Elemental Powder Metallurgy. *J. Mater. Eng. Perform.* **2019**, *28*, 5772–5792. [CrossRef]
25. Ivasishin, O.M.; Anokhin, V.M.; Demidik, A.N.; Savvakin, D.G. Cost-Effective Blended Elemental Powder Metallurgy of Titanium Alloys for Transportation Application. *Key Eng. Mater.* **2000**, *188*, 55–62. [CrossRef]
26. Ivasishin, O.M.; Moxson, V.S. Low cost titanium hydride powder metallurgy. In *Titanium Powder Metallurgy, Science, Technology and Applications*; Qian, M., Froes, S.H., Eds.; Elsevier: Amsterdam, The Netherlands; Boston, MA, USA; Heidelberg, Germany; London, UK; New York, NY, USA; Oxford, UK; Paris, France; San Diego, CA, USA; San Francisco, CA, USA; Singapore; Sydney, Australia; Tokyo, Japan, 2015; Chapter 8; pp. 117–148.
27. Qian, M. Cold Compaction and Sintering of Titanium and its Alloys for Near-Net-Shape or Preform Fabrication. *Int. J. Powder Metall.* **2010**, *46*, 29–44.
28. Lee, C.S.; Oh, J.C.; Lee, S. Improvement of Hardness and Wear Resistance of (TiC, TiB)/Ti-6Al-4V Surface-Alloyed Materials Fabricated by High-Energy Electron-Beam irradiation. *Metall. Mater. Trans. A* **2003**, *34A*, 1461–1471. [CrossRef]
29. Savvakin, D.G.; Prikhodko, S.V.; Stasyuk, O.O.; Markovsky, P.E. Influence of TiC content and Strengthening Heat Treatment on Microstructure and Hardness of Metal Matrix Composite Based on Ti-6Al-4V Alloy. In preparation.
30. Kolsky, H. Stress waves in solids. *J. Sound Vib.* **1964**, *1*, 88–110. [CrossRef]
31. Sreenivasan, P.R.; Ray, S.K. Mechanical Testing at High Strain Rates. In *Encyclopedia of Materials: Science and Technology*, 2nd ed.; Elsevier: Amsterdam, The Netherlands; Boston, MA, USA; Heidelberg, Germany; London, UK; New York, NY, USA; Oxford, UK; Paris, France; San Diego, CA, USA; San Francisco, CA, USA; Singapore; Sydney, Australia; Tokyo, Japan, 2001; pp. 5269–5271.
32. Chen, W.; Song, B. *Split Hopkinson (Kolsky) Bar: Design, Testing and Applications*; Springer: Berlin/Heidelberg, Germany; New York, NY, USA, 2011. [CrossRef]
33. Semiatin, S.L. An Overview of the Thermomechanical Processing of α/β Titanium Alloys: Current Status and Future Research Opportunities. *Metall. Mater. Trans. A* **2020**, *51*, 2593–2625. [CrossRef]
34. Zhong, Z.; Zhang, B.; Jin, Y.; Zhang, H.; Wang, Y.; Ye, J.; Liu, Q.; Hou, Z.; Zhang, Z.; Ye, F. Design and anti-penetration performance of TiB/Ti system functionally graded material armor fabricated by SPS combined with tape casting. *Ceram. Int.* **2020**, *46 Pt A*, 28244–28249. [CrossRef]
35. Wanjara, P.; Drew, R.A.L.; Root, J.; Yue, S. Evidence for Stable Stoichiometric Ti_2C at the Interface in TiC Particulate Reinforced Ti Alloy Composites. *Acta Mater.* **2000**, *48*, 1443–1450. [CrossRef]
36. Roger, J.; Gardiola, B.; Andrieux, J.; Viala, J.-C.; Dezellus, O. Synthesis of Ti matrix composites reinforced with TiC particles: Thermodynamic equilibrium and change in microstructure. *J. Mater. Sci.* **2017**, *52*, 4129–4141. [CrossRef]
37. Ohring, M. *Materials Science of Thin Films, Deposition and Structure*; Academic Press: Cambridge, MA, USA, 2002; 794p. [CrossRef]
38. Kyogoku, H.; Komatsu, S.; Shinohara, K.; Jinushi, H.; Toda, T. Microstructures and Mechanical Properties of Sintered Ti-4% Fe Alloy Compacts by Injection Moldings. *J. Jpn. Soc. Powder Powder Metall.* **1994**, *41*, 1075–1079. [CrossRef]
39. Krivoglaz, M.A. *X-ray and Neutron Diffraction in Nonideal Crystals*; Springer: Berlin/Heidelberg, Germany; New York, NY, USA, 1996; 494p. [CrossRef]

40. Biswas, N.; Ding, J.L. Numerical Study of the Deformation and Fracture Behavior of Porous Ti6Al4V Alloy under Static and Dynamic Loading. *Int. J. Impact Eng.* **2015**, *82*, 89–102. [CrossRef]
41. Kumar, P.; Ravi Chandran, K.S.; Cao, F.; Koopman, M.; Zak Fang, Z. The Nature of Tensile Ductility as Controlled by Extreme-Sized Pores in Powder Metallurgy Ti-6Al-4V Alloy. *Metall. Mater. Trans. A* **2016**, *47A*, 2150–2161. [CrossRef]
42. Markovsky, P.E. High-strength structural states in titanium alloys subjected to intense thermal action. *Met. Adv. Technol.* **2009**, *31*, 511–535. (In Russian)
43. Bignona, M.; Bertrand, E.; Tancret, F.; Rivera-Díaz-del-Castillo, P.E.J. Modelling martensitic transformation in titanium alloys: The influence of temperature and deformation. *Materialia* **2019**, *7*, 100382. [CrossRef]
44. Bignona, M.; Bertrand, E.; Rivera-Díaz-del-Castillo, P.E.J.; Tancret, F. Martensite formation in titanium alloys: Crystallographic and compositional effects. *J. Alloy. Compd.* **2021**, *872*, 159636. [CrossRef]
45. Prikhodko, S.V.; Kakoulli, I. Electron microscopy in conservation. In *The Encyclopedia of Archaeological Sciences*; Varela, S.L.L., Ed.; John Wiley & Sons: Hoboken, NJ, USA, 2018. [CrossRef]
46. Paramore, J.D.; Butler, B.G.; Dunstan, M.K.; Rhee, H.; El Kadiri, H.; Whittington, W.R.; Mujahid, S.H. The role of microstructure and strain rate on the mechanical behavior of Ti6Al-4V produced by powder metallurgy. *Int. J. Refract. Met. Hard Mater.* **2020**, *92*, 105268. [CrossRef]
47. Roebben, G.; Bollen, B.; Brebels, A.; Van Humbeeck, J.; Van der Biest, O. Impulse excitation apparatus to measure resonant frequencies, elastic moduli, and internal friction at room and high temperature. *Rev. Sci. Instrum.* **1997**, *68*, 4511. [CrossRef]
48. Clyne, T.W.; Campbell, J.E. Chapter 6—Compressive Testing. In *Testing of the Plastic Deformation of Metals*; Cambridge University Press: Cambridge, UK, 2021; pp. 107–122. [CrossRef]
49. Collings, E.; Welsch, G. *Materials Properties Handbook: Titanium Alloys*; ASM International: Novelty, OH, USA, 1994; 1176p.
50. Gui, G.; Gao, L.Y.G.; Shinohara, K.; Bollen, B.; Toda, T. Preparation, Mechanical Properties, and High-Temperature Wear Resistance of Ti–Al–B alloy. *Materials* **2019**, *12*, 3751. [CrossRef]
51. Balasubramanian, G.; Puria, I.K. Heat conduction across a solid-solid interface: Understanding nanoscale interfacial effects on thermal resistance. *Appl. Phys. Lett.* **2011**, *99*, 013116. [CrossRef]

materials

MDPI

Article

Third-Generation Thermodynamic Descriptions for Ta-Cr and Ta-V Binary Systems

Enkuan Zhang [1], Xinpei Xu [1], Yun Chen [2] and Ying Tang [1,*]

[1] School of Materials Science and Engineering, Hebei University of Technology, Tianjin 300130, China; ek_zhang@163.com (E.Z.); xu1020774690@163.com (X.X.)
[2] Department of Mechanical and Electrical Engineering, Xiamen University, Xiamen 361005, China; yun.chen@xmu.edu.cn
* Correspondence: ying_tang@hebut.edu.cn

Abstract: The third-generation thermodynamic descriptions for Ta-Cr and Ta-V binary systems were performed to construct the reliable thermodynamic database for refractory high-entropy alloys (RHEAs) containing Laves phase. The third-generation Gibbs energy expressions of pure Cr and V in both solid and liquid phases were established, from which the thermodynamic properties and thermal vacancy can be well described. The thermodynamic descriptions of Ta-Cr and Ta-V over the whole composition and temperature regions were carried out based on the reviewed phase equilibria and thermodynamic data with the CALPHAD (CALculation of PHAse Diagrams) approach. Specifically, the thermodynamic parameters of C14 and C15 Laves phases were evaluated by combining the theoretically computed and experimentally measured thermodynamic properties as well as the semiempirical relations. The calculated phase diagrams and thermodynamic properties in Ta-Cr and Ta-V systems according to the present thermodynamic parameters had a nice agreement with the experimental data even down to 0 K, indicating the reliability of the present modeling.

Keywords: third-generation thermodynamic description; Ta-Cr system; Ta-V system; CALPHAD; Laves phase

Citation: Zhang, E.; Xu, X.; Chen, Y.; Tang, Y. Third-Generation Thermodynamic Descriptions for Ta-Cr and Ta-V Binary Systems. *Materials* **2022**, *15*, 2074. https://doi.org/10.3390/ma15062074

Academic Editors: Gregory N. Haidemenopoulos and Edward Bormashenko

Received: 30 November 2021
Accepted: 2 February 2022
Published: 11 March 2022

Publisher's Note: MDPI stays neutral with regard to jurisdictional claims in published maps and institutional affiliations.

1. Introduction

Refractory high-entropy alloys (RHEAs) demonstrate excellent mechanical properties as well as good corrosion resistance and oxidation resistance up to 1600 °C, making them considerable candidates for high-temperature applications [1–4]. Generally, RHEAs contain refractory elements and form a single solid solution phase with a body-centered cubic (BCC, A2) structure. The addition of the elements, such as Cr and V, which have strong interactions with other refractory elements, can contribute to the formation of Laves phases in RHEAs [5–7]. With a minor amount of precipitated Laves phase in A2 solution, the strength of RHEAs can be significantly improved [6,7]. To identify the precipitation of Laves phase in RHEAs, varieties of empirical criteria have been proposed [8–10]. With the rapid development of a computational approach, the process of identifying the formation of Laves phase in RHEAs can be sped up to meet unique requirements. Of varied computational methods, the CALPHAD (CALculation of PHAse Diagrams) approach, which can calculate the phase equilibrium in multicomponent alloys, is one of the most powerful methods for Laves phases containing RHEA design and selection [11–13]. An accurate thermodynamic database is the prerequisite in CALPHAD-type calculations.

To develop an accurate thermodynamic database over the whole temperature region with physical sense, the concept of a third-generation thermodynamic database has been proposed since 1995 [14]. Under this framework, several research groups [14–19] have made efforts to establish new thermodynamic models for pure elements by using the Debye or Einstein models. Recently, a kind of third-generation thermodynamic model [13], which enables one to describe the thermodynamic properties of pure elements down to 0 K as well as thermal vacancy concentration near the melting point, has been proposed by combining the Segmented Regression (SR) model [15], two-state model [20] and thermal vacancy description [18]. By applying such a model, a third-generation thermodynamic database for MoNbTaW RHEAs was developed [13]. To further develop the third-generation thermodynamic databases for Laves containing RHEAs, reliable thermodynamic descriptions for Laves phase are essential. Currently, the reports on third-generation thermodynamic descriptions of Laves phase are rather limited in the literature. Jiang et al. [21,22] established the third-generation Gibbs energy of Laves phase in the Ta-Cr and Cr-Nb systems by considering the experimental heat capacity data with the SR model. Contrastingly, such a strategy was not valid for those systems when the heat capacity data of Laves phase were absent. Thus, a universal strategy to obtain thermodynamic description for Laves phase is needed.

In this work, the Ta-Cr and Ta-V systems that contain the hexagonal C14 ($MgZn_2$-type) and cubic C15 ($MgCu_2$-type) Laves phases are selected as the target. Following our previous work [13], the third-generation Gibbs energy expressions of pure Cr and V will be established first. Then, the experimental phase equilibrium and thermodynamic properties in Ta-Cr and Ta-V systems will be reviewed and their discrepancies will be clarified. After that, a universal strategy to obtain reliable thermodynamic parameters in Laves phase will be proposed, and the thermodynamic modeling of the Ta-Cr and Ta-V binary systems will be performed via the CALPHAD approach.

2. Thermodynamic Model

There are four stable phases (i.e., liquid, A2, C14, C15) in Ta-Cr and Ta-V binary systems. The liquid and A2 are the solution phases, while C14 and C15 are the Laves phase. Their thermodynamic models are briefly introduced as follows.

2.1. A2 and Liquid Solution Phases

To describe the thermal vacancy contribution in the A2 phase, the substitutional model (i, j, Va), where Va denotes thermal vacancy, was used. The molar Gibbs energy of the A2 phase is as follows:

$$\begin{aligned} G_m^{A2} &= x_i G_i^{A2,df} + x_j G_j^{A2,df} + \tfrac{y_{va}}{1-y_{va}} G_{va} + RT\left(x_i \ln x_i + x_j \ln x_j + \tfrac{y_{va}}{1-y_{va}} \ln y_{va}\right) \\ &+ (1-y_{va}) x_i x_j \sum_{n=0} {}^{n}L_{i,j}^{A2} (x_i - x_j)^n + \tfrac{1}{1-y_{va}} \left(x_i y_{va} L_{i,va}^{A2} + x_j y_{va} L_{j,va}^{A2}\right) \end{aligned} \quad (1)$$

where R and T are the gas constant and absolute temperature, x_i is the mole fraction of element i, and y_{va} is the thermal vacancy concentration. $L_{i,j}$ is the interaction parameter between the elements i and j, and $L_{i,Va}$ is the one between the element i and thermal vacancy. G_{va} represents the Gibbs energy of thermal vacancy in the A2 phase. $G_i^{A2,\,df}$ in Equation (1) is the molar Gibbs energy of the defect-free element i, the expression of which will be developed by using the SR model [15]. For the liquid phase, its molar Gibbs energy expression is similar to that of the A2 phase but excludes the contribution from the thermal vacancy (to make $y_{va} = 0$ in Equation (1)). The Gibbs energy of pure element in a liquid state will be established by applying the two-state model [20].

2.2. C14 and C15 Laves Phases

A two-sublattice model $(i, j)_2\,(i, j)$ was employed to describe the solid solubility in C14 and C15 phases. Take the C15 phase in Ta-Cr binary, for example; its molar Gibbs energy is given as follows:

$$\begin{aligned} G^{C15} &= {}^{0}G_{Cr:Ta}^{C15} y'_{Cr} y''_{Ta} + {}^{0}G_{Cr:Cr}^{C15} y'_{Cr} y''_{Cr} + {}^{0}G_{Ta:Ta}^{C15} y'_{Ta} y''_{Ta} + {}^{0}G_{Ta:Cr}^{C15} y'_{Ta} y''_{Cr} \\ &+ 2RT\left(y'_{Cr} \ln y'_{Cr} + y'_{Ta} \ln y'_{Ta}\right) + RT\left(y''_{Cr} \ln y''_{Cr} + y''_{Ta} \ln y''_{Ta}\right) \\ &+ y'_{Cr} y''_{Cr} y''_{Ta} \, {}^{0}L_{Cr:Cr,Ta}^{C15} + y'_{Ta} y''_{Cr} y''_{Ta} \, {}^{0}L_{Ta:Cr,Ta}^{C15} + y'_{Cr} y'_{Ta} y''_{Cr} \, {}^{0}L_{Cr,Ta:Cr}^{C15} + y'_{Cr} y'_{Ta} y''_{Ta} \, {}^{0}L_{Cr,Ta:Ta}^{C15} \end{aligned} \quad (2)$$

where y_i' and y_i'' denote the site fractions in the first and second sublattice of element i. ${}^0G_{i:j}^{C15}$ is the end-member Gibbs energy when the first sublattice is occupied by i while the second sublattice is occupied by j. Its expression is given as

$$ {}^0G_{i:j}^{C15} = A + B \cdot T + 2 \cdot G_i^{A2,\mathrm{def}} + G_j^{A2,\mathrm{def}} \tag{3} $$

where A and B are the coefficients to be optimized. ${}^0L_{Cr:i,j}^{C15}$ and ${}^0L_{Ta:i,j}^{C15}$ are the interaction parameters between i and j in the second sublattice when the first is occupied by Cr or Ta. Similarly, ${}^0L_{i,j:Cr}^{C15}$ and ${}^0L_{i,j:Ta}^{C15}$ represent the interaction parameters between i and j in the first sublattice when the second sublattice is occupied by Cr or Ta, which are all also needed to be determined during the thermodynamic assessment.

3. Results

3.1. The Third-Generation Gibbs Energy Expressions for Pure Cr and V

The reliable thermodynamic properties of pure Cr and V are the essential inputs to establish their Gibbs energy expressions. Recently, the measured thermodynamic properties including heat capacity (C_p), heat content (H_T–H_{298}), enthalpy of fusion for pure Cr and V have been collected and reviewed by Obaied et al. [19] and Arblaster [23], respectively. The discrepancies of the measured data from different resources have been clearly clarified by them. Therefore, the reviewed thermodynamic properties are adopted in the present modeling. In addition, the thermal vacancy concentrations in pure Cr and V at their melting temperature were experimentally investigated by means of modulation and drop methods [24,25]. It is generally accepted that the effects of thermal vacancy on thermodynamic properties will become obvious with the temperature above $2/3T_m$ [18]. Thus, the expressions of defect-free Gibbs energy expressions ($G_i^{A2,df}$) in Equation (1) of pure Cr and V in their solid stable states (below T_m) were evaluated by fitting the experimental heat capacity from 0 K to $2/3T_m$. As for the expressions above T_m, the strategy making C_p and dC_p/dT curves continue at T_m was used. Then, the interaction parameter $L_{i,Va}$ was evaluated by considering the thermodynamic properties above $2/3T_m$ as well as the thermal vacancy concentration at melting point. The Gibbs energy of the A2 phase with thermal vacancy over the whole temperature range for Cr and V can be evaluated by combining the estimated $G_i^{A2,df}$ and $L_{i,Va}$.The molar Gibbs energies of pure Cr and V in a liquid state were obtained with a two-state model by considering the experimental entropy, enthalpy and heat capacity at and above T_m. The finally obtained third-generation Gibbs energies for solid and liquid Cr and V are listed in Table 1.

Figure 1a,d show the calculated C_p of pure Cr and V in their solid A2 and liquid states. The experimental C_p data [26–46] are also appended in the figure for comparison, showing a nice agreement with present calculations. Figure 1b,d show the calculated H_T-H_{298} of pure Cr and V from 298 to 4000 K together with the reported data [23,47–51]. An excellent agreement can be observed. The Gibbs energies of solid A2 and liquid phases in pure Cr and V from 0 K to 6000 K were calculated and displayed in Figure 1c,f, revealing the lattice stability over the whole temperature range.

With the present established Gibbs energies, the thermal vacancy can be the equilibrium concentration of thermal vacancy in pure Cr, and V can be predicted by the following relation [18]:

$$ y_{va} = \exp(-\frac{G_i^{A2,df} + L_{i,va}^{A2}(1 - y_{va})^2}{RT}) \tag{4} $$

Figure 2 shows the calculated thermal vacancy concentrations in solid Cr and V. As can be seen, the thermal vacancy concentrations increase with the increase in temperatures, and a dramatic increase can be observed near the melting points. Besides, the present predicted thermal vacancy concentrations are in good agreement with those from measurements [24,25] at melting points. The above results indicate the reliability of the established third-generation Gibbs energies for pure Cr and V.

Table 1. The third-generation Gibbs energy expressions for pure Cr and V in solid A2 and liquid phases.

Element	Phases	Gibbs Energy (J/mol)
Cr	A2	$G_{Cr}^{A2,\,df} = \begin{cases} -13245.18 - 2.89 \times 10^{-3} \cdot T^2 + 16.21T\ln(e^{432.5532/T} - 1) + 8.71T\ln(e^{251.1734/T} - 1) & 0 < T < 531.9 \\ -13745.71 + 14.84T + 1.26 \times 10^{-3}T^2 - 1.12 \times 10^{-6}T^3 - 2.52T\ln(T) + 16.21T\ln(e^{432.5532/T} - 1) + 8.71T\ln(e^{251.1734/T} - 1) & 531.9 < T < 2136 \\ -38598.14 + 188.46 \cdot T - 23.90T\ln(T) - 5.42 \times 10^{19}T^{-5} + 9.03 \times 10^{38}T^{-11} + 16.21T\ln(e^{432.5532/T} - 1) + 8.71T\ln(e^{251.1734/T} - 1) & T > 2136 \end{cases}$ $G_{VA}^{A2} = +30T;\ L_{Cr,VA}^{A2} = +72220 - 41.50T$
	Liquid	$G_{Cr}^{liq} = 14164.90 + 24.942 \cdot T \cdot \ln(1 - e^{-297.2382/T}) - 2.23 \times 10^{-3}T^2 - 8.314 \cdot T \cdot \ln(1 + e^{(45995.32 - 14.8855 \cdot T - 1.16 \times 10^{-1} \cdot \ln(T))/8.314T})$
V	A2	$G_{V}^{A2,\,df} = \begin{cases} -11655.70 - 2.50 \times 10^{-3}T^2 + 23.58T\ln(e^{289.5879/T} - 1) + 1.29T\ln(e^{90.48463/T} - 1) & 0 < T < 2201 \\ -42621.43 + 144.41T - 17.69T\ln(T) + 4.38 \times 10^{19}T^{-5} - 4.76 \times 10^{38}T^{-11} + 23.58T\ln(e^{289.5879/T} - 1) + 1.29T\ln(e^{90.48463/T} - 1) & T > 2201 \end{cases}$ $G_{VA}^{A2} = +30T;\ L_{V,VA}^{A2} = +130500 - 60.01T$
	Liquid	$G_{V}^{liq} = 11355.04 + 24.942 \cdot T \cdot \ln(1 - e^{-205.2036/T}) - 1.77 \times 10^{-3}T^2 - 8.314 \cdot T \cdot \ln(1 + e^{(53359.83 - 10.4607 \cdot T - 5.66 \times 10^{-1} \cdot \ln(T))/8.314T})$

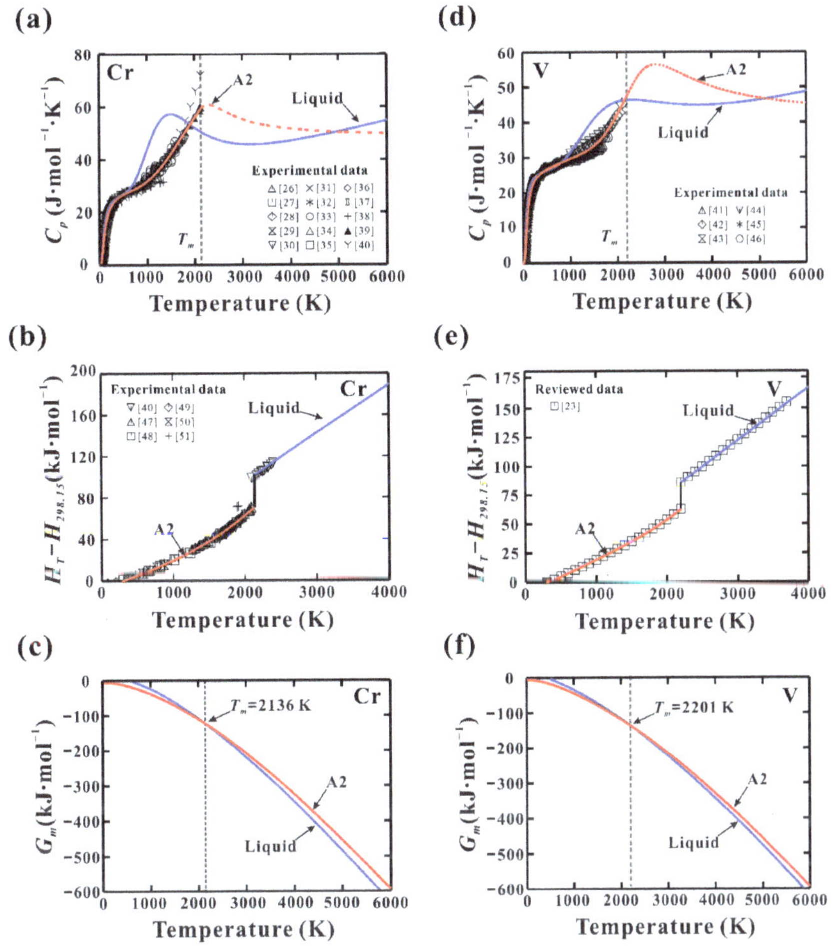

Figure 1. Calculated thermodynamic properties: (**a**,**d**) heat capacity (C_p); (**b**,**e**) heat content (H_T–H_{298}); (**c**,**f**) molar Gibbs energy of pure Cr and V in comparison with reported data [23,26–51].

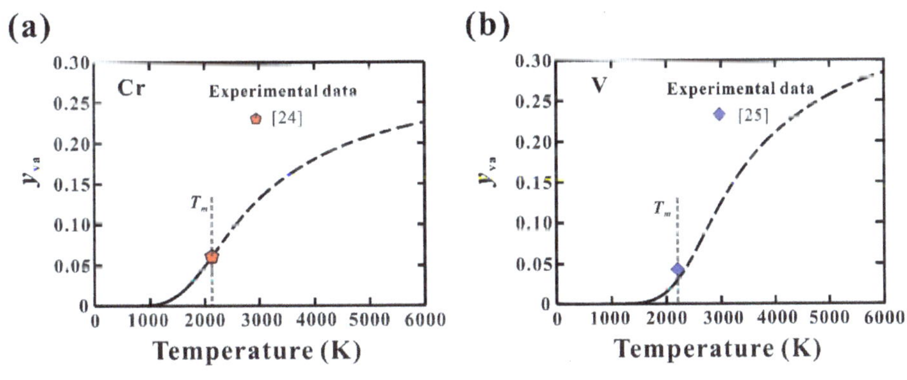

Figure 2. Calculated thermal vacancy concentrations in pure solid (**a**) Cr and (**b**) V along with measured data [24,25] at melting points.

3.2. Thermodynamic Descriptions of Ta-Cr and Ta-V Binary Systems

Ta-Cr and Ta-V binary systems were thermodynamically assessed by several research groups [52–56]. In these thermodynamic modeling, the Gibbs energy of pure elements were taken from SGTE (Scientific Group Thermodata Europe) data [57], which is also known as the second generation of thermodynamic description. In this description, the Gibbs energy was expressed as empirical temperature polynomials and valid only down to 298 K. In this section, the Ta-Cr and Ta-V binary systems were thermodynamically re-assessed by utilizing the above established third-generation Gibbs energy expressions for pure Cr and V as well as those for Ta from our previous work [13]. In addition, an effective strategy to obtain Gibbs energy for Laves phase was proposed and applied to C14 and C15 Laves phases in Ta-Cr and Ta-V systems.

Since the experimental phase equilibrium and thermodynamic properties in the Ta-Cr binary system have been recently reviewed by Jiang et al. [22], their reviewed data were adopted in the present modeling. As for the Ta-V binary system, most of the reported data have been reviewed by Danon and Servant [55] (2004). Thus, only the latest experimental data in the Ta-V binary system available in the literature will be briefly summarized and their discrepancies will be clarified. The greatest controversy in the Ta-V binary system is focused on the phase transition between C14 and C15 Laves phases. By reviewing the existing experimental data, Danon and Servant [55] concluded that C14 existed at a high temperature, while C15 existed at a low temperature, and there was a narrow (C15 + C14) two-phase region from 1400 to 1548 K. In addition, the eutectoid (C14 $\rightarrow$ C15+A2) and peritectoid (C14 + A2 $\rightarrow$ C15) reactions can be found in the V-rich and Ta-rich sides. After that, Pavlů et al. [56] pointed out that the C14 phase was more stable than C15 at 0 K based on their ab initio calculations. Thus, in their thermodynamic modeling, the C14 existed in two temperature regions, that is, from 0 to 626 K and from 1409 to 1703 K. Later, the measurements from Khan et al. [58] indicated that C14 was not found in four Ta-V alloys, which were annealed at 1473 K for 15 days. However, the recent phase equilibrium investigations in ternary Ta-V-Ni [59] and Ta-V-Co [60] systems displayed that the C14 phase existed in the Ta-V side at 1473 and 1573 K but was not found at 1173, 1273 and 1373 K after long-time annealing. Since no further experimental data confirm the existence of C14 below 626 K, as proposed by Pavlů et al. [56], the phase relationship between C14 and C15 recommended by Danon and Servant [55] was adopted in the present modeling.

During the thermodynamic modeling, the interaction parameters of the liquid and A2 phases were evaluated by considering the liquidus, solidus temperatures and invariant reactions. To obtain accurate thermodynamic descriptions for C14 and C15 Laves phases, the values of end-members in Equation (2) should be fixed initially. In this work, the values of ${}^0G_{Cr:Ta}$ and ${}^0G_{V:Ta}$ were evaluated by considering the experimental thermodynamic properties referenced to Cr2Ta and V2Ta alloys, which generally locate in the stable C14 or C15 phase region. The endmembers ${}^0G_{Cr:Cr}$, ${}^0G_{Ta:Ta}$ and ${}^0G_{V:V}$ are the Gibbs energies of pure elements with MgZn2-type (C14) and MgCu2-type (C15) crystal structures, which are the metastable ones, and their values cannot be directly obtained from the experimental investigation. Sluiter [61] computed the formation energies of a great deal of pure elements at 0 K in a variety of structures, including C14 and C15 by using electronic density functional theory (DFT). These theoretically predicted values can be served as start values for ${}^0G_{Cr:Cr}$, ${}^0G_{Ta:Ta}$ and ${}^0G_{V:V}$. To satisfy the Wagner–Schottky defects, there should be a constraint among the end-members when applying the two-sublattice model, that is, ${}^0G_{i:i} + {}^0G_{j:j} = {}^0G_{i:j} + {}^0G_{j:i}$ [62]. In the present work, the values of ${}^0G_{Ta:Cr}$ and ${}^0G_{Ta:V}$ are obtained by applying the above constraint. The interaction parameters in C14 and C15 Laves phase were adjusted to reproduce the invariant reactions. The thermodynamic parameters of Ta-Cr and Ta-V binary systems are summarized in Table 2.

Figure 3 shows the calculated phase diagrams of Ta-Cr and Ta-V binary systems according to the present thermodynamic descriptions. It shows that most of the experimental data [63–72] can be well reproduced. There are two invariant reactions between C14 and C15 Laves phases in Ta-Cr system, including a eutectoid reaction C14 $\rightarrow$ A2(Cr) + C15 and a peritectoid reaction (C14 + A2(Ta) $\rightarrow$ C15), as shown in the enlarged area in Figure 3a. A similar phase transition between C14 and C15 Laves phases in the Ta-V system can be observed. There is an azeotropic melting point in the Ta-V binary system, as shown in Figure 3b. The composition and temperature of this azeotropic melting point were calculated to be 12.4 at.% Ta and 2156 K, which were close to the measured ones [68–70]. Table 3 summarizes the calculated compositions and temperatures of the invariant reactions in Ta-Cr and Ta-V systems together with those from measurements [67–72] and previous calculations [53–56]. The present calculations agree reasonably with most of the measured data. Besides, the present results are quite close to those from previous assessments [53–56]. Figure 4 (a) presents the calculated activity of Cr over the whole composition range in the Ta-Cr system at 1472 K compared with the experimental data [73]. As can be seen, the calculations are within the uncertainty range of the experimental data. Figure 4b shows the calculated enthalpy of formation in the Ta-Cr system at 0 and1693 K, and the measured data [73] and DFT calculations [54,56,73–75] are also appended for a comparison. The calculated enthalpy formation of the C15 phase at 1693 K has a nice agreement with that from measurement [73]. In addition, the calculated enthalpy of formation at 0 K is more negative than that at 1693 K. Figure 4c gives the calculated enthalpy formation in the Ta-V system at 0 K. Since the theoretically computed formation enthalpy in the C15 phase from different sources is rather scattered, the present calculated result is reasonable.

Table 2. Summary of the thermodynamic parameters of Ta-Cr and Ta-V binary systems.

Systems	Phases	Thermodynamic Parameters (J/mol)
Ta-Cr	Liquid	${}^{0}L^{Liquid}_{Cr,Ta} = +5647.3 - 4.576T; {}^{1}L^{Liquid}_{Cr,Ta} = +10555.4 - 1.121T; {}^{2}L^{Liquid}_{Cr,Ta} = -21769.6 + 3.718T$
	A2	${}^{0}L^{A2}_{Cr,Ta} = +84411.6 - 30.246T; {}^{1}L^{A2}_{Cr,Ta} = +50274.2 - 21.825T; {}^{2}L^{A2}_{Cr,Ta} = -4321.6$
	C14	${}^{0}G^{C14}_{Cr:Cr} = +83400.0 + 3G^{A2,df}_{Cr}$ ${}^{0}G^{C14}_{Ta:Ta} = +29100.0 + 3G^{A2,df}_{Ta}$ ${}^{0}G^{C14}_{Cr:Ta} = -19550.7 - 5.770T + 2G^{A2,df}_{Cr} + G^{A2,df}_{Ta}$ ${}^{0}G^{C14}_{Ta:Cr} = +132050.7 + 5.536T + G^{A2,df}_{Cr} + 2G^{A2,df}_{Ta}$ ${}^{0}L^{C14}_{Cr,Ta:*} = +219623.9 - 74.407T; {}^{0}L^{C14}_{*:Cr,Ta} = +95266.4 - 77.887T$
	C15	${}^{0}G^{C15}_{Ta:Ta} = +33600.0 + 3G^{A2,df}_{Ta}$ ${}^{0}G^{C15}_{Cr:Cr} = +79200.0 + 3G^{A2,df}_{Cr}$ ${}^{0}G^{C15}_{Cr:Ta} = -26555.2 - 2.170T + 2G^{A2,df}_{Cr} + G^{A2,df}_{Ta}$ ${}^{0}G^{C15}_{Ta:Cr} = +139755.2 + 1.813T + G^{A2,df}_{Cr} + 2G^{A2,df}_{Ta}$ ${}^{0}L^{C15}_{Cr,Ta:*} = +51828.2 + 8.181T; {}^{0}L^{C15}_{*:Cr,Ta} = -61398.1 + 7.891T$
Ta-V	Liquid	${}^{0}L^{Liquid}_{V:Ta} = -5751.2; {}^{1}L^{Liquid}_{V:Ta} = -3341.9$
	A2	${}^{0}L^{A2}_{V:Ta} = +3731.5; {}^{1}L^{A2}_{V:Ta} = -10998.9$
	C14	${}^{0}G^{C14}_{Ta:Ta} = +29100.0 + 3G^{A2,df}_{Ta}$ ${}^{0}G^{C14}_{V:V} = +28800.0 + 3G^{A2,df}_{V}$ ${}^{0}G^{C14}_{V:Ta} = -18925.2 + 2G^{A2,df}_{V} + G^{A2,df}_{Ta}$ ${}^{0}G^{C14}_{Ta:V} = +76825.2 + G^{A2,df}_{V} + 2G^{A2,df}_{Ta}$ ${}^{0}L^{C14}_{V,Ta:*} = -11321.2; {}^{0}L^{C14}_{*:V,Ta} = -6751.6$
	C15	${}^{0}G^{C15}_{Ta:Ta} = +33600.0 + 3G^{A2,df}_{Ta}$ ${}^{0}G^{C15}_{V:V} = +33000.0 + 3G^{A2,df}_{V}$ ${}^{0}G^{C15}_{V:Ta} = -23389.9 + 3.004T + 2G^{A2,df}_{V} + G^{A2,df}_{Ta}$ ${}^{0}G^{C15}_{Ta:V} = +89989.9 - 3.004T + G^{A2,df}_{V} + 2G^{A2,df}_{Ta}$ ${}^{0}L^{C15}_{V,Ta:*} = -349.549 + 1.025T; {}^{0}L^{C15}_{*:V,Ta} = -7515.567$

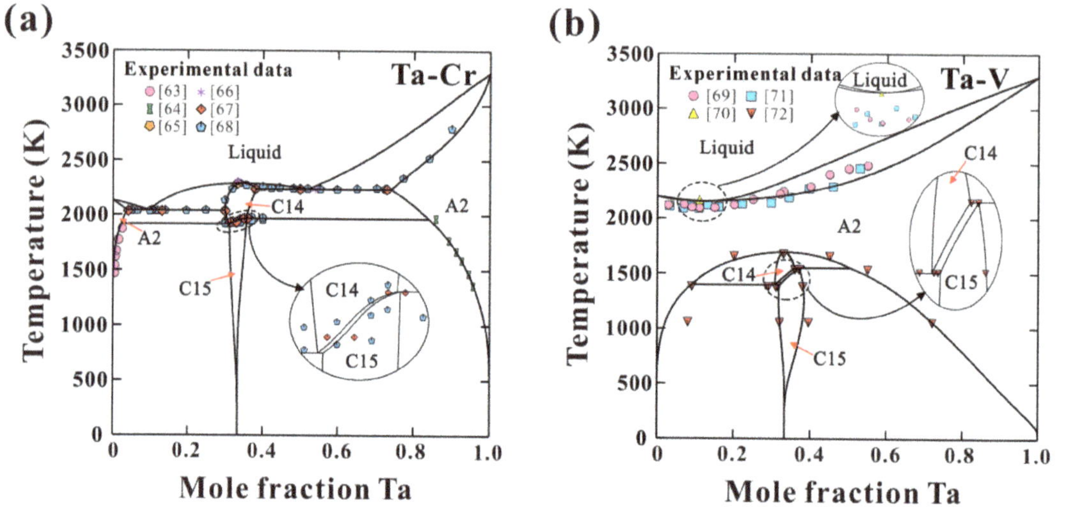

Figure 3. Calculated phase diagrams for (**a**) Ta-Cr and (**b**) Ta-V binary systems along with experimental data [63–72].

Table 3. Calculated temperatures and compositions of invariant reactions in Ta-Cr and Ta-V binary systems in comparison with the measured ones [67–72] and previous calculated ones [53–56].

Systems	Reactions	Temperature (K)	Composition (at. %Ta)			Ref.
Ta-Cr	Liquid → C14	2293.0 ± 20	–	33.30	–	[67] (Exp.)
		2309.7	–	34.90	–	[53] (Cal.)
		2304	–	34.22	–	[54] (Cal.)
		2290.0	–	34.87	–	This work (Cal.)
	Liquid → A2(Cr) + C14	2033.0 ± 20	~13.00	~4.00	~30.00	[67] (Exp.)
		2040.0 ± 10	~10.50	~3.50	~30.00	[68] (Exp.)
		2044.7	11.53	4.53	30.03	[53] (Cal.)
		2065.1	9.88	3.89	30.65	[54] (Cal.)
		2041.9	9.59	3.68	29.94	This work (Cal.)
	Liquid → A2(Ta) + C14	2238.0 ± 20	~50.0	~73.00	~38.00	[67] (Exp.)
		2223.4	49.41	72.75	37.90	[53] (Cal.)
		2239.2	49.96	73.32	37.70	[54] (Cal.)
		2239.0	52.12	74.13	38.21	This work (Cal.)
	C14 → A2(Cr) + C15	1933.0	~31.40	–	~33.00	[67] (Exp.)
		1917.2	30.96	2.81	32.23	[53] (Cal.)
		1903.3	30.97	1.87	31.11	[54] (Cal.)
		1921.0	30.83	2.17	31.15	This work (Cal.)
	C14 + A2(Ta) → C15	1968.0	~35.00	–	~36.00	[67] (Exp.)
		1982.9	35.17	80.58	35.55	[53] (Cal.)
		1991.2	37.29	84.72	37.39	[54] (Cal.)
		1969.0	35.52	83.86	35.72	This work (Cal.)
Ta-V	Liquid → A2	2153.0	–	11.00	–	[70] (Exp.)
		2098.0	–	15.00	–	[69] (Exp.)
		–	–	12.00	–	[68] (Exp.)
		2099	–	12.89	–	[55] (Cal.)
		2099	–	12.89	–	[56] (Cal.)
		2156.0	–	12.40	–	This work (Cal.)

Table 3. Calculated temperatures and compositions of invariant reactions in Ta-Cr and Ta-V binary systems in comparison with the measured ones [67–72] and previous calculated ones [53–56].

Systems	Reactions	Temperature (K)	Composition (at. %Ta)			Ref.
	A2 → C14	1693.0	–	~33.00	–	[71] (Exp.)
		1702.2	–	32.70	–	[55] (Cal.)
		1703.3	–	32.62	–	[56] (Cal.)
		1693.0	–	~33.00	–	[72] (Exp.)
		1695.1	–	33.22	–	This work (Cal.)
	C14 + A2(Ta) → C15	1553.0	36.00	37.00	55.00	[72] (Exp.)
		1550.2	35.97	37.36	50.27	[55] (Cal.)
		1556.1	36.14	37.78	57.21	[56] (Cal.)
		1552.5	35.87	37.02	50.46	This work (Cal.)
	C14 → A2(V) + C15	1398.0	29.00	9.00	31.50	[72] (Exp.)
		1403.0	31.02	6.48	31.29	[55] (Cal.)
		1409.5	30.69	3.41	31.64	[56] (Cal.)
		1396.0	30.70	8.90	31.85	This work (Cal.)

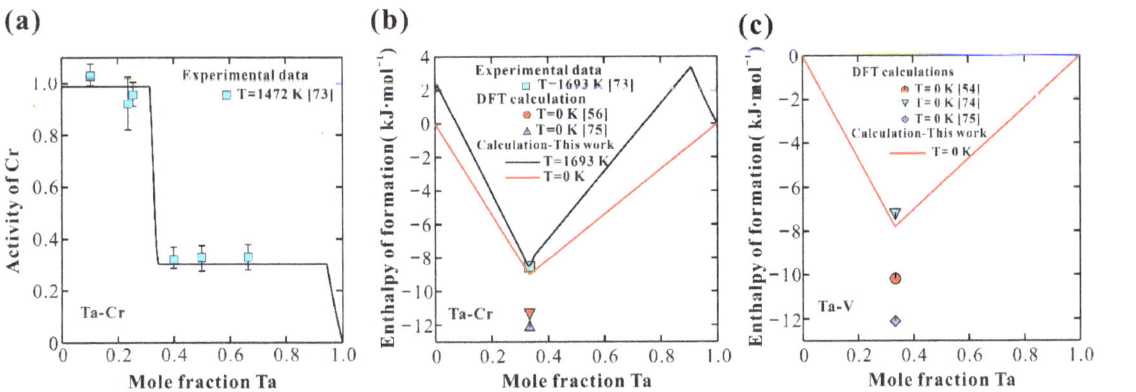

Figure 4. Calculated thermodynamic properties in Ta−Cr and Ta-V systems together with reported data [54,56,73–75]: (**a**) activity of Cr in Ta-Cr system at 1472 K; (**b**) enthalpy of formation in Ta-Cr system at 0 and 1693 K; (**c**) enthalpy of formation in Ta-V system at 0 K.

Figure 5 shows the calculated C_p of Cr2Ta and V2Ta alloys from 0 to 1400 K. Based on the present modeling, Cr2Ta and V2Ta alloys locate in the C15 single phase region below 1948.6 and 1446.7 K. Although the experimental C_p data were not considered during the modeling, the present calculations show a nice agreement with the reported data [76,77]. The calculations from previous thermodynamic assessments [52–56] are also presented in Figure 5 for a comparison. As shown in Figure 5a, the calculated C_p of Cr2Ta from [53,54] are quite close to the present calculated ones at high temperatures, while obvious deviations can be observed at low temperatures (below 298 K). The present calculations show better agreement with the measured ones. In addition, the values of C_p for Cr2Ta and V2Ta alloys, according to the previous assessments [53–56], decrease rapidly as the temperature decreases and become negative at 85 and 62 K, which is physically inaccurate. It also implies that the present thermodynamic description of C15 Laves phase in Ta-Cr and Ta-V binary systems is reliable. It is highly believed that the present strategy to establish the thermodynamic descriptions of Laves phase is reliable, which can be applied to develop the third-generation thermodynamic database for Laves phases containing RHEAs.

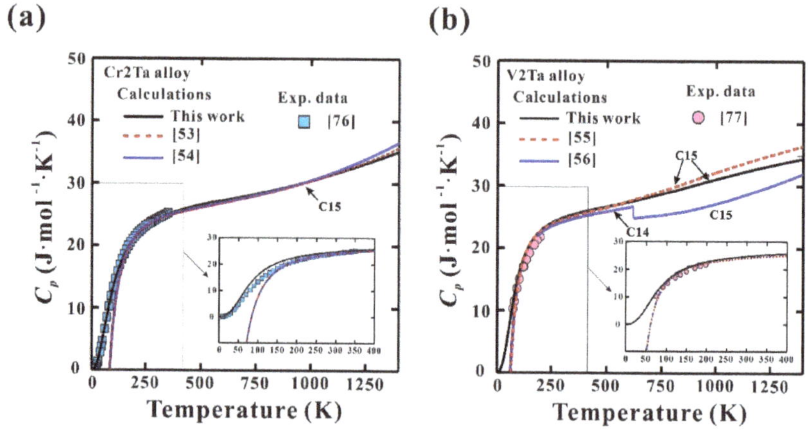

Figure 5. Calculated C_p for (**a**) Cr2Ta and (**b**) V2Ta alloys from 0 to 1400 K along with the reported data [76,77].

4. Conclusions

- The third-generation Gibbs energy expressions of pure Cr and V in both the liquid and A2 phases were established. By applying these expressions, the thermodynamic properties down to 0 K and thermal vacancy near the melting point can be well described. Besides, the lattice stability of Cr and V over the whole temperature range can be guaranteed.
- Based on the third-generation Gibbs energy expressions of pure elements, the Ta-Cr and Ta-V binary systems were thermodynamically assessed by considering the reviewed phase equilibria and thermodynamic data with the CALPHAD approach. A strategy to estimate the Gibbs energy of Laves phase was proposed by combining the theoretically computed and experimentally measured thermodynamic properties as well as semiempirical relation. Such a method was applied to C14 and C15 Laves phases in Ta-Cr and Ta-V binary systems. The calculated phase diagrams and thermodynamic properties showed nice agreement with the measured ones. Significant improvements can be observed at low temperatures compared with those from the second-generation thermodynamic descriptions, indicating the high reliability of the present thermodynamic descriptions.

Author Contributions: Methodology, Y.T.; validation, E.Z., Y.C. and X.X.; writing—original draft preparation, E.Z.; writing—review and editing, Y.T., X.X. and Y.C.; resources, Y.C. and X.X; supervision, Y.T.; funding acquisition, Y.T. All authors have read and agreed to the published version of the manuscript.

Funding: This research was funded by Natural Science Foundation of Hebei Province, China, grant number E2019202234, and National Joint Engineering Research Center for Abrasion Control and Molding of Metal Materials, grant number HKDNM2019019.

Institutional Review Board Statement: Not applicable.

Informed Consent Statement: Not applicable.

Data Availability Statement: The data presented in this study are available on request from the corresponding author.

Conflicts of Interest: The authors declare no conflict of interest.

References

1. Senkov, O.N.; Wilks, G.B.; Miracle, D.B.; Chang, C.P.; Liaw, P.K. Refractory high-entropy alloys. *Intermetallics* **2010**, *18*, 1758–1765. [CrossRef]
2. Senkov, O.N.; Miracle, D.B.; Chaput, K.J.; Couzinie, J.-P. Development and exploration of refractory high entropy alloys—A review. *J. Mater. Res.* **2018**, *33*, 3092–3128. [CrossRef]
3. Miracle, D.B.; Tsai, M.-H.; Senkov, O.N.; Soni, V.; Banerjee, R. Refractory high entropy superalloys (RSAs). *Scr. Mater.* **2020**, *187*, 445–452. [CrossRef]
4. Couzinié, J.-P.; Senkov, O.N.; Miracle, D.B.; Dirras, G. Comprehensive data compilation on the mechanical properties of refractory high-entropy alloys. *Data Brief* **2018**, *21*, 1622–1641. [CrossRef] [PubMed]
5. Qin, G.; Li, Z.; Chen, R.; Zheng, H.; Fan, C.; Wang, L.; Su, Y.; Ding, H.; Guo, J.; Fu, H. CoCrFeMnNi high-entropy alloys reinforced with Laves phase by adding Nb and Ti elements. *J. Mater. Res.* **2018**, *34*, 1011–1020. [CrossRef]
6. Yi, J.; Wang, L.; Xu, M.; Yang, L. Microstructure and mechanical properties of refractory high-entropy alloy HfMoNbTiCr. *Mater. Techol.* **2021**, *55*, 305–310. [CrossRef]
7. Yao, H.W.; Miao, J.W.; Liu, Y.M.; Guo, E.Y.; Huang, H.; Lu, Y.P.; Wang, T.M.; Li, T.J. Effect of Ti and Nb contents on microstructure and mechanical properties of $HfZrVTaMoWTi_xNb_y$ refractory high-entropy alloys. *Adv. Energy Mater.* **2021**, *23*, 2100225.
8. Stein, F.; Palm, M.; Sauthoff, G. Structure and stability of Laves phases. Part I. Critical assessment of factors controlling Laves phase stability. *Intermetallics* **2004**, *12*, 713–720. [CrossRef]
9. Yurchenko, N.; Stepanov, N.; Salishchev, G. Laves-phase formation criterion for high-entropy alloys. *Mater. Sci. Technol.* **2017**, *33*, 17–22. [CrossRef]
10. Yao, H.; Qiao, D.; Miao, J.; Wang, J.; Guo, E.; Lu, Y. Criteria for laves-phase formation in refractory high-entropy alloys. *Phil. Mag. Lett.* **2021**, in press. [CrossRef]
11. Gao, M.C.; Carney, C.S.; Doğan, Ö.N.; Jablonksi, P.D.; Hawk, J.A.; Alman, D.E. Design of refractory high-entropy alloys. *JOM* **2015**, *67*, 2653–2669. [CrossRef]
12. Raturi, A.; Aditya, J.; Gurao, N.P.; Biswas, K. ICME approach to explore equiatomic and non-equiatomic single phase BCC refractory high entropy alloys. *J. Alloys Compd.* **2018**, *806*, 587–595. [CrossRef]
13. Zhang, E.K.; Tang, Y.; Wen, M.W.; Obaied, A.; Roslyakova, I.; Zhang, L.J. On phase stability of Mo-Nb-Ta-W refractory high entropy alloys. *Int. J. Refract. Met. Hard Mater.* **2022**, *103*, 105780. [CrossRef]
14. Chase, M.W.; Ansara, I.; Dinsdale, A.; Eriksson, G.; Grimvall, G.; Hoglund, L.; Yokokawa, H. Group 1: Heat capacity models for crystalline phases from 0 K to 6000 K. *Calphad* **1995**, *19*, 437–447.
15. Roslyakova, I.; Sundman, B.; Dette, H.; Zhang, L.J.; Steinbach, I. Modeling of Gibbs energies of pure elements down to 0 K using segmented regression. *Calphad* **2016**, *55*, 165–180. [CrossRef]
16. Bigdeli, S.; Mao, H.H.; Selleby, M. On the third-generation Calphad databases: An updated description of Mn. *Phys. Status Solidi B* **2015**, *252*, 2199–2208. [CrossRef]
17. Hao, L.Y.; Ruban, A.; Xiong, W. CALPHAD modeling based on Gibbs energy functions from zero kevin and improved magnetic model: A case study on the Cr-Ni system. *Calphad* **2021**, *73*, 102268. [CrossRef]
18. Tang, Y.; Zhang, L.J. Effect of thermal vacancy on thermodynamic behaviors in BCC W close to melting point: A thermodynamic study. *Materials* **2018**, *11*, 1648. [CrossRef]
19. Obaied, A.; Bocklund, B.; Zomorodpoosh, S.; Zhang, L.J.; Otis, R.; Liu, Z.-K.; Roslyakova, I. Thermodynamic re-assessment of pure chromium using modified segmented regression model. *Calphad* **2020**, *69*, 101762. [CrossRef]
20. Ågren, J. Thermodynamics of supercooled liquids and their glass transition. *Phys. Chem. Liq.* **1988**, *18*, 123–139. [CrossRef]
21. Jiang, Y.X.; Zomorodpoosh, S.; Roslyakova, I.; Zhang, L.J. Thermodynamic re-assessment of binary Cr-Nb system down to 0 K. *Calphad* **2018**, *62*, 109–118. [CrossRef]
22. Jiang, Y.X.; Zomorodpoosh, S.; Roslyakova, I.; Zhang, L.J. Thermodynamic re-assessment of the binary Cr–Ta system down to 0 K. *Int. J. Mater. Res.* **2019**, *110*, 797–807.
23. Arblaster, J.W. Thermodynamic properties of vanadium. *J. Phase Equilib. Diffus.* **2017**, *38*, 51–64. [CrossRef]
24. Chekhovskoi, V.Y. The energy of formation of vacancies and their concentration in chromium. *Sov. Phys. Solid State* **1979**, *21*, 1089–1090.
25. Chekhovskoii, V.Y.; Tarasov, V.D.; Grigoreva, N.V. Contribution of equilibrium vacancies to vanadium caloric properties. *High Temp.* **2011**, *49*, 826–831. [CrossRef]
26. Anderson, C. The heat capacities of chromium, chromic oxide, chromous chloride and chromic chloride at low temperatures. *J. Am. Chem. Soc.* **1937**, *59*, 488–491. [CrossRef]
27. Estermann, I.; Friedberg, S.A.; Goldman, J.E. The specific heats of several metals between 1.8 and 4.2 K. *Phys. Rev.* **1952**, *87*, 582–588. [CrossRef]
28. Kemeny, T.; Fogarassy, B.; Arajs, S.; Moyer, C.A. Heat-capacity studies of chromium-rich antiferromagnetic chromium-iron alloys. *Phys. Rev. B* **1979**, *20*, 2975–2979. [CrossRef]
29. Beaumont, R.H.; Chihara, H.; Morrison, J.A. An anomaly in the heat capacity of chromium at 38.5 °C. *Phil. Mag.* **1960**, *5*, 188–191. [CrossRef]
30. Kohlhaas, R.; Braun, M. Die thermodynamischen funktionen des reinen eisens: Wärmeinhalt und spezifische wärme austenitischer eisenlegierungen und stähle. *Arch. Eisenhuttenwes* **1965**, *34*, 391–399. [CrossRef]

31. Armstrong, L.D.; Grayson-Smith, H. High temperature calorimetry: II. Atomic heats of chromium, manganese, and cobalt between 0° and 800 °C. *Can. J. Res.* **1950**, *28*, 51–59. [CrossRef]
32. Kalishevich, G.I.; Krentsis, R.V. The standard specific heats, entropies, and enthalpies of silicon and chromium and its alloys. *Zh. Fiz. Khim.* **1965**, *39*, 2999–3001.
33. Krauss, F. Die messung der spezifischen wärme von metallen bei hohen temperature. *Z. Metallkd.* **1958**, *49*, 386–392.
34. Martin, D.; Newton, R.F.; Girmes, W.R.; Blankenship, F.F. Activities in the chromium-nickel system. *J. Phys. Chem.* **1958**, *62*, 980–982.
35. Simon, F.; Simson, V.C.; Ruhemann, M. An investigation on the specific heat at low temperature, the specific heat of the halides of ammonia between −70 °C and room temperature. *Z. Phys. Chem.* **1927**, *129*, 339–348.
36. Clusius, K.; Franzosini, P. Ergebnisse der tieftemperaturforschung. XXXII. Atom-und elektronenwarme des chroms zwischen 14 K und 273 K. *Z. Naturforsch.* **1962**, *17*, 522–525. [CrossRef]
37. Rayne, J.A.; Kemp, W.R.G. The heat capacities of chromium and nickel. *Phil. Mag.* **1956**, *1*, 918–925. [CrossRef]
38. Jaeger, F.; Rosenbohm, E. The exact measurement of the specific heats of metals at high temperatures. XVII. Calorimetric retardation phenomena in the case of cerium and chromium. *Proc. Acad. Sci. Amst.* **1934**, *37*, 489–497.
39. Touloukian, Y.S.; Saxena, S.C.; Hestermans, P. *Thermophysical Properties of Matter-the TPRC Data Series, 11*; Ho, C.Y., Ed.; Thermophysical and Electronic Properties Information Analysis Center: Boulder, CO, USA, 1975; pp. 61–62.
40. Lin, R.; Frohberg, M. Enthalpy measurements of solid and liquid chromium by levitation calorimetry. *High Temp. High Press.* **1988**, *20*, 539–544.
41. Bendick, W.; Pepperhoff, W. The heat capacity of Ti, V. and Cr. *J. Phys. F Met. Phy.* **1982**, *12*, 1085–1090. [CrossRef]
42. Cezairliyan, A.; Righini, F.; McClure, J.L. Simultaneous measurements of heat capacity, electrical resistivity and hemispherical total emittance by a pulse heating technique: Vanadium, 1500 to 2100 K. *J. Res. Natl. Bur. Stand. A Phys. Chem.* **1974**, *78A*, 143–147. [CrossRef] [PubMed]
43. Clusius, K.; Franzosini, P.; Piesbergen, U. Ergebnisse der tieftemperaturforschung. XXXII. Die atom und elektronwarme des vanadins und niobs zwischen 10 and 273 K. *Z. Naturforschg.* **1960**, *15A*, 728–734. [CrossRef]
44. Filippov, L.P.; Yurchak, P. High temperature investigation of the thermal properties of solids. *Inzh. Fiz. Zh.* **1971**, *21*, 561–577. [CrossRef]
45. Kohlhaas, R.; Braun, M.; Vollmer, O. Die atomwarme von titan, vanadin und chrom im berich hoher temperature. *Z. Naturforsch.* **1965**, *20A*, 1077–1079. [CrossRef]
46. Takahashi, Y.; Nakamura, J.-I.; Smith, J.F. Laser flash calorimetry. III. Heat capacity of vanadium from 80 to 1000 K. *J. Chem. Thermodyn.* **1982**, *14*, 977–982. [CrossRef]
47. Conway, J.B.; Hein, R.A. Enthalpy measurements of solid materials to 2400 °C by means of a drop technique. In *Advances in Thermophysical Properties at Extreme Temperatures and Pressures*; Cratch, S., Ed.; The American Society of Mechanical Engineers: New York, NY, USA, 1965; pp. 152–155.
48. Lucks, C.; Deem, H. *Thermal Conductivities, Heat Capacities, and Linear Thermal Expansion of Five Materials. Wright Air Development Center*; Report No. WADC-TR-55-496; Wright-Patterson Air Force Base: Dayton, OH, USA, 1956.
49. Schimpff, H. On the heat capacity of metals and intermetallic compounds. *Z. Phys. Chem. Neue Folge* **1910**, *71*, 257–300. [CrossRef]
50. Schübel, P. Metallographische mitteilungen aus dem institut für physikalische chemie der universitat gottingen. lxxxvii. Über die warmekapazitat von metallen und metallverbindungen zwischen 18–600 °C. *Z. Anorg. Chem.* **1914**, *87*, 81–119. [CrossRef]
51. Umino, S. On the latent heat of fusion of several metals and their specific heats at high temperatures. *Sci. Rep. Tohuku Imp. Univ.* **1926**, *15*, 597–617.
52. Kaufman, L. Coupled thermochemical and phase diagram data for tantalum based binary alloys. *Calphad* **1991**, *15*, 243–259. [CrossRef]
53. Dupin, N.; Ansara, I. Thermodynamic assessment of the Cr-Ta system. *J. Phase Equilib.* **1993**, *14*, 451–456. [CrossRef]
54. Pavlů, J.; Vřešťál, J.; Šob, M. Re-modeling of Laves phases in the Cr–Nb and Cr–Ta systems using first-principles results. *Calphad* **2009**, *33*, 179–186. [CrossRef]
55. Danon, C.A.; Servant, C.A. Thermodynamic evaluation of the Ta-V system. *J. Alloys Compd.* **2004**, *366*, 191–200. [CrossRef]
56. Pavlů, J.; Vřešťál, J.; Chen, X.-Q.; Rogl, P. Thermodynamic modeling of Laves phases in the Ta–V system: Reassessment using first-principles results. *Calphad* **2011**, *35*, 103–108. [CrossRef]
57. Dinsdale, A.T. SGTE data for pure elements. *Calphad* **1991**, *15*, 317–425. [CrossRef]
58. Khan, A.U.; Broz, P.; Niu, H.Y.; Bursik, J.; Grytsiv, A.; Chen, X.-Q.; Giester, G.; Rogl, P. The system Ta-V-Si: Crystal structure and phase equilibria. *J. Solid State Chem.* **2012**, *187*, 114–123. [CrossRef]
59. Wang, C.P.; Liang, Y.; Yang, S.; Zhang, J.; Huang, Y.; Han, J.; Lu, Y.; Liu, X.J. Phase equilibria in the Ni-V-Ta ternary system. *Metals* **2018**, *8*, 774. [CrossRef]
60. Ruan, J.J.; Wang, C.P.; Yang, S.Y.; Omori, T.; Yang, T.; Kimura, Y.; Liu, X.J.; Kainuma, R.; Ishida, K. Experimental investigations of microstructures and phase equilibria in the Co-V-Ta ternary system. *J. Alloys Compd.* **2016**, *664*, 141–148. [CrossRef]
61. Sluiter, M.H.F. Ab initio lattice stabilities of some elemental complex structures. *Calphad* **2006**, *30*, 357–366. [CrossRef]
62. Coughanowr, C.A.; Ansara, I.; Luoma, R.; Hamalainen, M.; Lukas, H.L. Assessment of the Cu-Mg System. *Z. Metallkd.* **1991**, *82*, 574–581.
63. Auld, J.; Ryan, N. The solid solubility of tantalum in chromium. *J. Less Common. Met.* **1961**, *3*, 221–225. [CrossRef]

64. Gebhardt, E.; Rexer, J. Precipitations in Ta-Cr solid solutions. *Z. Metallkd.* **1967**, *47*, 611–617.
65. Elliott, R.; Rostoker, W. The occurrence of Laves-type phases among transition elements trans. *ASM* **1958**, *50*, 617–633.
66. Kocherzhinskii, Y.; Petkov, V.; Shishkin, E. Phase equilibrium and crystal structure of the intermediate phase in the Ta-Cr system. *Metallofizika* **1973**, *46*, 75–80.
67. Venkatraman, M.; Neumann, J. The Cr-Ta (chromium-tantalum) system. *J. Phase Equilib.* **1987**, *8*, 112–116. [CrossRef]
68. Rudy, E. *Compendium of Phase Diagram Data. Air Force Materials Laboratory*; Report No. AFML-Tr-65-2. Part V; Wrighr Patterson AFB: Dayton, OH, USA, 1969.
69. Nefedov, A.P.; Sokolovskaya, E.M.; Grigorev, A.T.; Chechernikov, V.I.; Sokolova, I.G.; Guzei, L.S. Solid-state phase transformations in vanadium-tantalum alloys. *Vestn. Moskov. Univ. Ser. II Khim.* **1965**, *5*, 42–47.
70. Smith, J.F.; Carlson, O.N. The Ta-V (tantalum-vanadium) system. *Bull. Alloy Phase Diagr.* **1983**, *4*, 284–289. [CrossRef]
71. Eremenko, V.N.; Tretyachenko, L.A.; Yakhimovich, R.I. Fusion diagram of the tantalum-vanadium system. *Russ. J. Inorg. Chem.* **1960**, *5*, 1110–1112.
72. Savitskii, E.M.; Efimov, J.V. Superconducting metallic compounds and their alloys. *Monatsh. Chem.* **1972**, *103*, 270–287.
73. Feschotte, P.; Kubaschewski, O. Thermochemical properties of the laves phase Cr_2Ta. *Trans. Faraday Soc.* **1964**, *60*, 1941–1946. [CrossRef]
74. Chihi, T.; Fatmi, M.; Ghebouli, B. Ab initio calculations for properties of laves phase V_2M (M = Zr, Hf, Ta) compounds. *Am. J. Phys.* **2013**, *2*, 88–92. [CrossRef]
75. Kirklin, S.; Saal, J.E.; Meredig, B.; Thompson, A.; Doak, J.W.; Aykol, M.; Rühl, S.; Wolverton, C. The Open Quantum Materials Database (OQMD): Assessing the accuracy of DFT formation energies. *NPJ Comput. Mater.* **2015**, *1*, 15010. [CrossRef]
76. Hafstrom, J.W.; Knapp, G.S.; Aldred, A.T. Electronic and lattice properties of $V_2Hf_{1-x}Ta_x$ C15 superconductors. *Phys. Rev. B* **1978**, *17*, 2892–2900. [CrossRef]
77. Martin, J.F.; Müller, F.; Kubaschewski, O. Thermodynamic properties of $TaCr_2$ and $NbCr_2$. *Trans. Faraday Soc.* **1970**, *66*, 1065–1072. [CrossRef]

materials

MDPI

Article

Nb/Sn Liquid-Solid Reactive Diffusion Couples and Their Application to Determination of Phase Equilibria and Interdiffusion Coefficients of Nb-Sn Binary System

Jiali Zhang, Jing Zhong, Qin Li * and Lijun Zhang *

State Key Lab of Powder Metallurgy, Central South University, Changsha 410083, China; jializhang@csu.edu.cn (J.Z.); zhongjingjogy@gmail.com (J.Z.)
* Correspondence: qinli333@csu.edu.cn (Q.L.); lijun.zhang@csu.edu.cn (L.Z.); Fax: +86-731-8871-0855 (L.Z.)

Abstract: Nb_3Sn plays an irreplaceable role in superconducting parts due to its stable performance under high field conditions. Accurate phase equilibria and interdiffusion coefficients are of great significance for designing novel Nb_3Sn superconductors. However, the related experimental information is still in a state of scarcity because of the difficulty in fabrication of Nb-Sn alloys caused by the large difference in melting points of Nb and Sn. In this paper, a simple but pragmatic approach was first proposed to prepare the Nb/Sn liquid-solid reactive diffusion couples (LSDCs) at 1100 °C and 1200 °C, of which the phase identification of the formed layer and the measurement of composition-distance profiles were conducted. The formed layer in Nb/Sn LSDCs was confirmed to be Nb_3Sn compound. While the measured composition profiles were employed to determine the phase equilibria according to the local equilibrium hypothesis and the interdiffusion coefficients with an aid of the latest version of HitDIC software. The determined phase equilibria of Nb_3Sn, (Nb) and liquid show good agreement with the assessed phase diagram. While the calculated interdiffusion coefficients and activation energy for diffusion in Nb_3Sn are consistent with both experimental and theoretical data in the literature. Moreover, the growth of the formed Nb_3Sn layer in Nb/Sn LSDCs was also found to be diffusion controlled. All the obtained phase equilibria and interdiffusion coefficients are of great value for further thermodynamic and kinetic modeling of the Nb-Sn system. Furthermore, it is anticipated that the presently proposed approach of fabricating liquid-solid reactive diffusion couple should serve as a general one for various alloy systems with large differences in melting points.

Keywords: liquid-solid diffusion couple; Nb-Sn system; Nb_3Sn; phase equilibria; diffusion; HitDIC

Citation: Zhang, J.; Zhong, J.; Li, Q.; Zhang, L. Nb/Sn Liquid-Solid Reactive Diffusion Couples and Their Application to Determination of Phase Equilibria and Interdiffusion Coefficients of Nb-Sn Binary System. *Materials* **2022**, *15*, 113. https://doi.org/10.3390/ma15010113

Academic Editor: Elena Pereloma

Received: 29 November 2021
Accepted: 20 December 2021
Published: 24 December 2021

1. Introduction

Due to its high superconducting critical temperature Tc = 18 K, high superconducting critical magnetic field Hc = 27T (4 K) and stable performance [1], Nb_3Sn superconductor has been of great importance in high field working environments since the 1960s [2]. Tokamak device, used in nuclear-fusion research for magnetic confinement of plasma, is the core unit of both the International Thermonuclear Experimental Reactor (ITER) project and the China Fusion Engineering Test Reactor (CFETR) project. Nb_3Sn acts as the vital toroidal field coil in Tokamak devices [3]. However, as one of A15 type intermetallic compounds in the cubic lattice system [4], Nb_3Sn owns the poor plasticity and toughness, and can be easily damaged, resulting in destruction of superconductivity. In order to improve the properties/performance of Nb_3Sn-based parts and even explore novel Nb_3Sn-based alloys, accurate phase equilibria and diffusion coefficients [5,6] are the prerequisites for understanding the Nb-Sn and its related systems. Unfortunately, a very large difference in melting points between Nb (2468 °C) and Sn (231.89 °C) makes it extremely difficult to fabricate the Nb-Sn alloys. To overcome this problem, researchers have made many attempts, and developed several techniques for preparing Nb-Sn alloys, including the powder-in-tube process [7], the bronze process [8], the internal tin process [9], the restacked rod process [10], the powder metallurgy method [11], and so on. Even so, the experimental data on phase equilibria and diffusion coefficients of the Nb-Sn binary system are still in a state of scarcity.

Up to now, only about 10 groups of researchers have performed experimental measurement of phase equilibria of the binary Nb-Sn system [12–20]. Based on the limited experimental phase equilibrium information, Toffolon [18] performed a CALPHAD (CALculation of PHAse Diagram) [21] thermodynamic assessment of Nb-Sn system in 1998. During the assessment, Toffolon [18] adopted the conclusion of Massalski [17] and Shunk [22] that Nb_3Sn would not be stable below 796 °C. However, on account of the enthalpy of formation of Nb_3Sn determined by drop calorimetry and other new literature data [13–15], it was believed that Nb_3Sn can exist stably down to the room temperature. Then in 2002, Toffolon [19] updated the thermodynamic assessment of binary Nb-Sn system accordingly. Although the calculated phase equilibria of binary Nb-Sn system according to the thermodynamic descriptions by Toffolon [19] are in general agreement with the limited experimental data [12–16,20], it is still necessary to provide more accurate experimental phase equilibrium data for further validation. While for the diffusion coefficients in Nb-Sn binary, there is only one report from Wallach [23], who obtained the interdiffusion coefficients of Nb_3Sn from the ternary Cu-Nb-Sn system measured by two different methods, i.e., the Nb-bronze multi-layered thin-film composites and the multi-filamentary composites. Moreover, the data from Wallach [23] only cover the low-temperature range between 350 °C and 800 °C. Again, accurate experimental diffusion coefficients of the binary Nb-Sn system, especially at high temperatures, are in urgent need.

Consequently, the major objectives of this paper are: (i) to fabricate the Nb/Sn liquid-solid reactive diffusion couples (LSDCs) at 1100 °C and 1200 °C. Here, a simple but pragmatic approach is to be proposed for preparing the Nb/Sn LSDCs, of which the melting points of two end-members show a large difference; and (ii) to apply the prepared Nb/Sn LSDCs to determine the accurate phase equilibria and interdiffusion coefficients of the binary Nb-Sn system at the corresponding temperatures. The phase equilibria are to be measured based on the classic local equilibrium hypothesis, while the interdiffusion coefficients of the compound Nb_3Sn will be evaluated by using the HitDIC (High-throughput Determination of Interdiffusion Coefficients) software [24–27], based on the measured composition-distance profiles over the interdiffusion zones in Nb/Sn LSDCs.

2. Materials and Methods

The raw materials for preparing Nb/Sn LSDCs were Nb blocks (25 × 15 × 14 mm) with a purity of 99.99% (wt.%) and Sn particles (1~2 mm in diameter) with a purity of 99.99% (wt.%). The Nb block was fixed on the leveling workbench for blind hole processing. An Nb cap (25 × 15 × 3 mm) was first cut from the Nb block by means of the wire-electrode technique, and then a blind hole (17 × 9 × 6 mm) was dug by the drill bit in the center of the remaining Nb block. Considering that the high-speed rotation of the drill bit may generate a large amount of heat, causing the slight oxidation on the surface of the Nb block, it is thus necessary to polish the inner surface of the blind hole after machining to remove the oxide layer and level the inner surface. After that, Sn particles were filled into the hole, and the Nb block was welded and sealed with the Nb cap. The assembled Nb/Sn LSDC is schematically shown in Figure 1a.

Two identical Nb/Sn diffusion couples were put into the corundum boats equipped with quartz cotton. The corundum boats containing the samples were preheated for a few minutes at the furnace port and then subjected to annealing at 1100 °C for 2.33 h and 1200 °C for 2 h in a high-temperature vacuum tube furnace (GSL1700X, Hefei Kejing Materials Technology Co., Ltd., Hefei, China), respectively. There are three reasons for choosing the annealing temperatures as 1100 °C and 1200 °C. First, according to the assessed phase diagram [19], only three stable phases (bcc, liquid and Nb_3Sn) exist in the Nb-Sn binary system between 911 °C~2152 °C. Such simple phase equilibria are beneficial for the study on thermodynamic and diffusion properties related to the Nb_3Sn phase.

Second, no experimental studies on the phase equilibria in Nb-Sn binary system around 1100 °C. Third, the reported interdiffusion coefficients of Nb_3Sn phase only cover the low-temperature range (350 °C~800 °C), and more information on interdiffusion coefficients of Nb_3Sn phase at higher temperatures is highly needed. Moreover, the formation of Nb_3Sn phase was clearly observed in Nb-bronze multi-filamentary composites after annealing at 700 °C for only 1 h in the work of Ref. [28]. Therefore, the annealing time at 1100 °C and 1200 °C was chosen to be around 2 h for the formation of Nb_3Sn phase with sufficient thickness in Nb/Sn LSDCs. At both 1100 °C and 1200 °C, the Nb block keeps the solid state, while the Sn particles are melted into the liquid state. The schematic diagrams for top view and side view of Nb/Sn LSDC are displayed in Figure 1b,c. After annealing, the Nb/Sn LSDCs were quickly quenched in liquid nitrogen. Subsequently, each Nb/Sn LSDC was cut into five pieces, and metallographically polished for the subsequent analysis of interdiffusion zones over the Nb/Sn interface. The photos of the physical samples of the Nb/Sn LSDCs after quenching are presented in Figure 1d,e. The optical microscopy (OM), scanning electron microscope (SEM), electron probe micro-analysis (EPMA), and electron backscattered diffraction (EBSD) techniques were used to analyze the microstructure and composition of the Nb/Sn LSDCs.

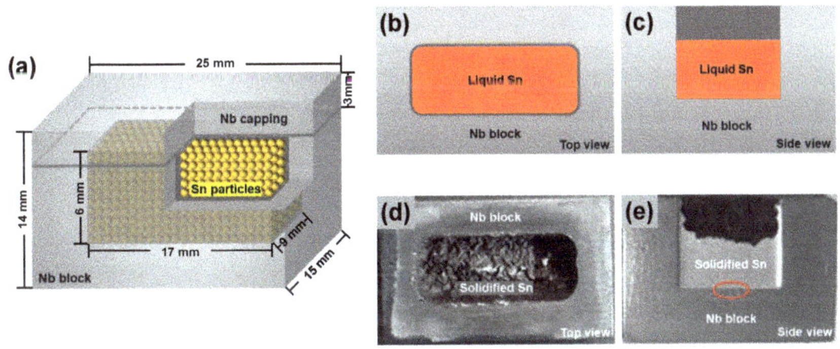

Figure 1. Geometry of Nb/Sn LSDC: (**a**) Schematic diagram for the assembled Nb/Sn LSDC; (**b**) top view and (**c**) side view of Nb/Sn LSDC during the heat treatment. Typical photos of physical sample for the Nb/Sn LSDC after annealing at 1200 °C for 2 h: (**d**) top view and (**e**) side view.

3. Results and Discussion

3.1. Phase Equilibria

According to the Nb-Sn binary phase diagram assessed by Toffolon [19], three phases are stable over the temperature range of 911~2152 °C, i.e., bcc, Nb_3Sn and liquid. The microstructure of the Nb/Sn LSDCs annealed at 1100 °C and 1200 °C taken from the position of the red circle in Figure 1e are respectively shown in Figure 2a,b. As can be observed in the figures, one apparent layer formed over the interdiffusion zone across the Nb/Sn interface at both 1100 and 1200 °C. The composition of the formed layer was measured by means of the EPMA technique, while its crystal structure (at point 1 in Figure 2b) was characterized by using the EBSD technique, as shown in Figure 3. Both EPMA and EBSD results indicated that the layer formed over interdiffusion zone is the Nb_3Sn phase, which corresponds well with the Nb-Sn phase diagram assessed by Toffolon [19]. Moreover, the average size of Nb_3Sn grains annealed at 1200 °C for 2 h was measured to be 4.36 ± 0.85 μm.

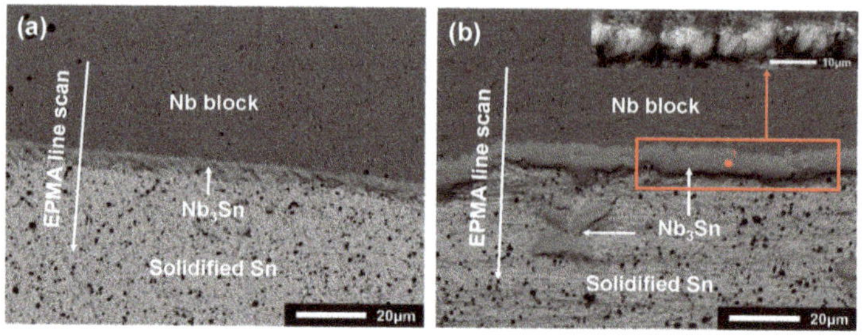

Figure 2. BSE images of Nb/Sn LSDCs annealed (**a**) at 1100 °C for 2.33 h, and (**b**) at 1200 °C for 2 h.

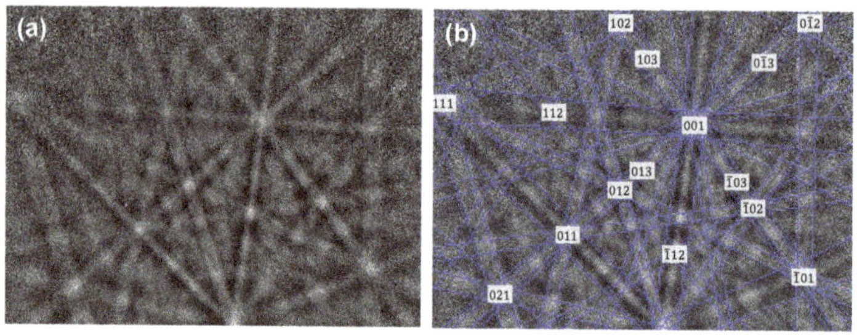

Figure 3. (**a**) EBSD pattern of Nb_3Sn phase and (**b**) its index calibration in Nb/Sn LSDC annealed at 1200 °C for 2 h. The pattern was acquired with the voltage of 20 kV, and the specimen tilting angle was set to 70°.

As also indicated in Figure 2, the EPMA scans of the compositions in Nb/Sn LSDCs across the entire diffusion zone perpendicularly to the diffusion interface were performed, and the obtained composition-distance profiles are displayed as scattered points in Figure 4a,b. Here, it should be noted that two EPMA runs were conducted for Nb/Sn LSDC at each temperature. Based on the measured composition-distance profiles, the average thickness of the Nb_3Sn phase layer can be evaluated to be 6.63 ± 1.86 μm at 1200 °C for 2 h and 3.43 ± 0.62 μm at 1100 °C for 2.33 h.

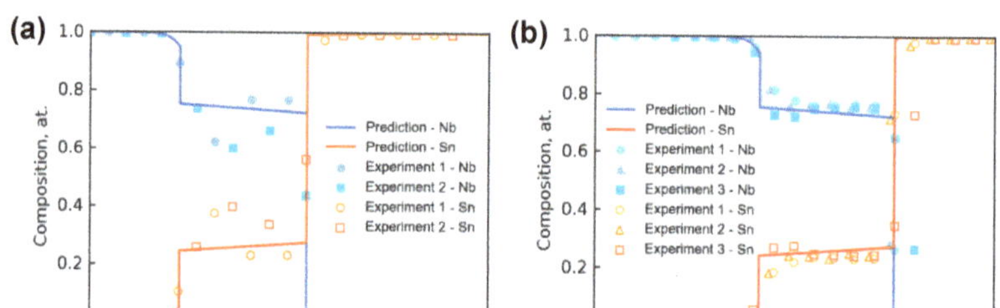

Figure 4. Experimental composition-distance profiles in the Nb/Sn LSDCs annealed (**a**) at 1100 °C for 2.33 h, (**b**) at 1200 °C for 2 h, compared with the corresponding model-predicted composition-distance profiles by HitDIC. Note that the caption "Experiment 1, 2 and 3" in the figures denotes the three separate measurements of composition profiles for each annealed Nb/Sn sample.

The measured concentration profiles of Nb/Sn LSDCs at two temperatures were then superimposed on the Nb-Sn binary phase diagram from Toffolon [19], as presented in Figure 5. It is worth mentioning that there should be no two-phase mixture regions in binary diffusion couples because they are thermodynamically forbidden. Thus, based on the local equilibrium hypothesis at the phase interfaces, the equilibrium compositions of phases could be obtained by extrapolating the measured concentration profiles to the phase interfaces. The concentration profiles of Nb_3Sn and liquid phases can be clearly seen in Figure 5, and their equilibrium compositions at 1100 and 1200 °C were determined and marked by solid circles in Figure 5 and summarized in Table 1. As can be clearly seen in the figure, the determined homogeneity range of Nb_3Sn compound at both 1100 °C and 1200 °C are in good agreement with the assessed phase equilibria by [19]. Moreover, it is astonishing to see a general agreement between the determined solubility limits of Nb in liquid (Sn) at both 1100 °C and 1200 °C and the assessed ones [19] exist, though there should be some further phase transformations from the remaining liquid to solids during quenching of LSDCs. For the solubility of Sn in (Nb) phase, the determined value at 1200 °C is in good consistency with the assessed one [19], as displayed in Figure 5. However, the solubility of Sn in (Nb) phase at 1100 °C cannot be accurately determined due to the relatively scattered composition profile of Nb/Sn LSDC at 1100 °C. Instead, a rough value between 3.1 at.% and 7.1 at.% was proposed for the solubility of Sn in (Nb) phase at 1100 °C, which is still close to the assessed one by [19]. Based on Figure 5 and the above analysis, the phase equilibria in binary Nb-Sn system at both 1100 °C and 1200 °C determined by the present LSDCs and predicted by thermodynamic assessment [19] should be reliable.

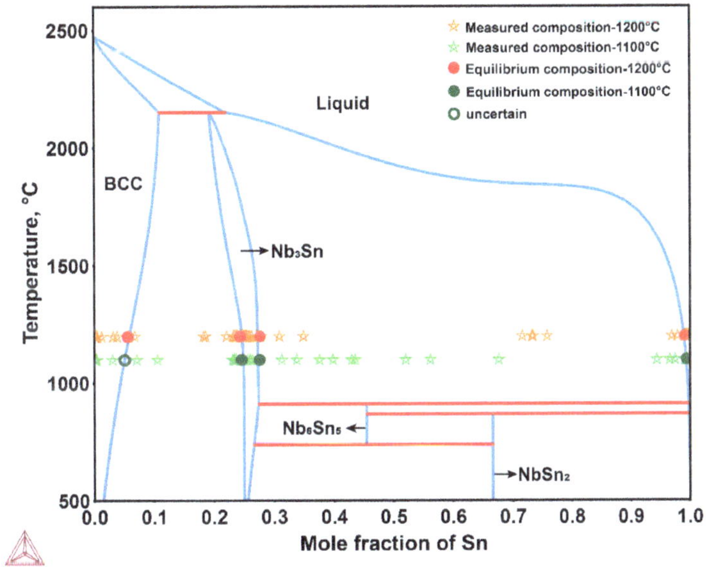

Figure 5. Comparison between the experimental composition profiles measured in this study and the Nb-Sn binary phase diagram calculated according to the thermodynamic parameters assessed by Toffolon [19].

Table 1. Equilibrium compositions for different phases measured from the Nb/Sn LSDCs.

Temperature (°C)	Equilibrium Phases		Equilibrium Compositions (at.% Sn)	
	Phase 1	Phase 2	Phase 1	Phase 2
1100	bcc	Nb_3Sn	3.1~7.1	24.6
	Nb_3Sn	liquid	27.6	99.4
1200	bcc	Nb_3Sn	5.7	24.4
	Nb_3Sn	liquid	27.6	99.1

3.2. Interdiffusion Coefficients

Based on the experimental composition-distance profiles yielded in the interdiffusion zones of Nb/Sn LSDCs, the interdiffusion coefficients were extracted using the latest version of HitDIC software [24–26] originally developed under the framework of pragmatic numerical inverse approach [29,30]. By simplifying the diffusion processes in reactive diffusion couple to be a one-dimensional moving boundary problem, the evolution of composition c_n for the n-th phase and interface position s_n between the n-th phase and the $(n + 1)$-th phase described as

$$\frac{\partial c_n}{\partial t} = \frac{\partial}{\partial r}\left(D_n \frac{\partial c_n}{\partial r}\right) \quad (1)$$

$$D_{n-1}\frac{\partial c_{n-1}}{\partial r}\bigg|_{r=s_n^-(t)} - D_n\frac{\partial c_n}{\partial r}\bigg|_{r=s_n^+(t)} = (c_{n+1} - c_n)\frac{ds_n}{dt} \text{ at } r = s_n(t) \quad (2)$$

where D_n are the interdiffusion coefficients to be recovered by means of the numerical inverse method. That is, with the setting of a diffusion couple, i.e., initial compositions of terminal alloys, diffusion time, and component-distance profile, an inverse routine based on optimization is used to iteratively evaluate a suggested set of interdiffusion coefficients for all the solution phases and intermetallic compounds. At each iteration, deviation between the prediction and experimental observations can be evaluated, while suggestions for an alternative set of interdiffusion coefficients may be provided according to the employed optimization methods, i.e., the genetic algorithm. The optimization processes can be terminated once the satisfactory prediction result towards the experimental observation is obtained, and the related set of interdiffusion coefficients can therefore be taken as the determined results. Among the calculation processes, the local equilibrium assumption is applied for adjacent phases in the reactive diffusion couple.

The measured composition-distance profiles were used to evaluate the interdiffusion coefficients for the Nb-Sn system at 1100 °C and 1200 °C, respectively. The predicted composition-distance profiles using HitDIC are also displayed as the solid lines in Figure 5. As can be seen in the plots, good consistency between the predicted composition-distance profiles and the experimental data exists for the two investigated temperatures. Here, only the interdiffusion coefficients of Nb_3Sn compound are to be evaluated. That is because (i) during the quenching, the liquid Sn near the interdiffusion zone and diffused Nb may form the $NbSn_2$ phase, resulting in the Sn content in the diffusion zone near the liquid side lower than the theoretical value, and (ii) the spatial resolution of EPMA makes the measured composition points close to (Nb) side insufficient for evaluating the interdiffusion coefficient of bcc phase especially at 1100 °C.

Figure 6a shows the presently extracted interdiffusion coefficients of Nb_3Sn from Nb/Sn LSDCs at 1100 °C and 1200 °C (i.e., 4.66×10^{-15} $m^2 \cdot s^{-1}$ and 1.85×10^{-14} $m^2 \cdot s^{-1}$), which generally follow the same trend as the low-temperature data from multi-layered thin-film composites and multi-filamentary composites [23]. Based on presently extracted interdiffusion coefficients at 1100 °C and 1200 °C, the pre-exponential factor and the activation energy for diffusion in Nb_3Sn compound can be obtained by fitting to the Arrhenius equation:

$$D = D_0 \exp\left(-\frac{Q}{RT}\right). \quad (3)$$

The obtained pre-exponential factor of the Nb_3Sn phase D_0 in Equation (4) is 1.80×10^{-6} $m^2 \cdot s^{-1}$, while the activation energy Q is 232.04 $kJ \cdot mol^{-1}$. The presently evaluated activation energy for Nb_3Sn compound shows very good agreement with the data from the kinetics measurements (i.e., 248.71 $kJ \cdot mol^{-1}$) by [31] and ab initio density functional theory calculations (i.e., 269.79 $kJ \cdot mol^{-1}$) by [32], validating the reliability of the presently obtained interdiffusion coefficients of Nb_3Sn compound.

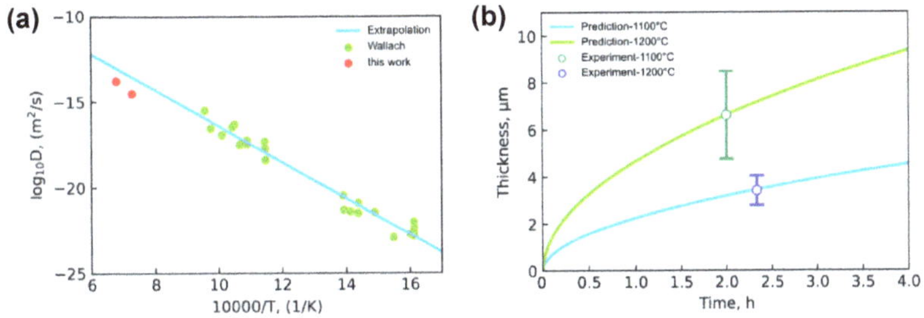

Figure 6. (**a**) Comparison between the presently evaluated interdiffusion coefficients of Nb_3Sn at 1100 °C and 1200 °C and the low-temperature data from Wallach [23]; (**b**) Model-predicted thickness of the Nb_3Sn layer as function of annealing time for the Nb/Sn LSDCs, compared with the experimental data in the present work.

Moreover, with interdiffusion coefficients calculated by HitDIC and the simultaneous diffusional growth of Nb_3Sn layer [33], the average thickness of the Nb_3Sn phase layer was determined as a function of the annealing time, which can be expressed as the following equation,

$$l = k\left(\frac{t}{t_0}\right)^n, \tag{4}$$

where l is the average thickness of the Nb_3Sn phase layer, t is the annealing time, and t_0 is the unit time. As shown in Figure 6b, we obtained k = 2.28 μm and n = 0.51 for the Nb/Sn LSDC annealed at 1100 °C, and k = 4.69 μm and n = 0.51 for the Nb/Sn LSDC annealed at 1200 °C, with the experimental thickness data locating precisely on the predicted profile. Such a fact indicates that the growth of the Nb_3Sn phase layer is controlled by diffusion.

It should be noted in Figure 6a that the interdiffusion coefficients of Nb_3Sn obtained in this work are a bit lower than the values extrapolated from low temperature data [23]. To explore the probable reasons for this case, the interface of the Nb/Sn liquid-solid reactive diffusion couple was further characterized by experiments. The element mapping characteristics over the diffusion zone were analyzed using the EPMA map analysis, as shown in Figure 7. The result shows that the Nb_3Sn phase layer formed across the Nb/Sn interface, while some free Nb_3Sn particles existed in the (Sn) region. The reason is considered that the Nb element diffused into melted Sn during the annealing process, and some Nb_3Sn particles formed and remained in the solidified Sn region after the quench. There are also some lath-like microstructures distributing in the solidified (Sn) region, and that might be due to the composition segregation in solidified liquid that resulted in the formation of the $NbSn_2$ phase. Because of the peritectic reactions, L + Nb_3Sn→$NbSn_2$ and L + $NbSn_2$→(Sn), a fraction of generated Nb_3Sn phase reacted with the remained liquid, which can make the measured thickness of the Nb_3Sn phase layer smaller than the actual one, thus lowering the interdiffusion coefficients of the Nb_3Sn phase.

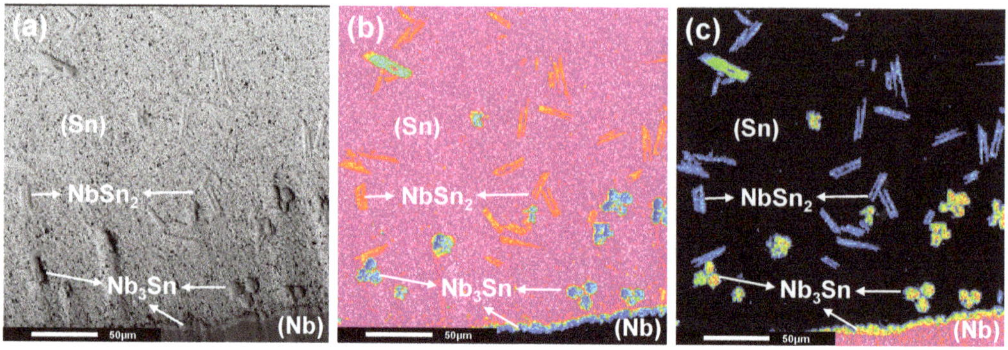

Figure 7. (**a**) BSE image, (**b**) mapping distribution of Sn, and (**c**) mapping distribution of Nb of Nb/Sn LSDC annealed at 1200 °C for 2 h.

4. Conclusions

A simple but pragmatic approach was proposed to fabricate the Nb/Sn LSDCs, which were subjected to annealing at 1100 °C and 1200 °C for 2.33 h and 2 h, respectively. It was found that apparent interdiffusion zones formed across the Nb/Sn interface in both LSDCs. The interdiffusion zone was then comprehensively characterized by EPMA and EBSD techniques. Both composition and crystal structure results confirm that the formed layer at the Nb/Sn interface is the Nb_3Sn phase.

The equilibrium homogeneity range of Nb_3Sn, solubility limits of (Nb) and liquid phases at 1100 °C and 1200 °C were determined, and the results are in general agreement with the Nb-Sn binary phase diagram assessed by [19].

The interdiffusion coefficients of the Nb_3Sn phase at 1100 °C and 1200 °C were evaluated by HitDIC software to be 3.22×10^{-15} $m^2 \cdot s^{-1}$ and 1.73×10^{-14} $m^2 \cdot s^{-1}$, respectively. The evaluated diffusion properties of compound Nb_3Sn, including the interdiffusion coefficients and activation energy, are consistent with the corresponding experimental and theoretical values in the literature. Moreover, the relation between the model-predicted thickness of the Nb_3Sn layer and the annealing time of Nb/Sn LSDCs was also observed to be in good agreement with the experimental data, indicating the growth of Nb_3Sn layer in Nb/Sn LSDCs is diffusion controlled.

The presently proposed approach for preparing Nb/Sn LSDCs is anticipated to be a universal one for various alloy systems with large differences in melting points. Furthermore, combined with the local equilibrium hypothesis and HitDIC software, the corresponding phase equilibria and diffusion coefficients can be accurately determined simultaneously.

Author Contributions: Conceptualization, L.Z.; methodology, L.Z., Q.L.; investigation, J.Z. (Jiali Zhang) and J.Z. (Jing Zhong); data curation, J.Z. (Jiali Zhang); Writing—original draft preparation, J.Z. (Jiali Zhang); Writing—review and editing, J.Z. (Jing Zhong), L.Z. and Q.L.; visualization, J.Z. (Jiali Zhang); supervision, L.Z.; project administration, L.Z. All authors have read and agreed to the published version of the manuscript.

Funding: This work was financially supported by Natural Science Foundation of Hunan Province (Grant No.: 2021JJ10062), China. Jiali Zhang acknowledges the financial support by the Fundamental Research Funds for the Central Universities of Central South University (Fund No. 2021ZZTS0583), China.

Institutional Review Board Statement: Not applicable.

Informed Consent Statement: Not applicable.

Data Availability Statement: Data sharing not applicable.

Acknowledgments: The authors would like to thank Jianbao Gao, Wang Yi, and Guangchen Liu from Central South University for their kind help on fixing the problems during the experimental tests.

Conflicts of Interest: The authors declare no conflict of interest.

References

1. Larbalestier, D.; Gurevich, A.; Feldmann, D.M.; Polyanskii, A. High-Tc superconducting materials for electric power applications. *Nature* **2001**, *414*, 368–377. [CrossRef] [PubMed]
2. Foner, S. High-field magnets and high-field superconductors. *IEEE Trans. Appl. Supercond.* **1995**, *5*, 121–140. [CrossRef]
3. Shikov, A.K.; Panstsyrnyi, V.I.; Vorob'Eva, A.; Dergunova, E.A.; Sud'Ev, S.; Mareev, K.A.; Belyakov, N.A.; Abdyukhanov, I.M.; Sergeev, V.V. Microstructure and properties of Nb_3Sn superconductors for the magnet system of the international thermonuclear experimental reactor. *Met. Sci. Heat Treat.* **2004**, *46*, 504–513. [CrossRef]
4. Wang, R.; Luo, Q. A15 structure. *Cryog. Supercond.* **1986**, *14*, 57–65.
5. Lu, Z.; Zhang, L. Thermodynamic description of the quaternary Al-Si-Mg-Sc system and its application to the design of novel Sc-additional A356 alloys. *Mater. Des.* **2017**, *116*, 427–437. [CrossRef]
6. Zhong, J.; Chen, L.; Zhang, L. High-throughput determination of high-quality interdiffusion coefficients in metallic solids: A review. *J. Mater. Sci.* **2020**, *55*, 10303–10338. [CrossRef] [PubMed]
7. Barsoum, M.W.; Ganguly, A.; Seifert, H.J.; Aldinger, F. The 1300 °C isothermal section in the Nb–Sn–C ternary phase diagram. *J. Alloy. Compd.* **2002**, *337*, 202–207. [CrossRef]
8. Adam, E.; Beischer, P.; Marancik, W.; Young, M. Fabrication and properties of niobium-tin (Nb_3Sn) tapes by the "bronze process": Application to superconducting power transmission lines. *IEEE Trans. Magn.* **1977**, *13*, 425–428. [CrossRef]
9. Gregory, E.; Gulko, E.; Pyon, T. Development of Nb_3Sn wires made by the Internal-Tin Process. In *Advances in Cryogenic Engineering Materials*; Balachandran, U.B., Gubser, D.G., Hartwig, K.T., Reed, R.P., Warnes, W.H., Bardos, V.A., Eds.; Springer: Boston, MA, USA, 1998; Volume 44, pp. 903–909.
10. Ghosh, A.K.; Cooley, L.D.; Moodenbaugh, A.R.; Parrell, J.A.; Field, M.B.; Zhang, Y.; Hong, S. Magnetization studies of high J_c Nb_3Sn strands. *IEEE Trans. Appl. Supercond.* **2005**, *15*, 3494–3497. [CrossRef]
11. Qiu, Q.; Suo, H.; Cheng, J.; Zhang, Z.; Ji, Y.; Wang, Q. Fabrication and superconducting properties of Nb_3Sn superconducting bulks. *Chin. J. Rare Met.* **2020**, *44*, 57–62.
12. Ellis, T.G.; Wilhelm, H.A. Phase equilibria and crystallography for the niobium-tin system. *J. Less-Common Met.* **1964**, *7*, 67–83. [CrossRef]
13. van Vucht, J.H.N.; van Ooijen, D.J.; Bruning, H.A.C.M. Some investigations of the niobium-tin phase diagram. *Philips Res. Rep.* **1965**, *20*, 136–161.
14. Charlesworth, J.P.; Macphail, I.; Madsen, P.E. Experimental work on the niobium-tin constitution diagram and related studies. *J. Mater. Sci.* **1970**, *5*, 580–603. [CrossRef]
15. Matstakova, A.; Lazarev, B. Peculiarities of the constitution phase diagram of the system niobium-tin. *Phys. Met. Metallogr.* **1973**, *35*, 133–141.
16. Schiffman, R.A.; Bailey, D.M. Thermodynamics of the incongruently subliming niobium-tin system. *High Temp. Sci.* **1982**, *15*, 165–177.
17. Massalski, T.B.; Murray, J.L.; Bennett, L.H.; Baker, H. *Binary Alloy Phase Diagrams*, 2nd ed.; American Society for Metals Research: Geauga, OH, USA, 1986.
18. Toffolon, C.; Servant, C.; Sundman, B. Thermodynamic assessment of the Nb-Sn system. *J. Phase Equilib.* **1998**, *19*, 479. [CrossRef]
19. Toffolon, C.; Servant, C.; Gachon, J.; Sundman, B. Reassessment of the Nb-Sn system. *J. Phase Equilib.* **2002**, *23*, 134. [CrossRef]
20. Vieland, L.J. High-temperature phase equilibrium and superconductivity in the system niobium-tin. *RCA Rev.* **1964**, *25*, 366–378.
21. Gao, J.; Zhong, J.; Liu, G.; Yang, S.; Song, B.; Zhang, L.; Liu, Z. A Machine Learning Accelerated Distributed Task Management System (Malac-Distmas) and Its Application in High-Throughput CALPHAD Computations Aiming at Efficient Alloy Design. *Adv. Powder Mater.* 2021, in press. [CrossRef]
22. Hansen, M.; Shunk, F.A. *Constitution of Binary Alloys*, 2nd ed.; McGraw-Hill Book Company: New York, NY, USA, 1969.
23. Wallach, E.; Evetts, J. The development of microstructure in multifilamentary bronze route A15 composites. In *Advances in Cryogenic Engineering Materials*; Gubser, D.G., Hartwig, K.T., Reed, R.P., Warnes, W.H., Bardos, V.A., Balachandran, B., Eds.; Springer: Berlin/Heidelberg, Germany, 1986; pp. 911–923.
24. Zhong, J.; Chen, W.; Zhang, L. HitDIC: A free-accessible code for high-throughput determination of interdiffusion coefficients in single solution phase. *Calphad* **2018**, *60*, 177–190. [CrossRef]
25. Zhong, J.; Zhang, L.; Wu, X.; Chen, L.; Deng, C. A novel computational framework for establishment of atomic mobility database directly from composition profiles and its uncertainty quantification. *J. Mater. Sci. Technol.* **2020**, *48*, 163–174. [CrossRef]
26. Zhong, J.; Chen, L.; Zhang, L. Automation of diffusion database development in multicomponent alloys from large number of experimental composition profiles. *npj Comput. Mater.* **2021**, *7*, 35–47. [CrossRef]
27. HitDIC: HIgh-Throughput Determination of Interdiffusion Coefficients. Available online: https://hitdic.com/ (accessed on 16 December 2021).
28. Shi, F. Calculation formula of multi-core Nb_3Sn layer thickness. *Min. Metall. Eng.* **1981**, *4*, 66–68.

29. Chen, W.; Zhang, L.; Du, Y.; Tang, C.; Huang, B. A pragmatic method to determine the composition-dependent interdiffusivities in ternary systems by using a single diffusion couple. *Scr. Mater.* **2014**, *90–91*, 53–56. [CrossRef]
30. Chen, W.; Zhong, J.; Zhang, L. An augmented numerical inverse method for determining the composition-dependent interdiffusivities in alloy systems by using a single diffusion couple. *MRS Commun.* **2016**, *6*, 295–300. [CrossRef]
31. Agarwal, S.; Narlikar, A. Ordering of Nb_3Sn layer formed in the bronze process-I. *Solid State Commun.* **1985**, *55*, 563–566. [CrossRef]
32. Besson, R.; Guyot, S.; Legris, A. Atomic-scale study of diffusion in A15 Nb_3Sn. *Phys. Rev. B Condens. Matter Mater. Phys.* **2007**, *75*, 1–7. [CrossRef]
33. Ma, S.; Xing, F.; Zhang, L. A novel analytical approach to describe the simultaneous diffusional growth of multilayer stoichiometric compounds in binary reactive diffusion couples. *Scr. Mater.* **2020**, *191*, 111–115. [CrossRef]

materials

Article

Thermal Processing Map and Microstructure Evolution of Inconel 625 Alloy Sheet Based on Plane Strain Compression Deformation

Yuelin Song [1], Jiangkun Fan [1,2,3,*], Xudong Liu [1], Peizhe Zhang [1] and Jinshan Li [1,2]

1 State Key Laboratory of Solidification Processing, Northwestern Polytechnical University, Xi'an 710072, China; 2020200843@mail.nwpu.edu.cn (Y.S.); liu_xudong@mail.nwpu.edu.cn (X.L.); zhangpeizhexgd@163.com (P.Z.); ljsh@nwpu.edu.cn (J.L.)
2 Innovation Center, NPU·Chongqing, Chongqing 401135, China
3 National & Local Joint Engineering Research Center for Precision Thermoforming Technology of Advanced Metal Materials, Xi'an 710072, China
* Correspondence: jkfan@nwpu.edu.cn

Citation: Song, Y.; Fan, J.; Liu, X.; Zhang, P.; Li, J. Thermal Processing Map and Microstructure Evolution of Inconel 625 Alloy Sheet Based on Plane Strain Compression Deformation. *Materials* **2021**, *14*, 5059. https://doi.org/10.3390/ma14175059

Academic Editor: Frank Czerwinski

Received: 4 August 2021
Accepted: 31 August 2021
Published: 3 September 2021

Abstract: Plane strain compression tests were used to study the deformation behavior of an Inconel 625 alloy sheet at various temperatures and strain rates. The peak stress was selected to establish the constitutive equation, and the processing maps under different strains were drawn. The results show that the effective stress–strain curve of Inconel 625 has typical dynamic recrystallization (DRX) characteristics. With the increasing deformation temperature and the decreasing strain rate, the softening effect is significantly enhanced. The parameters of the constitutive equation are calculated, and the average error of the constitutive equation is 5.68%. Through the analysis of the processing map, a deformation temperature of 950–960 °C with a strain rate of 0.007–0.05 s^{-1} were determined as the unstable region, and obvious local plastic-rheological zones were found in the unstable region. The optimum deformation condition was found to be 1020–1060 °C/0.005–0.03 s^{-1}. Through electron backscattered diffraction (EBSD) characterization, it was found that both the increase of temperature and the decrease of strain rate significantly promote the recrystallization process. At a low strain rate, the main recrystallization mechanism is discontinuous dynamic recrystallization (DDRX). It is expected that the above results can provide references for the optimization of the rolling process and microstructure control of an Inconel 625 alloy sheet.

Keywords: Inconel 625; constitutive equation; processing map; recrystallization; microstructure evolution

1. Introduction

Due to the stable matrix elements and the large extent of the γ'' phase, Inconel 625 has excellent high-temperature strength and structural stability [1–4]. In addition, Inconel 625 has good machinability and weldability [5,6]. Therefore, Inconel 625 can be easily processed into various components such as plates, bars, pipes, wires, and strips [7]. However, due to its complex composition, strong deformation resistance, and narrow range of hot-forming parameters, it is easy to segregate during preparation, and there are some problems in rolling processes, such as uneven distribution of stress or temperature, and fracture [8].

In order to solve various problems in processing, some scholars have studied the deformation behavior of Ni-based superalloys recently [9–12]. Chen et al. [13] studied the hot-deformation behavior of GH4169 under different strain modes. They found that when the strain rate remains high in the first stage and then becomes low in the second stage, the true stress decreases with the sudden decrease of the strain rate. Compared with the equivalent constant strain rate, there is no difference in the final stress. Both constant-strain-rates and changed-strain-rates GH4169 have the same recrystallization mechanism, while the recrystallization grain size and volume fraction of the former is between those of the latter. Jia and Tao et al. have carried out uniaxial compression and high-speed compression experiments on cylindrical Inconel 625 respectively [14,15]. Under different temperatures and strain rates, the real stress–strain curve shows an obvious steady-state flow rule, and the adiabatic heating effect decreased with decreasing temperature and increasing strain rate. Moreover, many new models have been developed to describe the high-temperature rheological behavior of nickel base alloys, such as the improved Arrhenius model [16], constitutive model based on physical mechanism [17], artistic neural network (ANN) model [18], deep belief network (DBN) model [19], multigene genetic programming (MGGP) [20], and so on. All of those have good prediction effects with different accuracy.

Some scholars have also studied the microstructure evolution and recrystallization behavior of the alloy [21–23]. By combining T-EBSD with transmission electron microscopy (TEM), Sun Fei [24] accurately studied the microstructure evolution of recrystallization process of U720Li disc superalloy. They found that with the increasing rate of strain, the slip dislocations tended to gather into the wall and form subgrain boundaries. Then, the dislocation wall was connected with the local primary γ' and combined to form subgrains. Kumar et al. [25] found that as for the advanced P/M Ni-based superalloy, the DRX reached the maximum at a higher deformation temperature. While at a lower deformation temperature, the DRX was limited and the structure highly deformed. Sun et al. [26] developed the effect of microstructure on recrystallization evolution of Inconel 625. Through high-resolution EBSD and transmission electron microscopy, they found that the δ phase can promote the progress of CDRX. In addition, Lamellar carbides (NbC) have a pinning effect on dislocations, which limits the growth of DRX grains in DRX process. Jiang et al. [27] simulated the microstructure evolution of the 690 alloy pipe-forming process by using the finite element method (FEM). The results show that the established numerical model can accurately simulate the dynamic recrystallization, subdynamic recrystallization, and grain-growth behavior of the 690 alloy pipe-forming process. Sun et al. [28] developed a model for predicting the microstructure evolution of nickel base alloy during hot deformation by 2D cellular automaton (CA). This method uses deterministic or stochastic element evolution rules, which are not limited in time and space and do not need to establish complex differential equations. The method can easily simulate the nucleation position, orientation, and growth process of grains, and provides great help for the study of recrystallization.

At present, the research on the hot deformation of nickel base superalloys mainly focuses on the uniaxial compression of cast cylindrical specimens, but there is little on the plane strain compression of rolled plates. Limited by the single deformation mode of the former, it is difficult to reflect the deformation characteristics of the alloy in multiple modes, and it is even more difficult to optimize the processing technology. With the increasing demand for plate and strip, the simulation of rolling behavior has become a part that cannot be ignored. However, the cylinder compression experiment cannot reflect the deformation behavior of the actual sheet. Due to the effect of friction, it is prone to uneven deformation and a "bulging" phenomenon. In addition, when the strain is large, it may lead easily to the abnormal increase of flow stress, which has an impact on the measurement of flow stress and the observation of microstructure [29]. In contrast, the geometric profile of the deformation zone during plane strain compression is closer to that of flat rolling. This method can well reflect the deformation state, heat conduction, or other information in the rolling process, and the measurement of rheological stress is more accurate [30]. In order to study the rolling deformation behavior and microstructure evolution of the Inconel 625 plate more accurately and systematically, the plane strain compression of a rolled rectangular plate at minimal strain rate was carried out, and some novel results were obtained. This complements the related work of observing a rolled Inconel 625 plate under different compression conditions and minimal strain-rate range. It is conducive to a more accurate and in-depth understanding of the rolling process of Inconel 625 sheets, which is of great significance for guiding production, reducing cost, and subsequent research.

2. Experimental

2.1. Plane Strain Isothermal Compression Test

An Inconel 625 industrial hot-rolled sheet was selected in this experiment, and its chemical composition is shown in Table 1.

Table 1. Chemical composition of the Inconel 625 used in this work (wt.%).

Ni	Cr	Mo	Nb	Fe	Si	Al	Ti	Mn	C	S
61.00	21.50	9.00	3.60	2.00	0.20	0.20	0.20	0.20	0.05	0.001

After annealing at 1150 °C for 50 min, the sheet was processed into a rectangular specimen with a size of 20 mm × 15 mm × 10 mm, and the original microstructure of the sample is shown in Figure 1. It can be seen that the annealed sample contains a large number of equiaxed and uneven grains, accompanied by the precipitation of white Nb-rich particles, and that twinning occurs in the grain. The plane strain compression test was carried out using a Thermecmastor-Z dynamic thermomechanical simulation testing machine (Fuji Electronics Industry Co., Ltd., Osaka, Japan). This equipment uses high-frequency heating and electric heating to heat the sample at the same time, and the thermal deformation is controlled by oil pressure. The data were collected by computer and program controller, and then the real stress–strain curve and processing map were drawn using the Origin 2021b software (OriginLab Corporation, Northampton, MA, USA). The hot-pressing experiments were carried out at 950–1100 °C and strain rates of 0.001–0.05 s^{-1}, and the strain of all samples was 0.7. The compression equipment and compression mode of this experiment are shown in Figure 2, where RD, TD, and ND represent rolling direction, transverse direction, and normal direction. In the actual compression process, the width of the platen is narrow, and there is graphite gasket lubrication between the platen and the sample. Therefore, the compressive stress P and strain ε_y in plane strain compression can be converted into effective stress σ and strain ε, according to Equations (1) and (2).

$$\sigma = \frac{\sqrt{3}}{2}P \quad (1)$$

$$\varepsilon = \frac{2\sqrt{3}}{3}\varepsilon_y \quad (2)$$

Figure 1. The original microstructure of annealed Inconel 625 characterized by BSE.

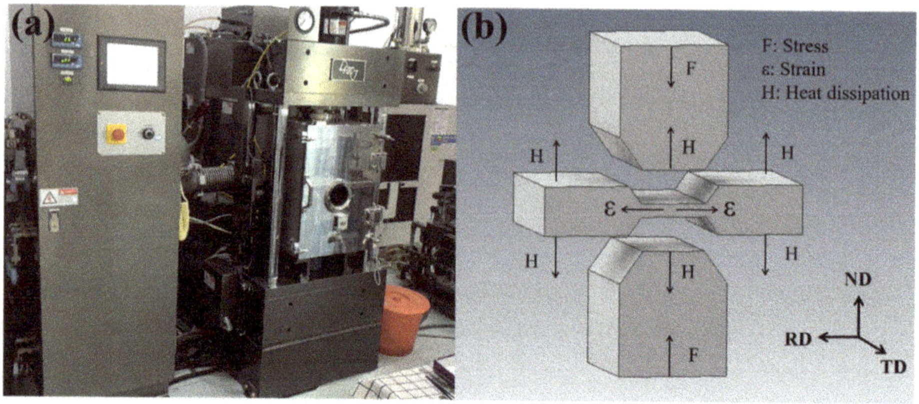

Figure 2. Experimental instruments and compression mode: (**a**) Thermecmastor-Z dynamic thermomechanical simulation testing machine, (**b**) schematic diagram of plane strain compression.

2.2. Microstructure Analysis

Firstly, the 80 # SiC sandpaper was used to remove the scratch and heat treatment oxide scale on the surface of the sample. Then Samples were ground up to 2000 # sandpaper and polished on an MP-1A metallographic polishing machine (Caikon Optical Instrument Co., Ltd., Shanghai, China), in which the disc was diameter 230 mm and the maximum power was 400 W. After ultrasonic cleaning, vibration polishing was carried out on a Buehler Vibromet2 polisher (Buehler, Lake Bluff, IL, USA) for 8 h. As Inconel 625 is a corrosion-resistant alloy, aqua regia was prepared as the corrosion solution, in which the ratio of hydrochloric acid and nitric acid was 3:1. After corrosion, the sample was cleaned with anhydrous ethanol and observed on an Olympus GX71 metallurgical microscope (Olympus Corporation, Tokyo, Japan).

The microstructure of the Inconel 625 was observed and analyzed by a ZEISS Gemini 500 field emission scanning electron microscope (Carl Zeiss Jena, Oberkochen, Germany), and the effect of thermal pressure deformation on tissue grain size was investigated by characterization with EBSD. Among the characteristics, the accelerating voltage was 20 kV and the step size was 0.7 μm. Finally, the results were analyzed using the Channel 5 5.0.9.0 software (Oxford Instruments, Oxford, UK).

3. Results and Discussion

3.1. The Effective Stress–Strain Curves and Flow Behavior

Figure 3 shows the effective stress–strain curves of Inconel 625 under different hot-pressing parameters. In the initial stage of alloy deformation, the stress increases in a nearly straight line and reaches the peak value rapidly. This phenomenon indicates that an obvious work-hardening effect occurs during the deformation process. Significantly, at low strain rate, the curve appeared to have jagged fluctuations after reaching the peak value. Meanwhile at high strain rate, the curve still rises gently after yielding until it enters a stable flow stage. The main reason for this is that at low strain rate, dynamic strain aging occurs. That is, many interstitial atoms gather near the dislocation to form a Cottrell atmosphere, which plays the role of pinning the dislocation. When the dislocation starts to move, the stress decreases. This process repeats continuously, which results in serrated undulations in the stress–strain curve [12,31]. At the initial stage of deformation, the hardening effect plays a dominant role and the flow stress increases. When the stress reaches the peak value, the DRX of Inconel 625 is obviously strengthened, and dynamic softening begins to take the lead. The flow stress gradually decreases and then remains stable. At this time, the dynamic balance between work hardening and dynamic softening is maintained.

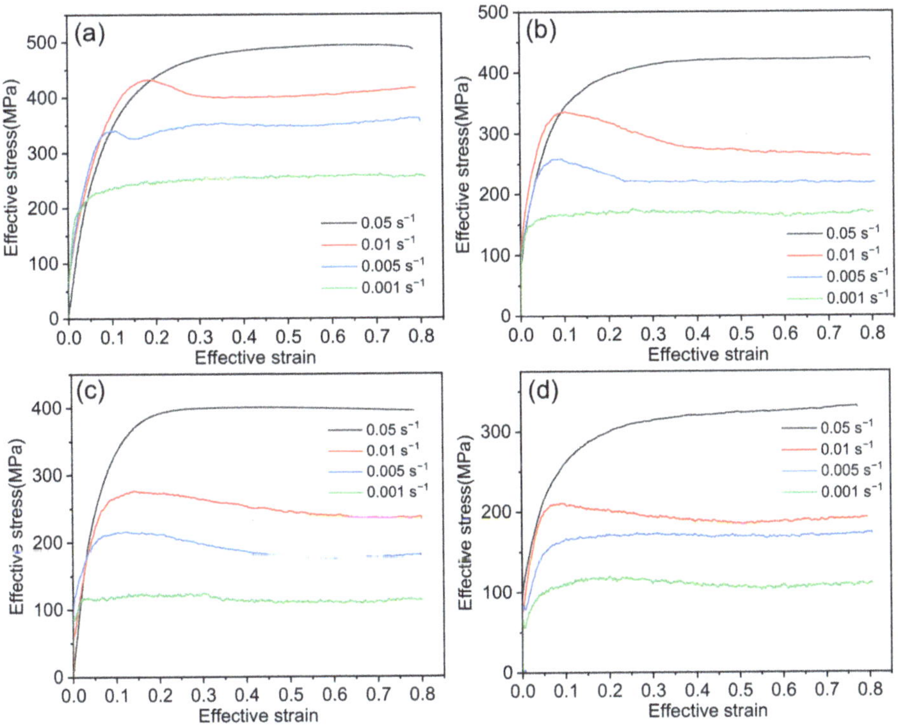

Figure 3. Stress–strain curves of Inconel 625 of 0.7 at (**a**) 950 °C, (**b**) 1000 °C, (**c**) 1050 °C, (**d**) 1100 °C.

Additionally, the peak stress decreases with increasing strain rate. It is well-known that flow stress is essentially an interactive system of work hardening and recrystallization softening. At the same deformation temperature, with the increasing strain rate, the time needed to reach the same strain will be shorter. The recrystallized grains do not have enough time to nucleate and grow, and the DRX process is incomplete, thus the softening effect is relatively reduced. At the same time, the rate of dislocation formation is also increasing, and the high density of dislocations causes dislocation movement to be impeded. In comparison, the work-hardening effect is relatively enhanced, and the flow stress increases continuously. On the curve, it is shown as an increase in the peak stress.

In addition, it can be observed that with the increasing of deformation temperature, the peak stress decreases gradually. This is because, at the same strain rate, the diffusion rate of grain boundary increases with deformation temperature. Under the same deformation conditions, more dislocation will be activated, which promotes the recrystallization softening process. As a result, the peak stress is constantly reduced. Under the same strain rate, the degree of work hardening of the alloy is approximately the same, and the DRX behavior will soften the material [14]. Therefore, in a certain temperature range, a rise in temperature can promote the DRX process of Inconel 625.

3.2. The Establishment and Solution of Constitutive Model

It can be seen from the above that the flow stress is closely related to the deformation temperature and the strain rate $\dot{\varepsilon}$. Through the constitutive equation of materials, we can know the relationship of thermal deformation parameters in the process of thermal deformation, such as flow stress, deformation temperature, strain rate, and strain [32–34]. Therefore, the constitutive model can predict the rheological behavior of Inconel 625, and provide theoretical support for selecting suitable processing equipment and forming parameters. Combining with the modified Arrhenius equation of hyperbolic sine form, we usually use the Zener–Hollomon parameter (the temperature compensation deformation rate factor) proposed by Sellars and Tegart to describe the relationship between thermal deformation parameters.

$$Z = \dot{\varepsilon}exp(Q/RT) = A_1\sigma^n \quad (\alpha\sigma \leq 0.8) \tag{3}$$

$$Z = \dot{\varepsilon}exp(Q/RT) = A_2exp(\beta\sigma) \quad (\alpha\sigma \geq 1.2) \tag{4}$$

$$Z = \dot{\varepsilon}exp(Q/RT) = A[sinh(\alpha\sigma)]^n \ (Full\ stress\ level) \tag{5}$$

$$\alpha = \beta/n \tag{6}$$

where σ represents the flow stress (MPa), $\dot{\varepsilon}$ represents the strain rate (s^{-1}), $\boldsymbol{R}$ is the molar gas constant, 8.314 J/(mol*K), and T is the thermodynamic temperature (K). When α, N, A, and Q are known, the above equation can be used to describe the rheological properties of Inconel 625 during hot deformation. Through the study of thermal deformation data, the relationship between σ and Z can be described by exponential relation at a low stress level, such as Equation (3). Meanwhile, the relationship between σ and Z can be described by power exponential relation at a high stress level, such as Equation (4). In the actual production, the hot deformation of Inconel 625 is under the full stress level, and the relationship between σ and Z is shown in Equation (5).

McQueen et al. [35] pointed out that for the material whose softening behavior is mainly DRX, the peak stress (σ_p) or steady-state stress (σ_s) is usually selected as σ for calculation and analysis. From the above analysis, Inconel 625 shows obvious DRX characteristics during hot deformation, so σ in the constitutive equation is marked σ_p. The peak stress under different deformation temperature and strain rate is shown in Figure 4.

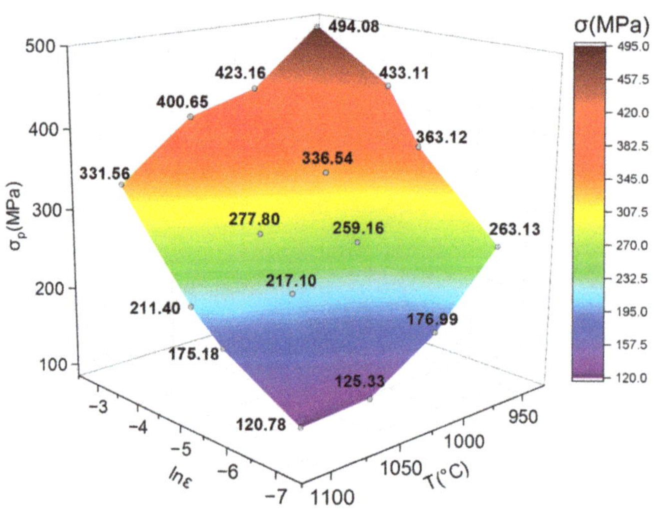

Figure 4. The peak stress at different deformation temperature and strain rate.

According to the flow stress model of the material during plastic deformation at high temperature, the correlation constants n_1, β, α can be solved. Take the logarithm for both sides of Equations (3) and (4):

$$ln\dot{\varepsilon} = lnA_1 - \frac{Q}{RT} + n_1 ln\sigma \tag{7}$$

$$ln\dot{\varepsilon} = lnA_2 - \frac{Q}{RT} + \beta\sigma \tag{8}$$

It can be seen from the above equation that there is an obvious linear relationship between $ln\dot{\varepsilon}$ and n_1, β. Plotting with $ln\dot{\varepsilon}$-$ln\sigma$ and $ln\dot{\varepsilon}$-σ as the coordinate axis, we can get Figure 5a,b.

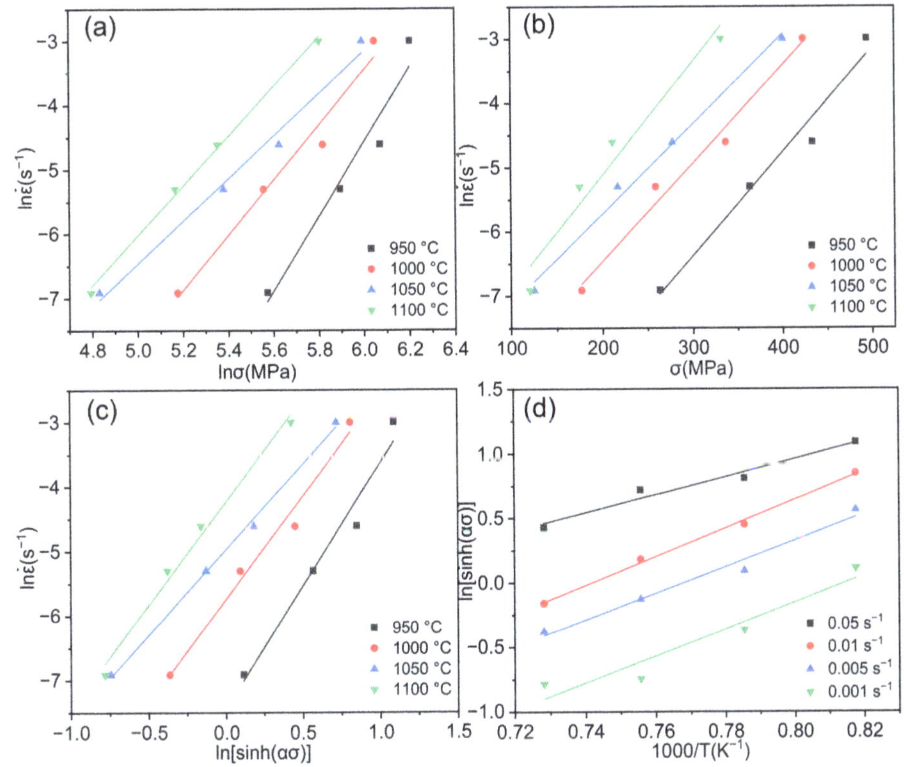

Figure 5. Relationship diagram of thermal deformation parameters. (**a**) $ln\dot{\varepsilon}$ and $ln\sigma$, (**b**) $ln\dot{\varepsilon}$ and σ, (**c**) $ln\dot{\varepsilon}$ and $ln[sinh(\alpha\sigma)]$, (**d**) $ln[sinh(\alpha\sigma)]$ and $1000/T$.

Then the slope is obtained by linear regression with the least square method. Calculating and taking the average, we find that β = 0.016, n_1 = 4.31, Then we can get that $\alpha = \beta/n_1$ = 0.016/4.31 = 0.0037.

For full stress level, the Equation (5) can also be expressed as:

$$\dot{\varepsilon} = A[sinh(\alpha\sigma)]^n exp\left(-\frac{Q}{RT}\right) \tag{9}$$

After obtaining the logarithm on both sides of Equation (9), the partial differential can be expressed as:

$$Q = R\{\frac{\partial ln\dot{\varepsilon}}{\partial ln[sinh(\alpha\sigma)]}\}_t \{\frac{\partial ln[sinh(\alpha\sigma)]}{\partial\left(\frac{1}{T}\right)}\}_{\dot{\varepsilon}} \tag{10}$$

It can be seen from Equation (10) that when the deformation temperature is fixed, $ln\dot{\varepsilon}$ and $ln[sinh(\alpha\sigma)]$ have a linear relationship, so let n_2 be the slope of this relationship. When the strain rate is constant, $ln[sinh(\alpha\sigma)]$ has a linear relationship with T^{-1}, and let K be the slope of the relationship. The above is represented by Figure 5c,d.

After calculating, n_2 = 3.21, K = 9.67, and $\boldsymbol{Q} = \boldsymbol{R} \times n_2 \times K$ =8.314 × 3.21× 9.67 = 258.22 kJ/mol.

Take the logarithm on both sides of Equation (5) to get:

$$lnZ = ln\dot{\varepsilon} + \frac{Q}{RT} \quad (11)$$

$$lnZ = lnA + nln[sinh(\alpha\sigma)] \quad (12)$$

From (12), we can see that there is a linear relationship between lnZ and $ln[sinh\ (\alpha\sigma)]$, and lnA is the intercept of this line. Calculate lnZ by Equation (11), and use the least square method and linear regression method to draw the corresponding linear relationship diagram as shown in Figure 6.

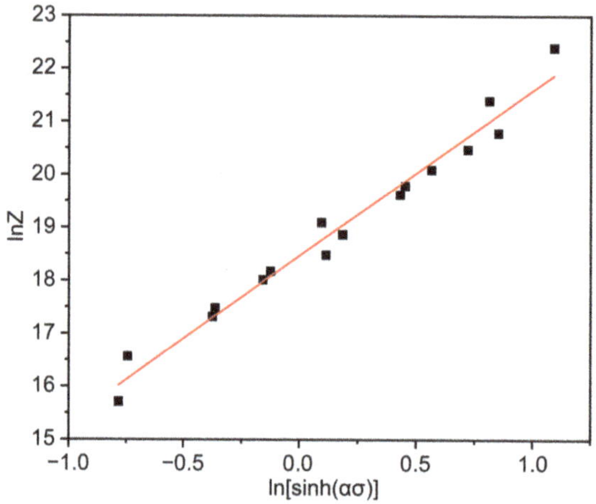

Figure 6. Relationship diagram of thermal deformation parameters lnZ and $ln[sinh(\alpha\sigma)]$.

According to its intercept, we can get lnA = 18.48, $A = 1.06 \times 10^8$.

Bring all parameters into Equation (5), we can get the constitutive equation of Inconel 625 as follows:

$$Z = \dot{\varepsilon}exp\left(2.58 \times 10^5/RT\right) = 1.06 \times 10^8[sinh(0.0037\sigma)]^{3.21} \quad (13)$$

It can also be expressed by flow stress as:

$$\sigma = 272.97ln\left\{\left(\frac{\dot{\varepsilon}exp\left(\frac{2.58\times10^5}{8.314T}\right)}{1.06 \times 10^8}\right)^{0.31} + \sqrt{\left(\frac{\dot{\varepsilon}exp\left(\frac{2.58\times10^5}{8.314T}\right)}{1.06 \times 10^8}\right)^{0.62} + 1}\right\} \quad (14)$$

The peak stresses of Inconel 625 obtained under the experimental conditions are compared with the calculated value by using the constitutive equation, as shown in Figure 7. The average error is 5.68%, which verifies the applicability of the constitutive equation under this deformation condition.

It is worth noting that compared with other studies [36,37], the deformation activation energy of this experiment is low. Deformation activation energy is usually considered to reflect the difficulty of plastic deformation, and there are many influencing factors, such as microstructure (initial grain size, dislocation pinning effect [38], and the concurrent dynamic precipitation [35]) and deformation conditions (temperature [39], strain rate [40], and accumulated deformation [41]). On the one hand, the deformation mode and stress state of plane strain compression are different from those of uniaxial compression. Moreover, this plane strain compression adopts lower heating rate and longer holding time, which can make the diffusion in the microstructure more sufficient and the alloy microstructure more uniform. This also reduces the deformation activation energy of the alloy. On the other hand, compared with the literature [36,37], the experiment in this paper is mainly carried out in the range of low strain rate, and the maximum strain rate is 0.05 s^{-1}, which is far lower than 10 s^{-1} in the above literature. High strain rate will cause a large number of dislocation entanglement, resulting in the increase of deformation activation energy. In addition, compared with the as-cast plate selected in the literature [36,37], the rolled plate was selected and annealed at 1150 °C for 50 min. The obtained equiaxed microstructure has low deformation resistance. Due to different initial conditions, heat treatment process, and deformation parameters, the results of the deformation activation energy will change.

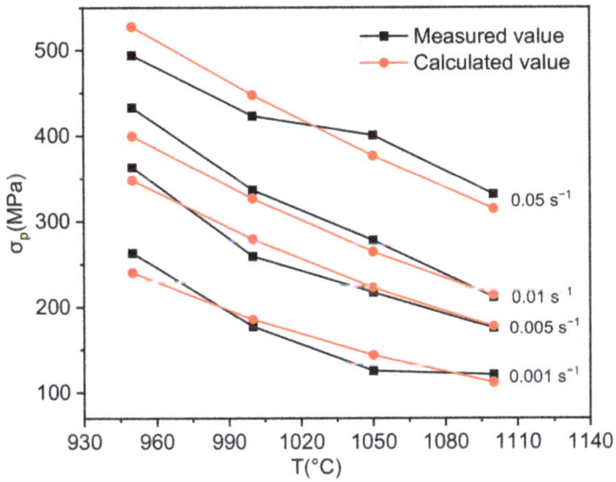

Figure 7. Comparison of measured and calculated peak stress of Inconel 625.

3.3. Thermal Processing Map

The hot workability of materials refers to the deformation ability of materials without damage in the process of plastic deformation. While the processing map reflects the processing advantages and disadvantages of materials under different processing conditions, which is related to the properties of materials [42]. During the hot working process, the work done by the external force (denoted by *P*) can be expressed by the plastic deformation (denoted by *G*) and the change of grain structure (denoted by J) [43]:

$$P = \sigma\dot{\varepsilon} = G + J = \int_0^{\dot{\varepsilon}} \sigma d\dot{\varepsilon} + \int_0^{\sigma} \dot{\varepsilon} d\sigma \quad (15)$$

The relationship between G and *J* is described by Equation (16), where *m* is strain rate sensitivity index.

$$m = \left(\frac{\partial J}{\partial G}\right)_{\varepsilon,T} = \frac{\partial P}{\partial G}\frac{\partial G}{\partial P} = \left[\frac{\partial ln\,\sigma}{\partial\dot{\varepsilon}}\right]_{\varepsilon,T} \quad (16)$$

When *G* equals *J*, *J* reaches the maximum J_{max}. In order to describe the power dissipation characteristics of material microstructure evolution, we use efficiency of power dissipations η to represent it.

$$\eta = \frac{J}{J_{max}} = \frac{P-G}{\sigma\cdot\frac{\dot{\varepsilon}}{2}} = 2\left(1 - \frac{\int_0^{\dot{\varepsilon}}\sigma d\dot{\varepsilon}}{\sigma\cdot\dot{\varepsilon}}\right) \quad (17)$$

When the strain is constant, the power dissipation diagram can be obtained from the contour map of temperature, strain rate and efficiency of power dissipations. However, according to the power dissipation diagram solely cannot reflect the actual processing situation. When the energy dissipation efficiency is high, instability may occur. Therefore, the power dissipation diagram has to be further improved.

Prasady et al. [44] considered that the dissipation function and strain rate satisfy the following inequality:

$$\frac{dD}{d\dot{\varepsilon}} \leqslant \frac{D}{\dot{\varepsilon}} \tag{18}$$

When system is unstable, the criterion of instability of material is obtained as follows:

$$\xi(\dot{\varepsilon}) = \frac{\partial lg\left(\frac{m}{m+1}\right)}{\partial lg\,\dot{\varepsilon}} + m \leqslant 0 \tag{19}$$

Plot the unstable region on the power dissipation diagram, and the processing map can be obtained. The processing map under different effective strains is shown in Figure 8.

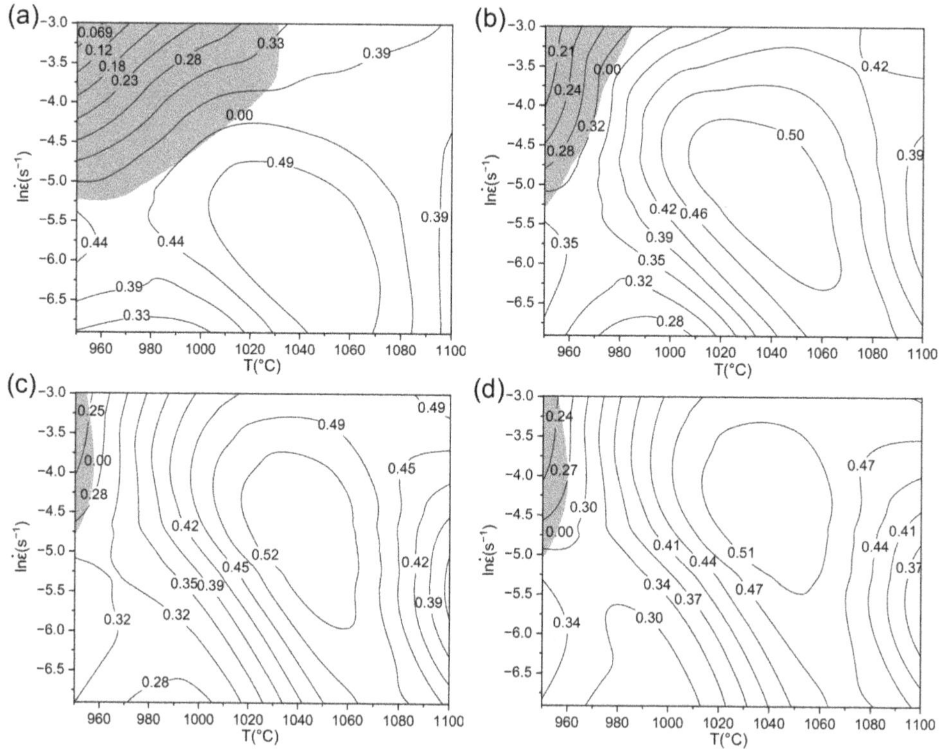

Figure 8. Thermal processing map of Inconel 625 under different effective strain. (**a**) ε = 0.2, (**b**) ε = 0.3, (**c**) ε = 0.5, (**d**) ε = 0.7.

The shadow area in the figure is the unstable region, and the value of isoline is η. The region corresponding to the highest energy dissipation rate is the optimal deformation region. It can be seen that the unstable region is mainly concentrated in the range of low temperature and high strain rate, and with the increasing of temperature, the proportion of energy used for microstructure evolution raises. This is because the increase of temperature is conducive to the progress of DRX. As can be seen from Figure 8a, when the strain is low, the area of unstable region is large. At this time, the flow stress is near the peak value, and the recrystallization behavior has not been fully carried out. With the increasing of strain, the area of the unstable region gradually decreases, the optimal deformation zone moves to the high temperature and high strain rate zone, and the shape of hot working diagram tends to be stable. On the whole, the effect of strain on the change of hot working diagram is not great, so Inconel 625 can be regarded as a strain insensitive material. When the true strain is 0.7, the flow curve is basically stable. At this time, the optimum deformation condition is found to be 1020–1060 °C/0.005–0.03 s^{-1}, and the maximum energy conversion can reach 0.51. The deformation temperature of 950–960 °C with the strain rate of 0.007–0.05 s^{-1} are determined as the unstable region.

The specimens in the unstable region at the conditions of 950 °C/0.01 s^{-1} was selected for optical observation under metallographic microscope, and the results are shown in Figure 9. It can be seen that a banded recrystallization dense area appears in the sample. This is because under low temperature and high strain rate, the heat generated by local plastic deformation is too fast to be transmitted to the low temperature area, resulting in an instantaneous sharp rise in local temperature. The deformation resistance of this area decreases, and local large deformation occurs. All this promotes local DRX process, leading to the generation of local plastic-rheological zones. The above phenomena illustrate that the deformation of the alloy is very uneven during hot pressing in the unstable region, which indicates the characteristics of local instability. If the instability is further increased, adiabatic shear bands and microcracks may develop and cause sample failure. This phenomenon further verifies the guidance of the processing map. In the actual machining process, the unstable region should be avoided.

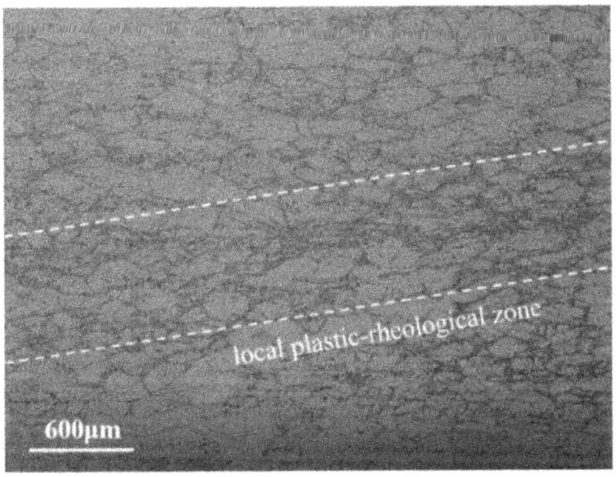

Figure 9. Microstructures of Inconel 625 under in the unstable region.

3.4. Microstructure Evolution of Inconel 625 during Thermal Deformation

The Inconel 625 samples after the plane strain compression test were characterized by EBSD, and the results are shown in Figure 10. The misorientation angle of 2–15° is defined as a low angle grain boundary (LAGBs), and the misorientation angle greater than 15° is a high angle grain boundary (HAGBs). Figure 11 shows the quantitative statistics of grain sizes and misorientation angles at different strain rates. The noise below 1.5 μm and grains connected to the picture boundary are excluded through the subset function. The twin boundary is not included in the statistics, and the grain size is expressed by the equivalent diameter. In addition, Figure 12 shows the volume fraction of different grains.

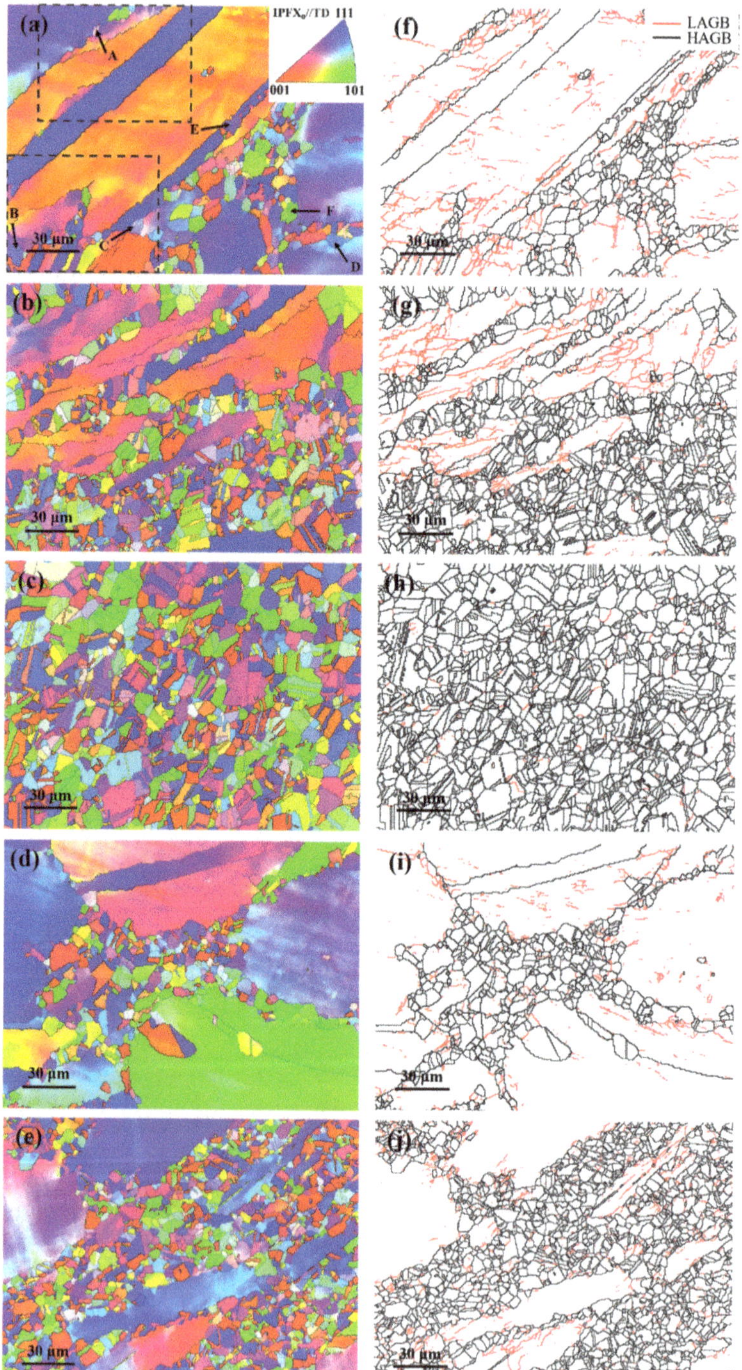

Figure 10. IPF maps and grain misorientations distribution of Inconel 625 under different hot-deformation conditions: (**a**,**f**) 1000 °C, 0.05 s^{-1}, (**b**,**g**) 1000 °C, 0.01 s^{-1}, (**c**,**h**) 1000 °C, 0.005 s^{-1}, (**d**,**i**) 950 °C, 0.01 s^{-1}, (**e**,**j**) 1050 °C, 0.01 s^{-1}.

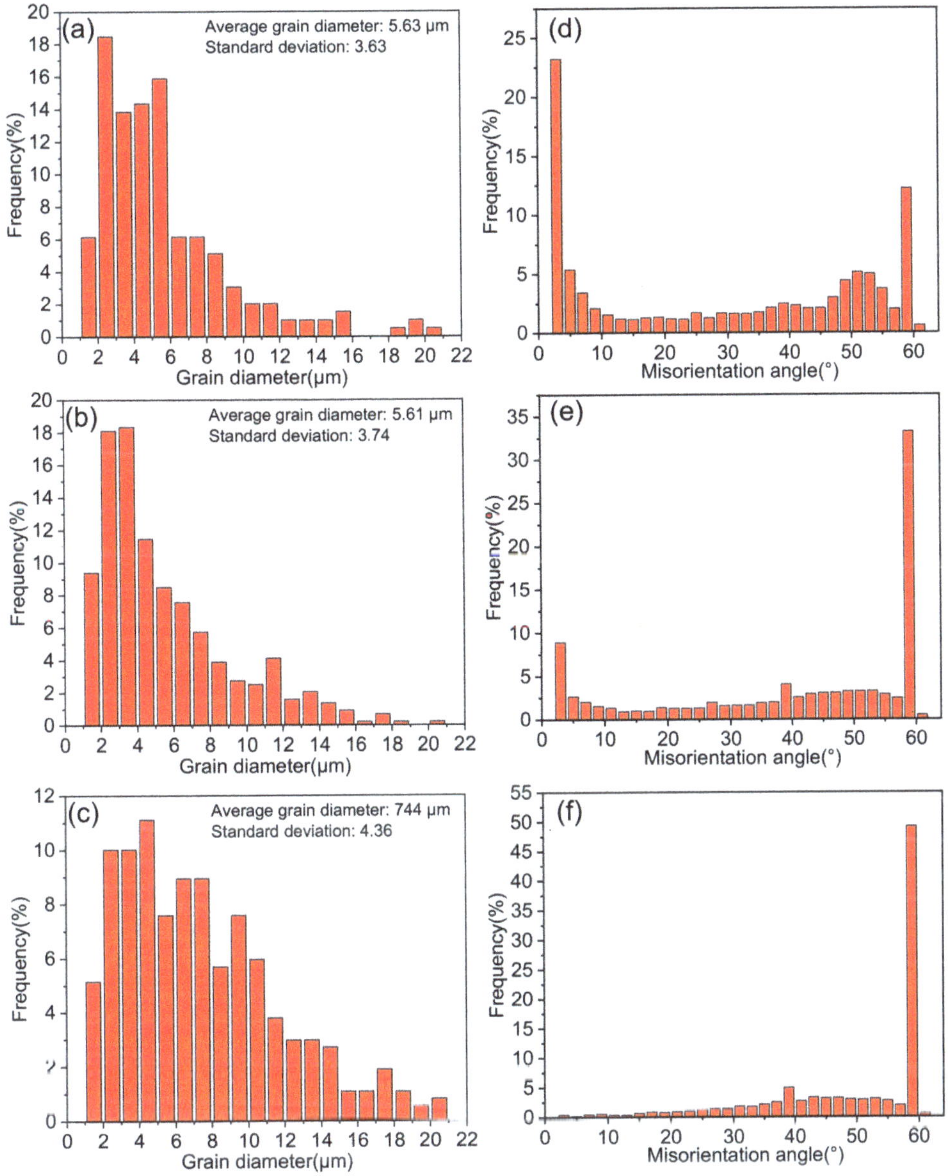

Figure 11. Grain sizes and misorientation angles under different strain rates: (**a**,**e**) 1000 °C, 0.05 s^{-1}, (**b**,**d**) 1000 °C, 0.01 s^{-1}, (**c**,**f**) 1000 °C, 0.005 s^{-1}.

From the Figures 10, 11a and 12 we can see under the temperature of 1000 °C and the strain rate of 0.05 s^{-1}, the recrystallized grain is fine, and the coarse original grains are dominant. As shown in Figures 10, 11b and 12, with decreasing strain rate, the original coarse grains decrease gradually, and the grains with size of 0–5 μm increase rapidly. The fine recrystallized grains form uniformly, and twins appear in the grains. And Figure 10c shows when the strain rate further decreases, the recrystallized grains have enough time to grow. The original coarse grains almost disappeared, and the size of twin and recrystallized grains increases, as shown in Figures 11c and 12. Therefore, whether the strain rate is too high or too low, it is unbeneficial for Inconel 625 to obtain fine grain structure during plane strain compression, which is unfavorable for the macro properties of the alloy. In the actual production, selection of the appropriate strain rate should be considered. The effect of deformation temperature on recrystallization structure is similar to that of strain rate. From Figure 10b,d,e, it can be found that the recrystallized grains of Inconel 625 are fine at low temperature. When the temperature rises, both the proportion and the size of recrystallized grains increase. This is consistent with the conclusion obtained from the stress–strain curve. As a result, the increasing temperature can promote the DRX behavior of the alloy.

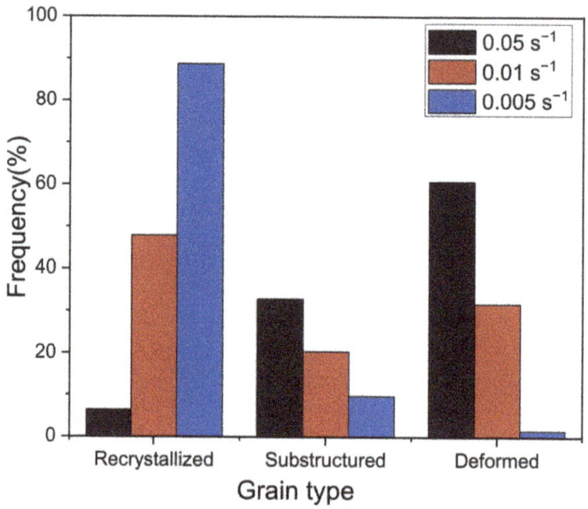

Figure 12. The volume fraction of different grains.

In addition, it can be seen from Figure 10a that when the strain rate is low, some grain boundaries bend into a stepped shape due to stress concentration. These stepped grain boundaries hinder the movement of dislocations, resulting in the increasing of dislocation density. The nucleated DRX grains attached to these grain boundaries for nucleation preferentially. During the hot-pressing process, fine recrystallized grains surround the initial coarse grains and form chain structure. This is a typical sign of discontinuous dynamic recrystallization (DDRX). In the process of DDRX, there is a big difference between the initial grain size and the steady-state recrystallization grain size. The former is consumed by the latter in the form of alternate nucleation and grain growth. Through the analysis in Figure 11d–f, it can be seen that the distribution of the misorientation angle of Inconel 625 has bimodal characteristics. When the angle is less than 5° or about 60°, the frequency is the highest. In addition, with the decreasing strain rate, the low angle grain boundary decreases and the high angle grain boundary (mainly 60°) increases. This is related to the formation of annealing twins during thermal deformation. It is generally believed that the formation of annealing twins can reduce the boundary energy of growing grains and increase the mobility of grain boundaries [45]. Therefore, annealing twins can promote the dynamic recrystallization process.

In order to explore the recrystallization mechanism of the Inconel 625 alloy at low strain rate, we take the cumulative orientation difference along arrows A–E in Figure 10a, and the results are shown in Figure 13. In addition, the partial enlarged view of the dotted line is shown in Figure 14.

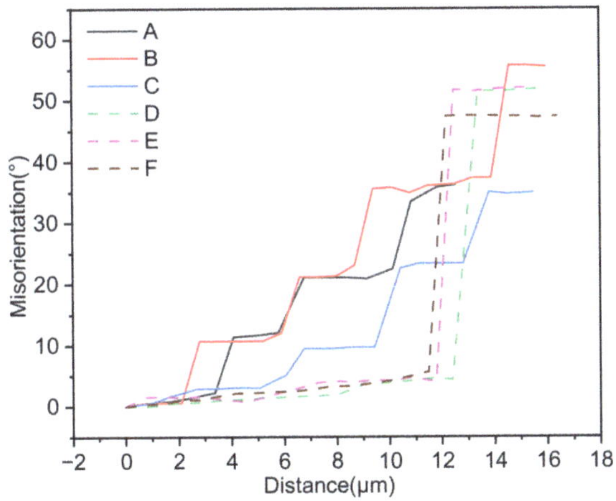

Figure 13. Cumulative misorientation along the arrows of Figure 10a.

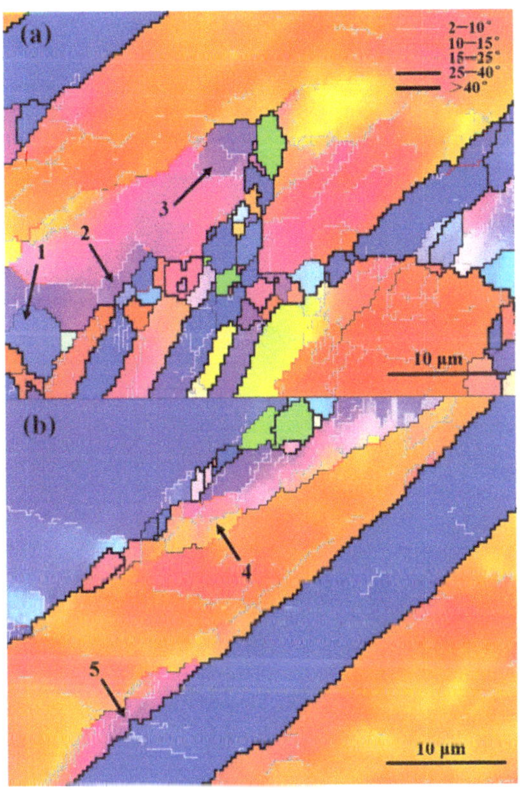

Figure 14. Partial enlarged view of the dotted line in Figure 10: (**a**) the lower part, (**b**) the upper part.

It can be found from Figure 13 that the cumulative misorientation at A–C increases in a step shape, indicating that the dynamic recrystallization grain here is formed by the evolution from a small angle grain boundary to a large angle grain boundary. And the grain boundary distribution in Figure 14 also proves this phenomenon, in which grains 1–5 are typical CDRX grains. On the other hand, the cumulative misorientation at D-F rises sharply when crossing the grain boundary, while in other regions it is flat; and the cumulative misorientation is much greater than the discrimination value of the large angle grain boundary by 15°. This phenomenon shows that the recrystallized grains here are formed by nucleation and growth through grain boundary bow, that is, discontinuous dynamic recrystallization, which shows that DDRX occurs. From the statistics of microstructure, we find that the number of recrystallized grains produced by CDRX is very small. To further illustrate this phenomenon, the changes in the fractions of different misorientation angle scopes under different deformation conditions are shown in Figure 15.

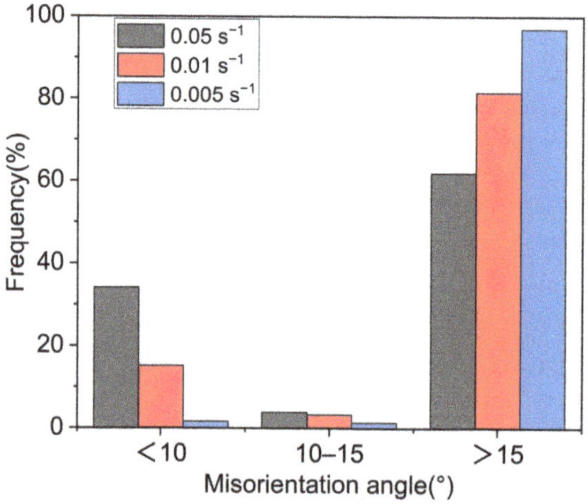

Figure 15. The changes in the fractions of different misorientation angles scopes at 1000 °C.

As can be seen from Figure 15, with the decreasing of strain rate, the LAGBs decrease gradually and remain in a small proportion. Under the above conditions, the proportion of LAGBs with an misorientation of 10–15° is always very low, and the proportion does not change significantly. It is well-known that the CDRX mechanism usually leads to an increase of 10–15° misorientation rate [46]. Therefore, under plane strain compression at high temperature and low strain rate, the DRX mechanism of the Inconel 625 alloy is DDRX based on grain boundary bulging and HAGBs migration. Meanwhile, the CDRX with sub-crystal merging and rotation is not dominant and plays an auxiliary role.

4. Summary and Conclusions

In this paper, a plane strain compression experiment on an Inconel 625 sheet under different thermal deformation conditions was carried out. By analyzing the stress–strain curves and microscopic characterization, we came to the following conclusions:

(1) Recrystallization occurs during hot pressing, and with the increasing of deformation temperature and the decreasing of strain rate, the recrystallization softening is more significant, so the stress level decreases. When the strain rate is higher, the curve is smooth. While the strain rate is low, the stress–strain curve shows high-frequency periodic oscillation.

(2) Calculating the related parameters, we get the constitutive equation of plane compression strain of an Inconel 625 sheet as $Z = \dot{\varepsilon} exp\left(2.58 \times 10^5/RT\right) = 1.06 \times 10^8 [sinh(0.0037\sigma)]^{3.21}$, and the average error is 5.68%.

(3) The processing map of the Inconel 625 sheet under different strains was drawn. It was found that the Inconel 625 is a strain-insensitive material. The unstable region is: deformation temperatures at 950–960 °C, and strain rates at 0.007–0.05 s^{-1}. The best deformation region is: deformation temperatures at 1020–1060 °C, and strain rates at

0.005–0.03 s^{-1}. Under the condition of instability, obvious local plastic-rheological zones can be observed, in which the fine DRX grains are densely distributed.

(4) Through EBSD characterization, it was found that increasing temperature and decreasing strain rate will promote the recrystallization behavior, and both the proportion and the size of recrystallized grains increase. Under the condition of high temperature and low strain rate, DDRX is the main deformation mechanism, while CDRX plays an auxiliary role.

Author Contributions: Y.S.: Investigation, Writing-original draft, Visualization, Formal analysis, Data curation, Methodology. J.F.: Investigation, Visualization, Formal analysis, Data curation, Writing—review & editing. X.L. and P.Z.: Investigation, Data curation. J.L.: Supervision, Funding acquisition. All authors have read and agreed to the published version of the manuscript.

Funding: This research was funded by the National Science and Technology Major Project, grant number 2017-VI-0014-0086 and MJ-2018-G-48.

Institutional Review Board Statement: Not applicable.

Informed Consent Statement: Not applicable.

Data Availability Statement: All data are available from the corresponding author on reasonable request.

Conflicts of Interest: The authors declare no conflict of interest.

References

1. Gao, Y.B.; Ding, Y.T.; Meng, B.; Ma, Y.J.; Chen, J.J.; Xu, J.Y. Research progress in evolution of precipitated phases in Inconel 625 superalloy. *Mater. Eng.* **2020**, *48*, 13–22.
2. Zhang, Q.; Wang, J.W.; Shi, Y.S.; Wei, Q.S. Effects of solution treatment on microstructure and tensile properties of hot isostatic pressed Inconel 625 alloy. *Heat Treat. Met.* **2013**, *38*, 65–69.
3. Nguejio, J.; Szmytka, F.; Hallais, S.; Tanguy, A.; Nardone, S.; Martinez, M.G. Comparison of microstructure features and mechanical properties for additive manufactured and wrought nickel alloys 625. *Mater. Sci. Eng.* **2019**, *764*, 138214. [CrossRef]
4. Yu, L.J.; Marquis, E.A. Precipitation behavior of Alloy 625 and Alloy 625 plus. *J. Alloys Compd.* **2019**, *811*, 151916. [CrossRef]
5. Fei, J.X.; Liu, G.L.; Patel, K. Effects of Machining Parameters on Finishing Additively Manufactured Nickel-Based Alloy Inconel 625. *Manuf. Mater. Process.* **2020**, *4*, 32–54.
6. Satish, G.J.; Gaitonde, V.N.; Kulkarni, V.N. Traditional and non-traditional machining of nickel-based superalloys: A brief review. *Mater. Today Proc.* **2021**, *44*, 1448–1454. [CrossRef]
7. Chen, S.; Ma, J.X. Trend analysis of nickel-based superalloy based on journal literature. *Mater. Rev.* **2014**, *28*, 116–124.
8. Wang, J.Z.; Li, K.Y.; Liu, A.J.; Ding, K.L. Microstructure and hardness of hot rolled Inconel 625 after heat treatment. *Met. Heat Treat.* **2018**, *43*, 96–100.
9. Liu, X.D.; Fan, J.K.; Zhang, P.Z.; Xie, J.; Chen, F.L.; Liu, D.G.; Yuan, R.G.; Tang, B.; Kou, H.C.; Li, J.S. Temperature dependence of deformation behavior, microstructure evolution and fracture mechanism of Inconel 625 superalloy. *J. Alloys Compd.* **2021**, *869*, 159342. [CrossRef]
10. Chen, F.L.; Li, K.D.; Tang, B.; Liu, D.G.; Zhong, H.; Li, J.S. Deformation behavior and microstructural evolution of Inconel 625 superalloy during the hot compression process. *Metals* **2021**, *11*, 824. [CrossRef]
11. Parida, A.K.; Maity, K. FEM analysis and experimental investigation of force and chip formation on hot turning of Inconel 625. *Def. Technol.* **2019**, *15*, 853–860. [CrossRef]
12. Liu, X.D.; Fan, J.K.; Li, K.D.; Song, Y.L.; Liu, D.G.; Yuan, R.H.; Wang, J.; Tang, B.; Kou, H.C.; Li, J.S. Serrated flow behavior and microstructure evolution of Inconel 625 superalloy during plane-strain compression with different strain rates. *J. Alloys Compd.* **2021**, *881*, 160648. [CrossRef]
13. Chen, M.S.; Zou, Z.H.; Lin, Y.C.; Li, K.K. Hot deformation behaviors of a solution-treated Ni-based superalloy under constant and changed strain rates. *Vacuum* **2018**, *155*, 531–538. [CrossRef]
14. Jia, Z.; Gao, Z.X.; Ji, J.J.; Liu, D.X.; Guo, T.B.; Ding, Y.T. Study of the Dynamic Recrystallization Process of the Inconel625 Alloy at a High Strain Rate. *Materials* **2019**, *12*, 510. [CrossRef] [PubMed]
15. Tao, L.; Cheng, M.; Zhang, W.H.; Song, G.S.; Zhang, S.H. High-temperature and high-speed hot deformation behavior of Inconel 625. *J. Mater. Heat Treat.* **2012**, *33*, 55–59.
16. Lin, Y.C.; Min, H.E.; Chen, M.S.; Wen, D.X.; Chen, J. Effects of Initial Phase (Ni3Nb) on Hot Tensile Deformation Behaviors and Material Constants of a Ni-Based Superalloy. *Trans. Nonferrous Met. Soc. China* **2016**, *26*, 107–117. [CrossRef]
17. Shi, X.N.; Liu, Y.Y.; Ning, Y.Q.; Zhang, J.Y.; Wang, M.T. Physically-Based Constitutive Model of Nickel-Based Superalloy ATI 718Plus. *Chin. J. Rare Met.* **2019**, *43*, 613–620.

18. Zhang, M.; Liu, G.; Wang, H.; Hu, B. Modeling of thermal deformation behavior near γ' solvus in a Ni-based powder metallurgy superalloy. *Comput. Mater. Sci.* **2019**, *156*, 241–245. [CrossRef]
19. Lin, Y.C.; Li, J.; Chen, M.S.; Liu, Y.X.; Liang, Y.J. A deep belief network to predict the hot deformation behavior of a Ni-based superalloy. *Neural Comput. Appl.* **2018**, *29*, 1015–1023. [CrossRef]
20. Lin, Y.C.; Nong, F.Q.; Chen, X.M.; Chen, D.D.; Chen, M.S. Microstructural evolution and constitutive models to predict hot deformation behaviors of a nickel-based superalloy. *Vacuum* **2017**, *137*, 104–114. [CrossRef]
21. Liu, Y.H.; Liu, Z.Z.; Wang, M. Gradient microstructure evolution under thermo-mechanical coupling effects for a nickel-based powder metallurgy superalloy-Dynamic recrystallization coexist with static recrystallization. *J. Mater. Process. Technol.* **2021**, *294*, 117142. [CrossRef]
22. Li, H.Z.; Yang, L.; Wang, Y.; Tan, G.; Qiao, S.C.; Huang, Z.Q.; Liu, M.X. Thermal deformation and dynamic recrystallization of a novel HEXed P/M nickel-based superalloy. *Mater. Charact.* **2020**, *163*, 110285. [CrossRef]
23. Jia, D.; Sun, W.R.; Xu, D.S.; Yu, L.X.; Xin, X.; Zhang, W.H.; Qi, F. Abnormal dynamic recrystallization behavior of a nickel-based superalloy during hot deformation. *J. Alloys Compd.* **2019**, *787*, 196–205. [CrossRef]
24. Sun, F. Integrated TEM/transmission-EBSD for recrystallization analysis in nickel-based disc superalloy. *Prog. Nat. Sci. Mater. Int.* **2020**, *31*, 63–67. [CrossRef]
25. Kumar, S.; Raghu, T.; Bhattacharjee, P.P.; Rao, G.A.; Borah, U. Evolution of microstructure and microtexture during hot deformation in an advanced P/M nickel base superalloy. *Mater. Charact.* **2018**, *146*, 217–236. [CrossRef]
26. Jia, Z.; Sun, X.; Ji, J.J.; Wang, Y.J.; Wei, B.L.; Yu, L.D. Hot Deformation Behavior and Dynamic Recrystallization Nucleation Mechanisms of Inconel 625 during Hot Compressive Deformation. *Adv. Eng. Mater.* **2021**, *23*, 2001048. [CrossRef]
27. Jiang, H.; Dong, J.X.; Zhang, M.C.; Yao, Z.H. Phenomenological model for the effect of strain rate on recrystallization and grain growth kinetics in the 617B alloy. *J. Alloys Compd.* **2018**, *735*, 1520–1535. [CrossRef]
28. Liu, Y.X.; Lin, Y.C.; Zhou, Y. 2D cellular automaton simulation of hot deformation behavior in a Ni-based superalloy under varying thermal-mechanical conditions. *Mater. Sci. Eng. A* **2017**, *691*, 88–99. [CrossRef]
29. Hsu, Y.C.; Yang, T.S.; Sung, S.Y.; Chang, S.Y. Constructing the predictive models of friction coefficient using cylindrical compression testing. *Mater. Sci. Forum* **2006**, *540*, 745–750. [CrossRef]
30. Pan, H.B.; Tang, D.; Hu, P.P.; Wang, X.X. Study on plane strain physical compression technology. *Forg. Technol.* **2008**, *33*, 75–79.
31. Jiang, Z.W.; Zhu, L.L.; Yu, L.X.; Sun, B.A.; Cao, Y.; Zhao, Y.H.; Zhang, Y. The mechanism for the serrated flow induced by Suzuki segregation in a Ni alloy. *Mater. Sci. Eng. A* **2021**, *820*, 141575. [CrossRef]
32. Sadovskii, V.M.; Sadovskaya, O.V.; Petrakov, I.E. On the theory of constitutive equation for composites with different resistance in compression and tension. *Compos. Struct.* **2021**, *268*, 113921. [CrossRef]
33. Dai, Q.S.; Deng, Y.L.; Tang, J.G.; Wang, Y. Deformation characteristics and strain-compensated constitutive equation for AA5083 aluminum alloy under hot compression. *Trans. Nonferrous Met. Soc. China* **2019**, *29*, 2252–2261. [CrossRef]
34. Tan, G.; Li, H.Z.; Wang, Y.; Yang, L.; Qiao, X.C.; Huang, Z.Q.; Liu, M.X. Hot working characteristics of HEXed PM nickel-based superalloy during hot compression. *Trans. Nonferrous Met. Soc. China* **2020**, *30*, 2079–2723. [CrossRef]
35. McQueen, H.J.; Ryan, N.D. Constitutive analysis in hot working. *Mater. Sci. Eng. A* **2002**, *322*, 43–63. [CrossRef]
36. Guo, S.; Li, D.; Guo, Q.; Wu, Z.; Peng, H.; Hu, J. Investigation on hot workability characteristics of Inconel 625 superalloy using processing maps. *J. Mater. Sci.* **2012**, *47*, 5867–5878. [CrossRef]
37. Jia, Z.; Gao, Z.X.; Ji, J.J.; Liu, D.X.; Ding, Y.T. High-temperature deformation behavior and processing map of the as-cast Inconel 625 alloy. *Rare Met.* **2021**, *40*, 2083–2091. [CrossRef]
38. Huang, X.D.; Zhang, H.; Han, Y.; Wu, W.X.; Chen, J.H. Hot deformation behavior of 2026 aluminum alloy during compression at elevated temperature. *Mater. Sci. Eng. A* **2010**, *527*, 485–490. [CrossRef]
39. Langkruis, J.; Kool, W.H.; Zwaag, S. Assessment of constitutive equations in modelling the hot deformability of some overaged Al–Mg–Si alloys with varying solute contents. *Mater. Sci. Eng. A* **1999**, *266*, 135–145. [CrossRef]
40. Yin, X.Y.; Liu, X.K.; Ding, H.X.; Zhu, J. High temperature constitutive model modification and evolution law of deformation activation energy for HNi55-7-4-2 alloy. *Forg. Stamp. Technol.* **2021**, *46*, 221–228.
41. Shen, B.; Deng, L.; Wang, X. A new dynamic recrystallisation model of an extruded Al-Cu-Li alloy during high-temperature deformation. *Mater. Sci. Eng.* **2015**, *625*, 288–295. [CrossRef]
42. Zhang, B.X.; Zhao, X.; Zhang, B.H. Optimization of hot stamping process for deep blind hole cylinders based on thermal processing map. *Ordnance Mater. Sci. Eng.* **2021**, *44*, 17–24.
43. Ahamed, H.; Senthilkumar, V. Hot deformation behavior of mechanically alloyed Al6063/0.75Al_2O_3/0.75Y_2O_3 nano-composite-A study using constitutive modeling and processing map. *Mater. Sci. Eng. A* **2012**, *539*, 349–359. [CrossRef]
44. Prasad, Y.V.R.K.; Seshacharyulu, T. Modeling of hot deformation for microstructural control. *Int. Mater. Rev.* **1998**, *43*, 243–258. [CrossRef]
45. Coryell, S.P.; Findley, K.O.; Mataya, M.C.; Brown, E. Evolution of microstructure and texture during hot compression of a Ni-Fe-Cr superalloy. *Metall. Mater. Trans. A* **2012**, *43*, 633–649. [CrossRef]
46. Mandal, S.; Bhaduri, A.K.; Sarma, V.S. A study on microstructural evolution and dynamic recrystallization during isothermal deformation of a Ti-modified austenitic stainless steel. *Metall. Mater. Trans. A* **2011**, *42*, 1062–1072. [CrossRef]

materials

MDPI

Article

Modeling and Characterization of Complex Concentrated Alloys with Reduced Content of Critical Raw Materials

Beatrice-Adriana Șerban [1,†], Ioana-Cristina Badea [1,2,*,†], Nicolae Constantin [2], Dumitru Mitrică [1], Mihai Tudor Olaru [1,†], Marian Burada [1], Ioana Anasiei [1,2,*], Simona-Elena Bejan [1], Andreea-Nicoleta Ghiță [1] and Ana Maria-Julieta Popescu [3]

1 National R&D Institute for Non-Ferrous and Rare Metals, 102 Biruinței, 077145 Pantelimon, Romania; beatrice.carlan@imnr.ro (B.-A.Ș.); dmitrica@imnr.ro (D.M.); o.mihai@imnr.ro (M.T.O.); mburada@imnr.ro (M.B.); sbejan@imnr.ro (S.-E.B.); andreea.lupu9@yahoo.com (A.-N.G.)
2 Engineering and Management of Obtaining Metallic Materials Department, University Politehnica of Bucharest, 313 Splaiul Independentei, 060042 Bucharest, Romania; nctin2014@yahoo.com
3 Romanian Academy, "Ilie Murgulescu" Institute of Physical Chemistry, 202 Splaiul Independenței, 060021 Bucharest, Romania; popescuamj@yahoo.com
* Correspondence: cristina.banica@imnr.ro (I.-C.B.); ioana.anasiei@gmail.com (I.A.)
† These authors have equally contributed to this study.

Abstract: The continuous development of society has increased the demand for critical raw materials (CRMs) by using them in different industrial applications. Since 2010, the European Commission has compiled a list of CRMs and potential consumption scenarios with significant economic and environmental impacts. Various efforts were made to reduce or replace the CRM content used in the obtaining process of high-performance materials. Complex concentrated alloys (CCAs) are an innovative solution due to their multitude of attractive characteristics, which make them suitable to be used in a wide range of industrial applications. In order to demonstrate their efficiency in use, materials should have improved recyclability, good mechanical or biocompatible properties, and/or oxidation resistance, according to their destination. In order to predict the formation of solid solutions in CCAs and provide the optimal compositions, thermodynamic and kinetic simulations were performed. The selected compositions were formed in an induction furnace and then structurally characterized with different techniques. The empirical results indicate that the obtained CCAs are suitable to be used in advanced applications, providing original contributions, both in terms of scientific and technological fields, which can open new perspectives for the selection, design, and development of new materials with reduced CRM contents.

Keywords: complex concentrated alloys (CCAs); simulation; critical raw materials (CRMs); elaboration; properties

Citation: Șerban, B.-A.; Badea, I.-C.; Constantin, N.; Mitrică, D.; Olaru, M.T.; Burada, M.; Anasiei, I.; Bejan, S.-E.; Ghiță, A.-N.; Popescu, A.M.-J. Modeling and Characterization of Complex Concentrated Alloys with Reduced Content of Critical Raw Materials. *Materials* **2021**, *14*, 5263. https://doi.org/10.3390/ma14185263

Academic Editor: Lijun Zhang

Received: 24 August 2021
Accepted: 10 September 2021
Published: 13 September 2021

1. Introduction

Metallic materials have extended their application spectrum with protometallurgy, when meteorite iron was used for the manufacturing of weapons, household items, or tools. Later, human communities learned to exploit and process the underground ores and people were able to see the benefits that these activities bring. To simplify the quotidian activities and increase productivity, people began to alloy the subterranean materials discovered and developed metallurgy—an important contribution for humankind [1,2].

Precious metal alloys have had great importance in medical applications, with gold being used in dental applications for over 2000 years. However, for reasons of ancient religious dogmas, the medical applications were slowed down until 1500 A.D. Despite all these impediments, in 659 B.C., a metallic material with applicability in dental medicine was developed, containing 100 parts of mercury, 45 parts of silver, and 900 parts of tin [3].

Up to the present, the most used metallic materials for the manufacturing of medical implants are Co–Cr–Mo and Ti alloys. They have superior mechanical and biocompatible properties, so they can be used for the replacement of artificial joints (hip, knee, and shoulder prostheses). Although they have good biocompatible characteristics, titanium alloys (e.g., Ti_6Al_4V) cannot be used as contact surfaces between two materials due to the low wear resistance, which is associated with low shear strength and repassivation behavior of the oxide layer from the surface [4]. Alloys from the Co–Cr–Mo system (e.g., $Co_{28}Cr_6Mo$) have higher wear resistance and can be used as contact surfaces for joint prostheses, but there are identified situations where the implant has been rejected by the human body [5]. It has been observed that the level of chromium and cobalt metal ion increases in the blood of patients with implants from the analyzed alloys' systems. In vivo and in vitro tests were also performed to demonstrate the risk of cytotoxicity, DNA damage, hypersensitivity reactions to metal and pseudo-tumors [6]. The materials currently used in the manufacturing of implants have many disadvantages, so the development of new materials that are resistant to wear and prolong the life of implants remains a topical challenge [7].

Civilizational progress has led to the development of new materials with improved properties, based on exhaustible mineral resources. This has favored an irrational consumption of the existing natural resources, which can compromise the ability of future generations to meet their needs [8]. To prevent this and to promote the sustainable development-exploiting tendency of mining resources, the European Union has compiled a report containing the global situation of critical metals and materials [9]. This class of materials is widely used in high-tech products, and industrialized societies are dependent on them [10]. This study included the potential consumption scenarios and the necessity to move to a climate-neutral economy through the development of new technologies that decrease the critical material demand and the industrial ecological footprint. Table 1 presents the critical raw materials list, which was first published in 2020.

Table 1. Critical raw materials list [9].

Critical Raw Materials in 2020				
Antimony	Fluorspar	Magnesium	Scandium	Titanium
Baryte	Gallium	Natural graphite	Silicon	Strontium
Beryllium	Germanium	Natural rubber	Tantalum	
Bismuth	Hafnium	Niobium	Tungsten	
Borate	Heavy rare earth elements	Platinum group metals	Vanadium	
Cobalt	Light rare earth elements	Phosphate rock	Bauxite	
Coking Coal	Indium	Phosphorus	Lithium	

To reduce or totally replace the critical metals from the composition of advanced materials some changes have been made, but challenges arise when the performances of the obtained product depended on them. For example, silicon is necessary for semiconductors, titanium is a key element in biocompatible materials, and magnesium provides strength and corrosion resistance in light alloys [9].

To decrease the high demand for critical materials, various strategies have been developed to modify the established technologies or to design new materials with similar properties but with a lower content of CRMs [10].

Commonly, metallic alloys are based on the existence of a major element that influences the properties of the material with the addition of alloying elements, which improve the final structure of the metallic material [11]. Complex concentrated alloys (CCAs) [12] are a new family of materials that have a distinct design strategy, due to their multitude of attractive characteristics, such as high hardness, good wear resistance, fatigue resistance, superior thermal properties, low elasticity modulus values, and/or increased oxidation and corrosion resistance [13]. Complex concentrated complex alloys are extending the boundaries of high entropy alloys by accepting a smaller number of alloying elements, which can influence the formation of mixed structures based on solid solutions and intermetallic phases. CCAs have a higher configurational entropy value, have no single, dominant element, and can form structures based on disordered solid solutions [14,15].

The multi-principle nature of complex concentrated alloys and the unique structures they produce have a significant effect on reducing the critical raw material contents in biomaterials with specific properties [16]. The structural particularities of CCAs increase the recyclability rate, which is relevant for the current societal concerns regarding environmental protection and the need to reduce the industrial ecological footprint [17].

To better predict the structure and to develop the most appropriate compositions of CCAs, thermodynamic and kinetic criteria were calculated. In order to have a more complex selection process for CCA compositions, the CALPHAD modeling method was used, as well as for thermodynamic evaluations and empirical data [18]. The kinetic modules included in the simulation program were based on the solid-state phase transformations, especially considering the accuracy of the calculations performed and their applicability to multicomponent systems [19–21].

The focus of this work was on developing new complex concentrated alloys with advanced properties, such as increased oxidation resistance, low density, good surface and mechanical properties, and/or biocompatibility [22]. The constituent elements of the alloy system are already used in conventional alloys, but the main purpose was to reduce or totally replace, where it is possible, the critical raw material contents from CCA compositions, without having a negative influence on the material properties. Another important aspect was to design an inexpensive alloy that can be obtained using common manufacturing processes.

2. Materials and Methods

To obtain the required mechanical and physical characteristics, it was very important to identify an appropriate composition of the complex concentrated alloys. The constitutive elements have a strong influence on CCA properties, but it is important to meet the needs of the present society by reducing the content of the used critical raw materials. To reduce the alloy density, there can be used chemical elements such as Ti, Al, Mg, or Si. Good anti-corrosive and mechanical properties or improved tensile strength can be obtained by adding Ti and Nb; mixing them with Zr, the alloys were able to form ternary solid solutions and had improved thermal properties and superplasticity. Mn presence favored the formation of solid solutions structures and Fe improved wear resistance and machinability. Therefore, by introducing inexpensive elements to the alloy, the final cost of the alloy was reduced, which has a favorable economic impact.

The most suitable CCA compositions were selected using semi-empirical criteria [23], which were defined by the following equations:

a. Using Boltzmann's equations, the mixing entropy was determined:

$$\Delta S_{mix} = -R \cdot \sum c_i \cdot \ln c_i, \quad (1)$$

where R is the gas's constant (8.314 J/molK); c_i is the molar fraction of the element i.

b. The enthalpy of mixing (ΔH_{mix}) was calculated using the Miedema macroscopic model [24]:

$$\Delta H_{mix} = \sum 4 c_i c_j \cdot \Delta H_{ij}, \quad (2)$$

where ΔH_{ij} is the binary enthalpy of the alloy; i and j are the elements introduced to the mix.

c. The atomic size difference (δ) was calculated using Equation (3) [15]

$$\delta = 100 \cdot \sqrt{\sum c_i \cdot \left(1 - \frac{r_i}{\bar{r}}\right)^2}, \quad (3)$$

where r_i is the atomic radius of element i; $\bar{r}$ is the average atomic radius.

d. The derived parameter Ω, which is influenced by the mixing entropy and the mixing enthalpy [25], was calculated using the following equation:

$$\Omega = T_m \Delta S_{mix} / |\Delta H_{mix}|\,, \tag{4}$$

where T_m is the melting temperature calculated with $T_m = \sum c_i \times T_{mi}$, where T_{mi} is the melting temperature of element i.

e. The Allen electronegativity difference ($\Delta\chi$) was calculated using Equation (5) [26]:

$$\Delta\chi^{\mathrm{Allen}} = 100 \cdot \sqrt{\sum \mathrm{c_i} \cdot \left(1 - \frac{\chi_\mathrm{i}}{\overline{\chi}}\right)^2}\,, \tag{5}$$

where χ_i is the electronegativity after Pauling for element i; $\overline{\chi}$ is the electronegativity average.

f. The valence electron concentration (VEC) was calculated (Equation (6)) to determine the type of solid solution that is found in the alloy [27]:

$$VEC = \sum c_i \cdot VEC_i, \tag{6}$$

where VEC_i is the valence electron concentration of element i.

g. The geometrical parameter (Λ) depends on the mixing entropy (ΔS_{mix}) and the atomic size difference (δ) of the mixture [28], expressed as follows:

$$\Lambda = \Delta S_{\mathrm{mix}} / \delta^2, \tag{7}$$

The thermodynamics and kinetics of the systems were simulated with the MatCalc Pro edition software, version 6.02. The thermodynamic modeling was based on the CALPHAD method (CALculation of PHAse Diagrams) analysis and the kinetic modules were developed based on solid-state phase transformations, taking into account the applicability of these models for multicomponent systems.

The selected alloys were prepared by mixing the technically pure raw materials of Fe, Ta, Nb, Ti, Zr, and Mn to obtain a charge of 350 g for each alloy composition. The alloys were placed in an alumina–zirconia crucible and melted in an induction furnace—Linn MFG, 300 type—under an inert atmosphere. The heating rate was related to the induction coupling capacity and the power applied. In this case, a 20% coupling at 22 kW of power and a 14 kHz frequency allowed for a heating rate of 75 °C/min. The alloy melt was cast inside the furnace in a cylindrical copper mold and then cooled in the furnace under vacuum.

To obtain more homogeneity in the alloy, certain alloys were remelted in the same working conditions and the obtained samples were characterized by structural, physical, and chemical analyses.

The resulting alloys were chemically analyzed by inductively coupled plasma spectrometry (ICP-OES) using an Agilent 725 spectrometer (Agilent, Santa Clara, CA, USA) to determine the percentage of chemical elements. Optical microscopy was accomplished using a Zeiss Axio Scope A1m Imager microscope (Zeiss, Jena, Germany), with bright field, dark field, DIC, and polarization characteristics, and high-contrast ECEpiplan 109/509/1009 lenses. SEM-EDAX characterization was performed with a high-resolution scanning electron microscope, FEI Quanta 3D FEG (FEI Europe B.V., Eindhoven, Netherlands), which was equipped with an X-ray spectrometer (EDS). The configuration of the phases was inspected using an X-ray diffractometer, type BRUKER D8 ADVANCE (Bruker Corporation, Billerica, MA, USA), which had a DIFFRACplus XRD Commender (Bruker AXS) software, as well as the Bragg–Brentano diffraction method, Θ—Θ, coupled in a vertical configuration with the following parameters: CuKα radiation, 2Θ region of 20–124, 2Θ step of 0.020, time of 8.7 s/step, rotation speed of 15 rot/min. Using the reference intensity ratio (RIR) method, a semi-quantitative evaluation of the identified phase concentrations was performed. The method was based upon scaling all the diffraction data to the diffraction levels of the standard reference materials. The results offer a qualitative indication of phase proportion in the alloy.

The Vickers microhardness of the alloys were determined at 25 °C using a micro-indenter attachment (Anton Paar MHT10, Anton Paar GmbH, Graz, Austria), which had an applied load of 2 N and a slope of 0.6 N/s.

3. Results

3.1. CCAs Structure Design Depending on Concentration

The redistribution of the solid solutions during the solidification process was studied by means of the MatCalc simulation program. The analysis tools obtained the equilibrium and Scheill–Gulliver diagrams, which provided information regarding the slow diffusion effect and the theoretical estimation of the solidification temperature.

The thermodynamic and kinetic criteria were calculated for the FeTaNbTiZr and FeMnNbTiZr alloy systems, in which the concentrations of each element were varied, to identify the most appropriate compositions for potential biocompatible applications.

Figure 1 presents the calculated phase distribution for the FeTaNbTiZr alloy system, which was dependent on the content of each constituent element. The step calculation simulation was performed by increasing the selected element concentration while maintaining proportional values for the rest of the elements.

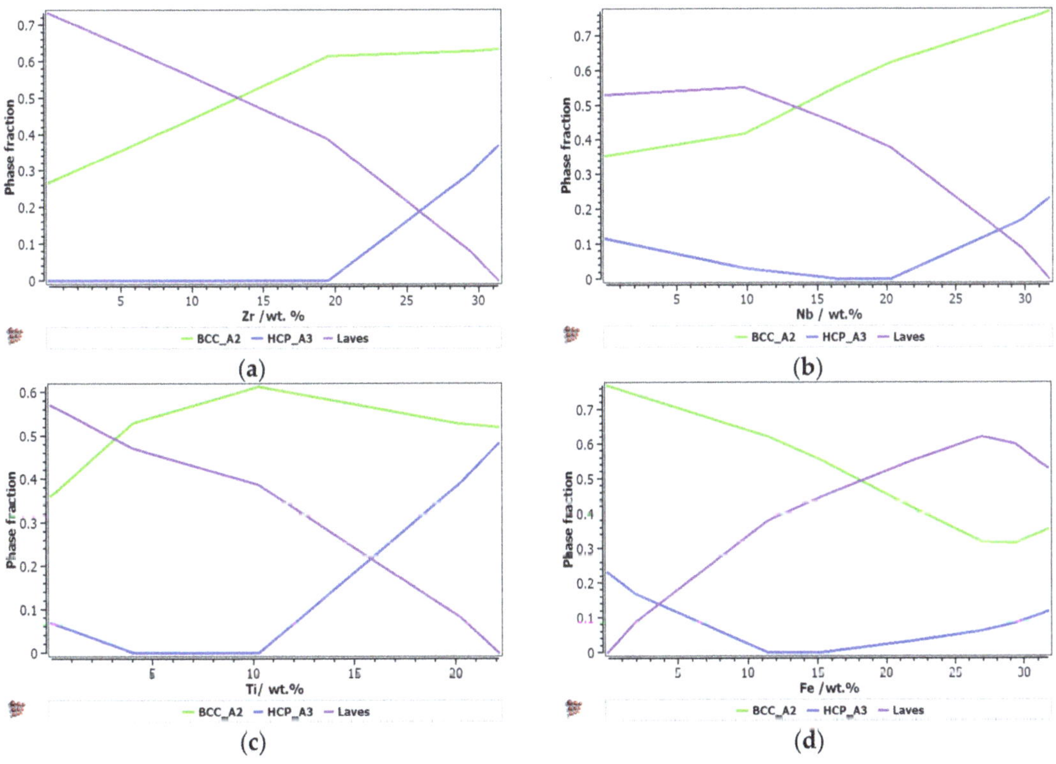

Figure 1. *Cont.*

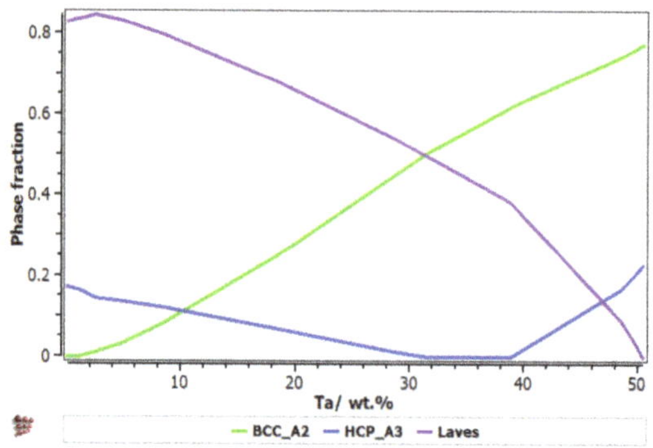

(e)

Figure 1. The proportions of phases that can be found in the FeTaNbTiZr alloy system at 200 °C, depending on the variation of (**a**) Zr; (**b**) Nb; (**c**) Ti; (**d**) Fe; (**e**) Ta.

It can be observed that the proportion of the BCC–A2 solid solution increased with the content of Zr, but the Laves intermetallic phase content decreased. At a percentage of 20%, Zr appeared to be A3 hexagonal, which is a brittle phase and specific to zirconium. Brittle phases are usually avoided in biocompatible alloys, so up to 20% of Zr is recommended to be used for alloy selection. The increase in the Nb content in the alloy had a similar result to Zr; with a proportion higher than 20%, and HCP–A3 phase starts to form. Up to 15% Nb, a significant proportion of the HCP–A3 phase can be observed, which was suppressed between 15 and 20% Nb. This concentration interval is recommended to be used for the intended application. Similar to Zr and Nb, Ti is important for stabilizing the BCC–A2 phase, up to 10%. When the titanium content exceeded 10%, the HCP–A3 phase started to form in the alloy structure. In Figure 1d, it can be observed that the presence of iron influenced the formation of phase structure by the inhibition of the BCC phase and the increase in the brittle HCP–A3 and Laves phases. Unlike the other elements, Ta had a constant influence on the formation of the BCC-A2 solid solution phase, up to a high concentration value of 50% (Figure 1e). The formation of the hardening HCP–A3 phase started at 40% Ta, which was a higher value compared to other elements. It was calculated the influence of brittle phases. As previously mentioned, the HCP–A3 phase was not desired in our configuration, and it was preferable, in this case, to have two-phase instead of three-phase structures. As a result, it was established that the optimal concentration interval for Ta was 30–40%.

For the second alloy system, FeMnNbTiZr, the influence of the alloying elements on the evolution of the constituent phases was analyzed (Figure 2). It can be observed that with an increase in the Ti percentage, there was a significant increase in the BCC–A2 solid solution phase and a decrease in the HCP–A3 hexagonal phase (Figure 2a). The Laves intermetallic phases had a maximum area of 0.4 M in the concentration interval of 8–15%Ti. A high percentage of 25–30%Ti is recommended for obtaining predominantly solid solutions in the alloy. The presence of Zr in the alloy at a percentage of up to 26% determined a decrease in the solid solution distribution and an increase in the content of Laves intermetallic phases (Figure 2b). A high concentration of Zr in CCA determined a decrease in the intermetallic compounds and an increase in the BCC–A2 and HCP–A3 solid solutions. When the Zr content was increased by more than 26%, the C14 Laves phase was reduced to a minimum. Nb has a high BCC–A2 phase stabilization capacity, higher than Zr and Ti, so at values over 27%, it can form exclusively solid solutions in the alloy (Figure 2c). In Figure 2d, it can be observed that the influence of Mn in the alloy system is significant. Up to 15% of Mn in the alloy formed a complex structure composed of the majority of the phases met in the system. At higher percentages, there was an accentuated stabilization of the main BCC–A2 phase, reaching a structure based mostly on the BCC–A2 solid solution at 32%. Similar to the FeTaNbTiZr alloy system, the influence of iron on the phase evolution in the FeMnNbTiZr system was significant, as it considerably decreased the BCC phase. An interval of up to 10 wt.% is recommended (Figure 2e).

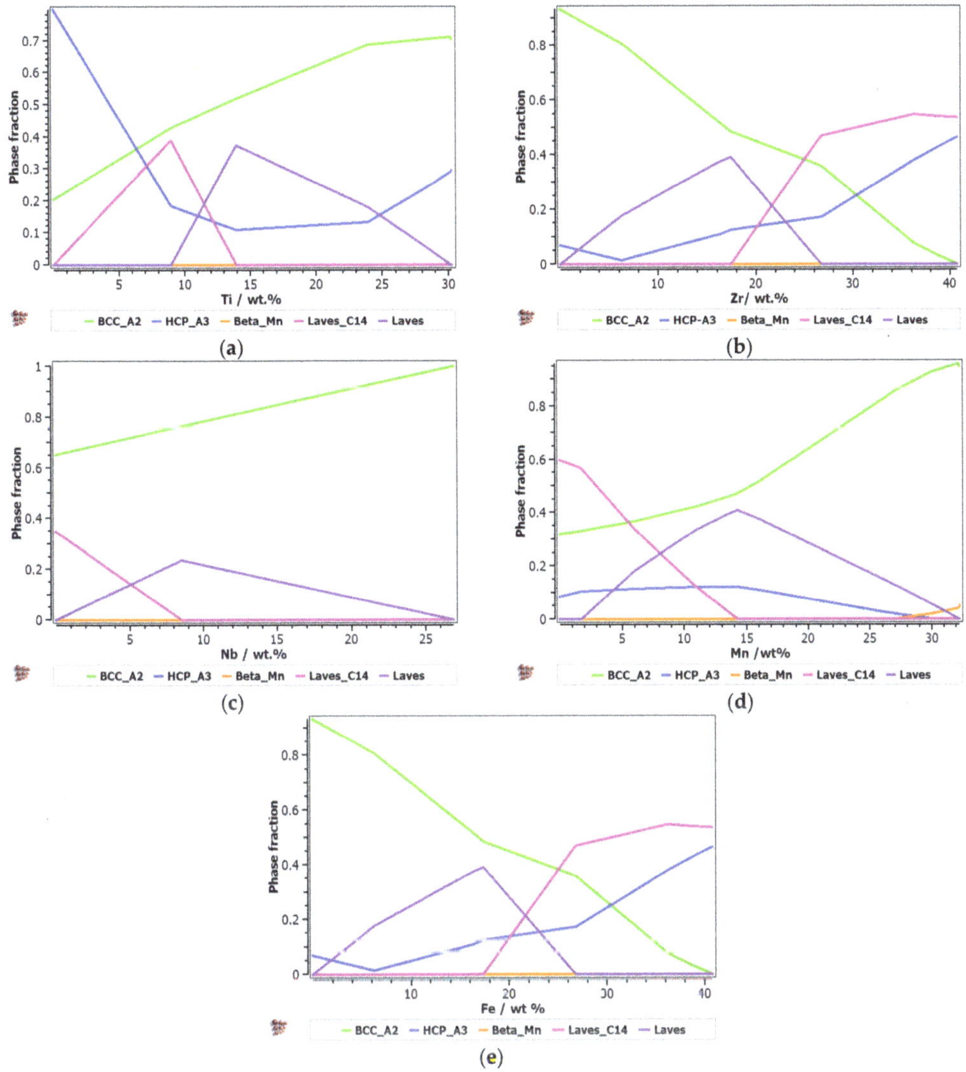

Figure 2. The proportions of phases that can be found in the FeMnNbTiZr alloy system at 200 °C, depending on the variations of (**a**) Ti; (**b**) Zr; (**c**) Nb; (**d**) Mn; (**e**) Fe.

3.2. Kinetic and Thermodynamic Criteria Calculation

To design the optimal composition of complex concentrated alloys with reduced contents of critical raw materials and potential biocompatible properties, for both selected alloy systems (FeTaNbTiZr and FeMnNbTiZr), calculations of the thermodynamic criteria were performed, varying the proportions of each element. The influence of each constituent element proportion on the calculated parameters is shown in Figures 3 and 4. The areas between the dotted lines represent the boundary intervals between which solid solutions can be formed, depending on the conditions of each parameter.

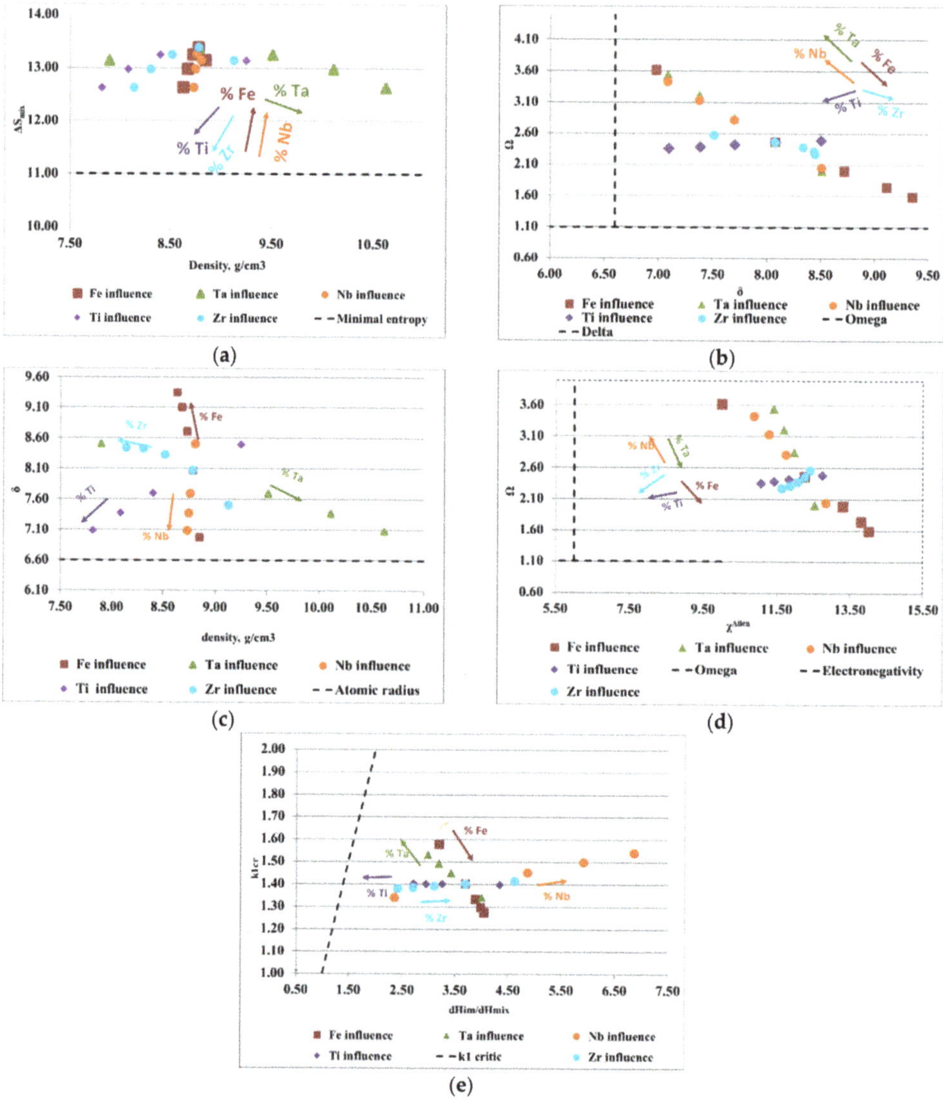

Figure 3. Graphical representation of the variations between (**a**) density, ρ, and the configurational entropy, ΔS_{mix}; (**b**) the atomic radius difference, δ, and the derived parameter, Ω; (**c**) density, ρ, and the atomic radius difference, δ; (**d**,**e**) the ratio between the enthalpy of intermetallic compounds (ΔH_{im}) and the mixing enthalpy (ΔH_{mix}) with the $k1_{cr}$ factor for the FeTaNbTiZr alloy system.

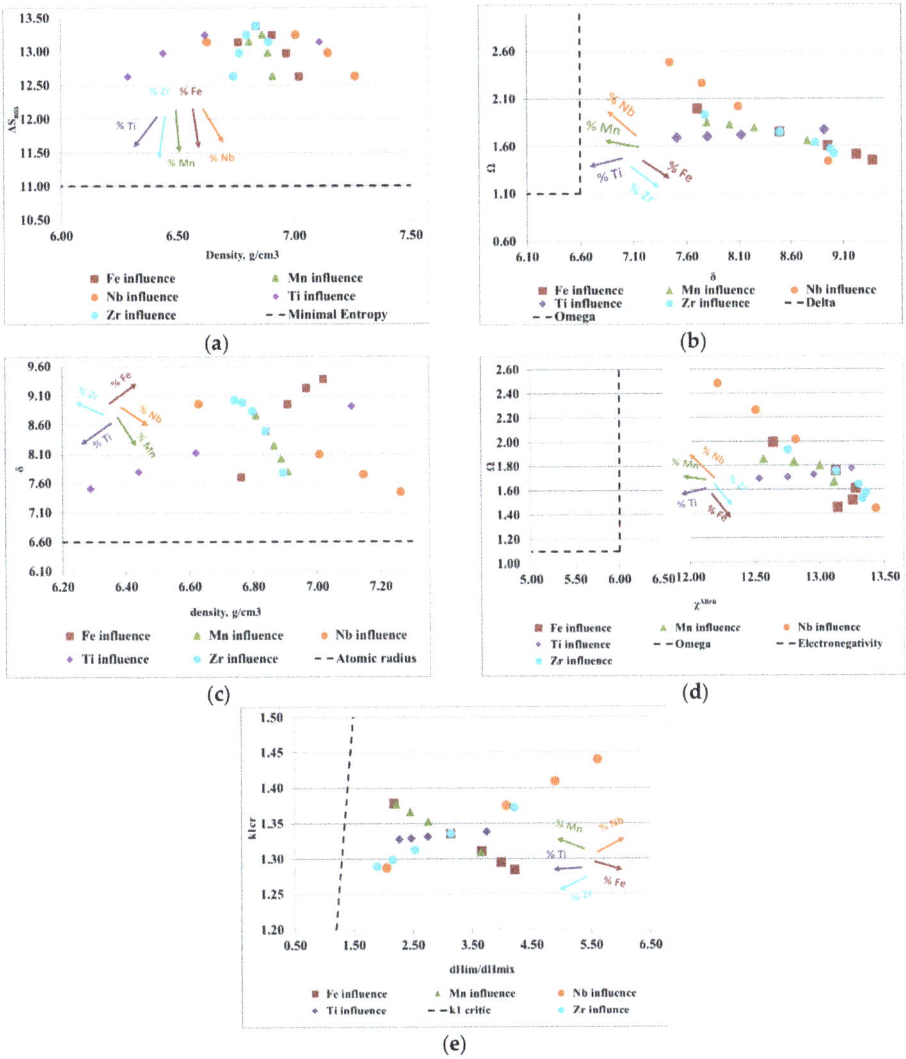

Figure 4. Graphical representation of the variations between (**a**) density, ρ, and the configurational entropy, ΔSmix; (**b**) the atomic radius difference, δ, and the derived parameter, Ω; (**c**) density, ρ, and the atomic radius difference, δ; (**d**) Allen electronegativity, χ^{Allen}, and the derived parameter, Ω; (**e**) the ratio between the enthalpy of intermetallic compounds (ΔHim) and the mixing enthalpy (ΔHmix) with the $k1_{cr}$ factor for the FeMnNbTiZr alloy system.

Figure 3a presents the ratio between the density of the alloy and the configurational entropy. It can be observed how the content of Ti, Zr, and Ta positively influenced the formation of solid solutions, while the increase in the content of Fe and Nb removed the alloy from the area considered optimal for the formation of solid solutions. Analyzing the variations of the derived parameter, Ω, with the atomic radius difference, δ (Figure 3b), it can be observed that an increase in the content of Ta, Nb, and Ti and a decrease in the content of Fe and Zr positively influenced the formation of solid solutions. Figure 3c illustrates the variations of the atomic radius difference, δ, with the density, ρ. Analyzing the trend of the chemical elements in the FeTaNbTiZr alloy system, it can be observed that increasing the content of Ta, Ti, and Nb, and decreasing the Fe and Zr content, positively influenced the formation of solid solution structures. Figure 3d presents the variations of the derived parameter Ω with Allen electronegativity, where it can be noticed that the elements Ti and Zr had a direct influence on the formation of solid solutions. On the other hand, maintaining low values of Fe and Ta had a positive influence, due to the positive characteristics they offered to the alloy; however, an increase in the concentrations of these elements can negatively influence the formation of solid solutions. Figure 3e shows the variations of the ratio between the enthalpy of the intermetallic compounds and the mixture enthalpy, considering the $k1_{cr}$ factor, where it can be estimated that an increase in the Ti and Ta contents had a positive influence on solid solutions formation, while Zr, Fe, and Nb are not good solid solution formers.

Figure 4a shows the ratio between the alloy density and mixing entropy, where Ti, Zr, Mn, Fe, and Nb have a positive influence on the formation of solid solutions. Analyzing the variation of the derived parameter, Ω, with atomic radius difference, δ (Figure 4b), it can be observed that an increase in Mn, Nb, and Ti and a decrease in Zr and Fe stimulates the formation of solid solutions. The variations of the atomic radius difference, δ, with the density, ρ (Figure 4c), provide information regarding the elements that are good solid solution formers, such as Mn, Ti, and Nb. Regarding the diagram (Figure 4d), where the influence of the Allen electronegativity on the derived parameter is presented, Ω, it can be noted that Ti, Mn, and Nb were good solid solution formers. Moreover, the fluctuation of the ratio between the enthalpy of intermetallic compounds (ΔH_{im}) and the mixing enthalpy (ΔH_{mix}) with the $k1_{cr}$ factor provides information regarding the elements such as Ti, Mn, and Nb that are desirable for the composition of complex concentrated alloys.

3.3. CCA Selection Using the CALPHAD Method and Kinetic and Thermodynamic Criteria

With the analysis of the CALPHAD modeling results for the FeTaNbTiZr alloy system, it was observed that a structure that was preponderantly based on solid solutions was obtained at high percentages of Ta [29], Nb, and Zr. Otherwise, the Ti had to be kept at relatively low percentages. By replacing the Ta with Mn in the FeMnNbTiZr alloy system, the most important influence on the formation of a structure composed of solid solutions was Zr [30].

The criteria calculation results provided information regarding the alloys' structures, which were influenced by the proportion of constituent elements, which can have a positive or negative influence on the formation of solid solutions. For the FeTaNbTiZr alloy system, the Ta and Nb had a favorable influence on the evolution of the Ω and δ parameters, while the Ti reduced the value of density and improved the $k1_{cr}$ critical factor. However, according to the European Commission list (Table 1), titanium is considered a critical metal, which implies that a sustainable alloy design has low Ti content, in terms of the FeTaNbTiZr alloy system. Although Ta increased the density of the alloy, it contributed substantially to the improvement of the selection criteria. Because Ta is also included in the critical raw material list (Table 1), a substitution with Mn was suggested, due to the similar solid solution formation behavior that both metals develop in combination with other elements, thus obtaining a new alloy system, FeMnNbTiZr.

The modeling results show a good capability of the selected systems to form structures based majorly on solid solution phases. It has been shown before that high percentages of Ta, Ti, Nb, Zr, and Mn favor the formation of BCC-type solid solution structures, while Fe has a negative influence on the formation of hard HCP–A3 and Laves phases. The results of the criteria simulations also show that Zr and Ta could have a negative impact on the final structure, as opposed to the CALPHADS modeling results. Suggestions made by the CALPHAD and criteria optimization were taken into account in the selection of the alloy compositions. However, considering the high cost of some elements (e.g., Ta and Nb) and their scarcity, the selection of the appropriate composition changed considerably. In this case, it was preferable to have higher contents of Fe and Mn, as well as lower contents of Ta, Nb, and Ti, which are considered critical metals. However, the simulation results show a favorable influence of the critical elements on the formation of solid solution structures. A compromise composition was selected in order to meet all the proposed requirements, after several preliminary trials. Thus, three new CCAs with low contents of critical raw materials and inexpensive obtaining processes were developed—$FeTa_{0.5}Nb_{0.5}TiZr_{0.5}$, $FeTa_{0.5}Nb_{0.5}Ti_{1.5}Zr_{0.5}$, and $FeMnNb_{0.5}TiZr_{0.5}$. The Ti presence was still significant, even though it is an unwanted critical element, so a comparison was done with a lower Ti content for the second alloy.

The thermodynamic and kinetic criteria, calculated for the selected complex concentrated alloys, are presented in Table 2.

Table 2. Thermodynamic and kinetic criteria for the formation of optimal CCAs.

Alloy	ΔS_{mix}, J/mol•K	ΔH_{mix}, kJ/mol	δ, %	VEC, %	$\Delta\chi^{Allen}$, %	T_m	Ω	ΔH_{im}	$k1_{cr}$	ρ, g/cm^3
$FeTa_{0.5}Nb_{0.5}Ti_{1.5}Zr_{0.5}$	12.42	−15.56	8.02	5.25	12.78	2201.78	1.76	−15.56	1.31	7.61
$FeTa_{0.5}Nb_{0.5}TiZr_{0.5}$	12.89	−16.41	8.52	5.43	13.28	2239.01	1.76	−16.41	1.31	8.06
$FeMnNb_{0.5}TiZr_{0.5}$	12.97	−15.38	8.23	5.88	12.92	1927.40	1.63	−15.38	1.33	6.65

Equilibrium and Scheil–Gulliver diagrams were calculated for the selected CCA compositions to determine the phase proportions identified in the alloy in specific solidification conditions. In the FeTaNbTiZr alloy system, it can be observed that the complex concentrated alloy with a high Ti content had a higher content of Laves intermetallic phases at room temperature (Figure 5). However, titanium made a considerable contribution to increasing the biocompatible properties of the alloy, so it was absolutely necessary to include it in the final composition. The Scheil–Gulliver diagrams (Figure 6) indicate high stability of the BCC–A2 phase, which was formed first during the solidification process. Thus, the Laves intermetallic phases mostly formed in the interdendritic area of the alloy.

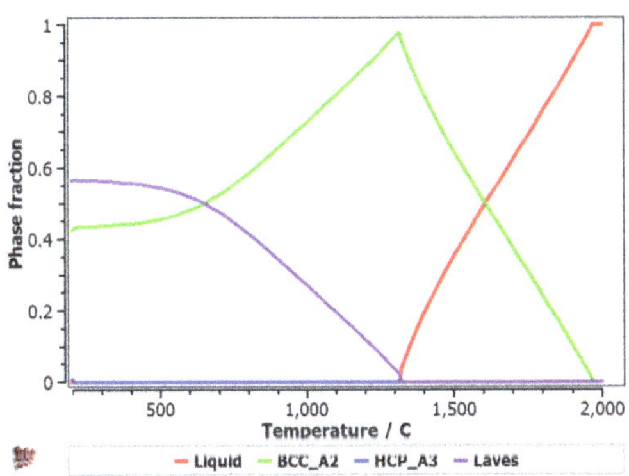

(a)

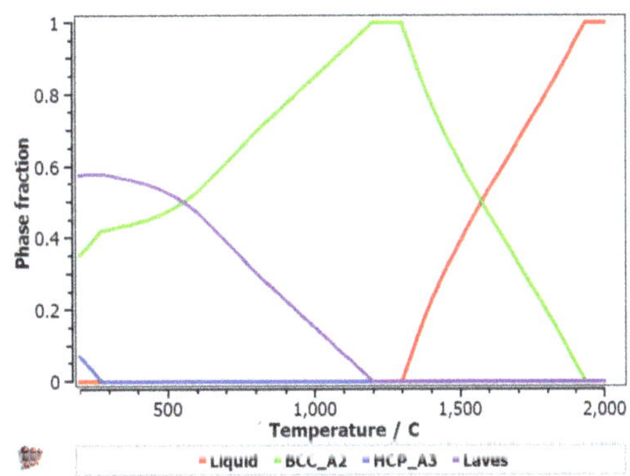

(b)

Figure 5. Equilibrium diagrams for the (a) $FeTa_{0.5}Nb_{0.5}TiZr_{0.5}$ and (b) $FeTa_{0.5}Nb_{0.5}Ti_{1.5}Zr_{0.5}$ alloys.

The equilibrium diagram of the $FeMnNb_{0.5}TiZr_{0.5}$ (Figure 7) indicates the high stability of the solid solution, based on a BCC–A2 structure. At low temperatures, the Laves intermetallic phases had high contents, which determined a complex multiphase structure with low percentages of the HCP–A3 and Beta–Mn phases.

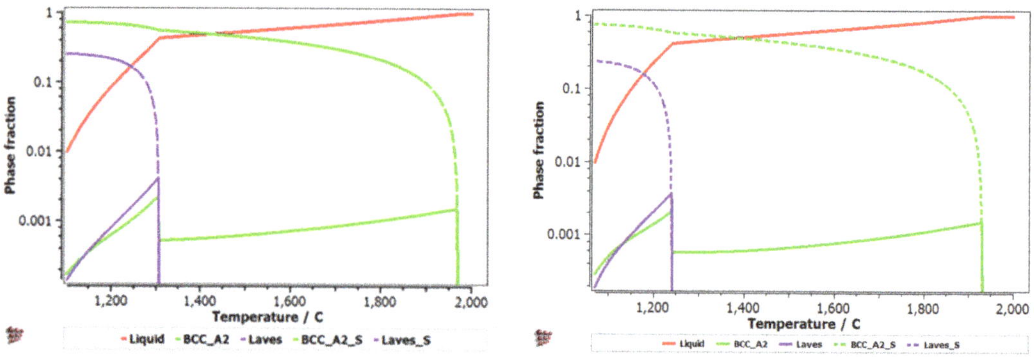

(a) (b)

Figure 6. Scheill–Gulliver diagrams for the (**a**) $FeTa_{0.5}Nb_{0.5}TiZr_{0.5}$ and (**b**) $FeTa_{0.5}Nb_{0.5}Ti_{1.5}Zr_{0.5}$ alloys.

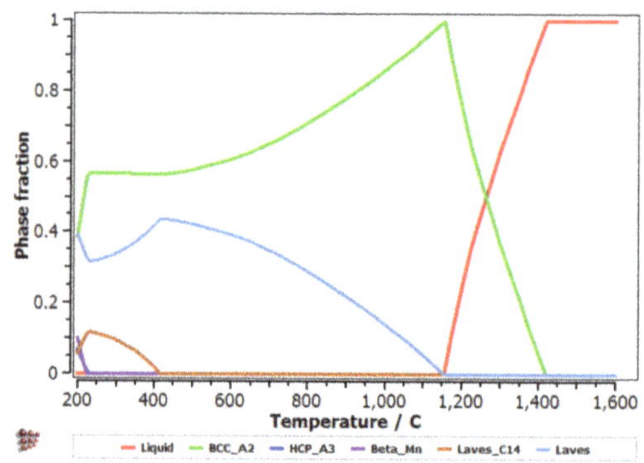

Figure 7. Equilibrium diagram of the $FeMnNb_{0.5}TiZr_{0.5}$ alloy.

The Scheill–Gulliver solidification curve (Figure 8) indicates a high degree of subcooling. The model of the solidification process of the alloy and considering Zr diffusion and its impact on changing the structure in the solidification area, a decrease in the subcooling degree can be observed.

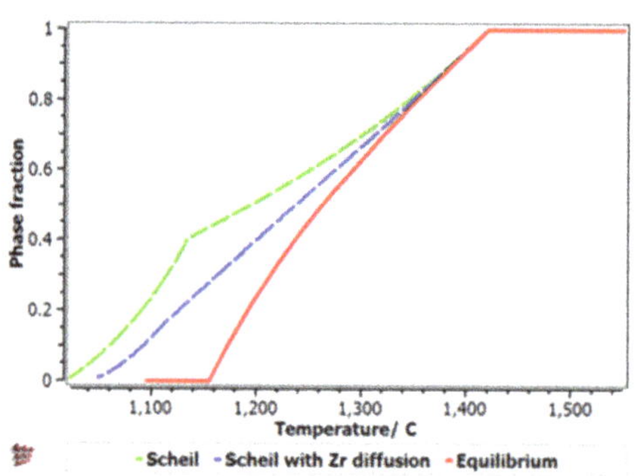

Figure 8. Scheil–Gulliver cooling diagram with Zr diffusion for the $FeMnNb_{0.5}TiZr_{0.5}$ alloy.

With the analysis of the equilibrium diagram, it can be observed that the high stability of the BCC–A2 phase determined a primary solidification at a temperature of 1420 °C, according to the Scheil–Gulliver solidification diagram (Figure 9). The process was followed by the formation of the Laves intermetallic phase at a temperature difference lower than 200 °C.

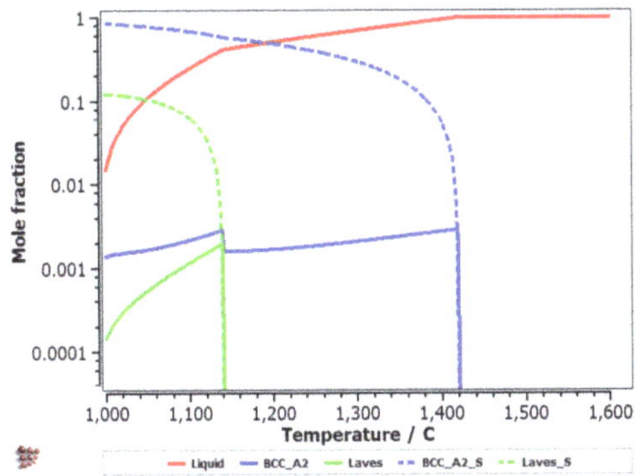

Figure 9. Scheill–Gulliver diagrams for $FeMnNb_{0.5}TiZr_{0.5}$.

3.4. Empirical Results

The alloys were examined in the as-cast state using chemical analysis, scanning electron microscopy (SEM), and X-ray diffraction (XRD).

The chemical and physical analysis results are presented in Table 3. The resulting alloy composition was found to be within 3% of the limits of the nominal values for the used elements [31].

Table 3. The chemical composition and density of alloys.

Alloy	Composition Tip	wt.%						Density g/cm³
		Ti	Fe	Mn	Nb	Ta	Zr	
$FeTa_{0.5}Nb_{0.5}Ti_{1.5}Zr_{0.5}$	Analyses	24.17	19.15		15.55	27.56	13.57	7.61
	Nominal	23.14	18.02	-	14.97	29.17	14.7	
$FeMnNb_{0.5}TiZr_{0.5}$	Analyses	17.89	21.73	22.36	22.04		15.97	6.65
	Nominal	19.09	22.28	21.92	18.53	-	18.18	
$FeTa_{0.5}Nb_{0.5}TiZr_{0.5}$	Analyses	18.83	21.21		15.27	29.74	14.95	8.06
	Nominal	16.73	19.51	-	16.22	31.62	15.92	
$FeTa_{0.5}Nb_{0.5}Ti_{1.5}Zr_{0.5}$ remelted	Analyses	19.2	18.42	-	15.8	25	12.4	7.61
	Nominal	22.16	20.28		16.39	27.52	13.65	7.61

The SEM analyses of the as-cast and remelted $FeTa_{0.5}Nb_{0.5}Ti_{1.5}Zr_{0.5}$ alloy showed different types of dendritic structures with variable phase compositions from the center to the borders (Figure 10a,b). The remelted alloy presented larger and branched dendrites compared to the as-cast sample, which presented small and well-dispersed dendrite fragments. Six separate phases were distinguished in both structures. According to the EDS analysis, the dendrite area (DR) contained more Nb, Ti, and Ta (Figure 10) in the melted (Table 4) and re-melted (Table 5) $FeTa_{0.5}Nb_{0.5}Ti_{1.5}Zr_{0.5}$ alloys. The interdendritic area contained two phases–the phase corresponding to ID1 contained Ti, Fe, Zr, and small quantities of Ta and Nb. The phase corresponding to ID2 contained Ti, Nb, Fe, and small quantities of Ta and Zr. Additionally, in the interdendritic area, the presence of a eutectic with fine lamellar morphology in an as-cast sample and with fine punctiform distribution in the remelted state (ID3) was observed, which, according to the EDS analysis on the component corresponding to the ID3 point, contained high percentages of Ti, Fe, Zr, and less Ta and Nb. In the case of the component corresponding to ID4, a predominant Ti and Fe composition was distinguished; the rest of the elements were at lower values but still significant. A small phase that was identified in the interdendritic area is indicated by ID5, with different composition gradients for the two samples. The as-cast sample showed a predominant Ti and Fe composition, and the re-melted sample showed a predominant Ti, Nb, and Ta composition.

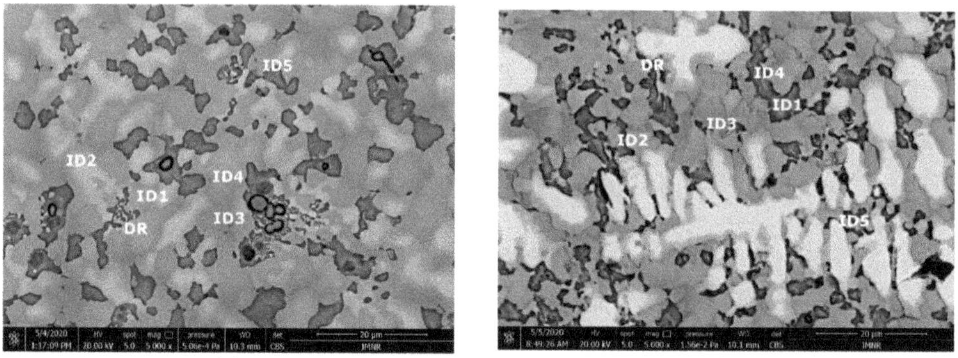

(a) (b)

Figure 10. SEM-EDS image of the $FeTa_{0.5}Nb_{0.5}Ti_{1.5}Zr_{0.5}$ alloy in as-cast (**a**) and re-melted (**b**) states.

Table 4. Phase compositions for the as-cast $FeTa_{0.5}Nb_{0.5}Ti_{1.5}Zr_{0.5}$ alloy.

Phase	Composition, wt.%				
	Zr	Nb	Ti	Ta	Fe
DR	4.17	30.30	43.39	14.38	7.76
ID1	14.16	4.01	45.47	2.35	34.01
ID2	6.88	15.23	51.36	5.53	20.99
ID3	11.98	7.18	45.62	3.74	31.48
ID4	9.76	14.75	41.87	9.03	24.59
ID5	11.59	9.03	41.48	5.66	32.24

Table 5. EDS compositions for the remelted $FeTa_{0.5}Nb_{0.5}Ti_{1.5}Zr_{0.5}$ alloy, in wt.%.

Phase	Composition, wt.%				
	Zr	Nb	Ti	Ta	Fe
DR	0.00	32.84	31.57	29.00	6.58
ID1	14.70	0.00	46.94	2.36	36.00
ID2	13.64	6.06	37.13	5.11	38.06
ID3	13.52	6.00	37.83	5.16	37.49
ID4	11.70	8.03	36.61	8.06	35.61
ID5	4.38	24.93	45.90	16.78	8.01

The EDS results (Figures 11 and 12) show large differences in terms of dendrite size and element distribution. The as-cast alloy presented a homogenous structure with uniform element distributions, besides small size segregations for Zr. The re-melted sample presented well-defined dendrites with a high concentration of Ta, and less Nb and Ti. The interdendritic area was characterized by a major presence of Fe and Zr.

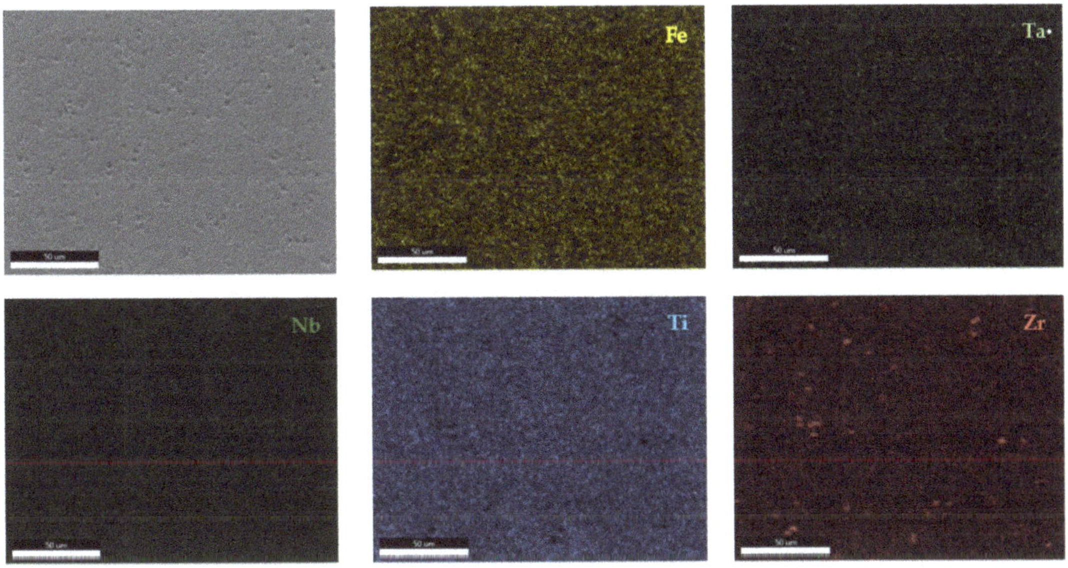

Figure 11. EDS mapping of the $FeTa_{0.5}Nb_{0.5}Ti_{1.5}Zr_{0.5}$ alloy.

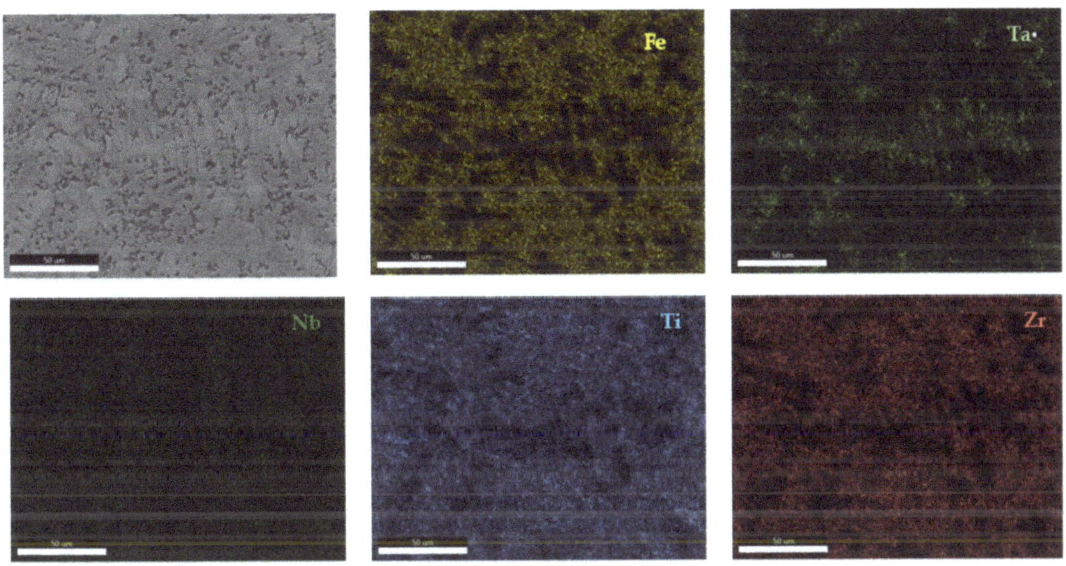

Figure 12. EDS mapping of the remelted $FeTa_{0.5}Nb_{0.5}Ti_{1.5}Zr_{0.5}$ alloy.

The XRD phase analysis for the melted (Figure 13) and remelted (Figure 14) $FeTa_{0.5}Nb_{0.5}Ti_{1.5}Zr_{0.5}$ alloys revealed similar structure configurations, showing three different BCC–A2 type phases, a Laves phase, a complex cubic phase, and a hexagonal compact phase. Phase proportion calculations revealed high contents of Laves–C14 and BCC–β1–A2 phases, which changed between the two samples from 40 wt.% and 20 wt.% to 32 wt.% and 31 wt%, respectively. The other phases, BCC—β2–A2 (14 wt.%), HCP–A3 (13 wt.%), complex cubic–D8a (8 wt.%), and BCC–A2 (5 wt.%) had minor proportions. The high number of phases suggests a less stable structure that needs special heat treatment processing to reach the equilibrium configurations.

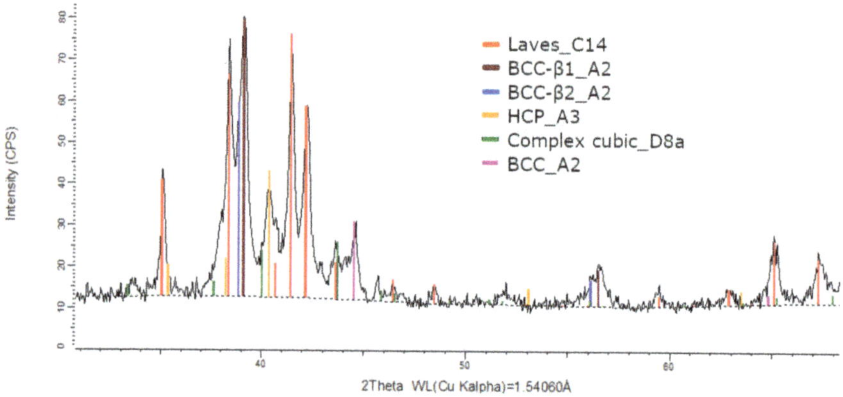

Figure 13. X-ray diffraction pattern for the $FeTa_{0.5}Nb_{0.5}Ti_{1.5}Zr_{0.5}$ alloy.

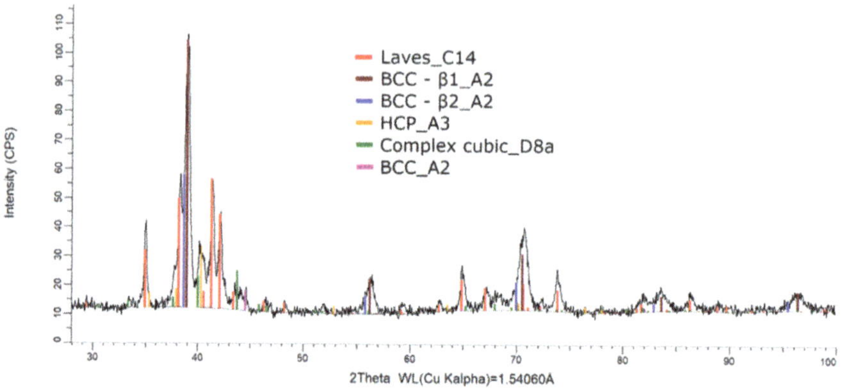

Figure 14. X-ray diffraction pattern for the remelted $FeTa_{0.5}Nb_{0.5}Ti_{1.5}Zr_{0.5}$ alloy.

The SEM analyses of the $FeTa_{0.5}Nb_{0.5}TiZr_{0.5}$ alloy revealed a dendritic structure (Figure 15) with larger branches than for the $FeTa_{0.5}Nb_{0.5}Ti_{1.5}Zr_{0.5}$ alloy. According to the EDS analysis (Table 6 and Figure 16), the dendritic area (DR) contained mostly Ta, followed by Nb and smaller quantities of Ti and Fe. The interdendritic area (ID2) contained Ta, Fe, Ti, and smaller quantities of Zr and Nb. A smaller phase was distinguished in the interdendritic area, with a major proportion of Ti and less Ta, Nb, Fe, and Zr (ID3). Additionally, in the interdendritic area, the presence of another phase that contained mostly Ti, Fe, Zr, and smaller quantities of Ta and Nb (ID1) was observed.

Table 6. EDS composition for the $FeTa_{0.5}Nb_{0.5}TiZr_{0.5}$ alloy, in wt.%.

Phase	Composition, wt.%				
	Zr	Nb	Ti	Ta	Fe
DR	-	28.95	21.93	41.02	8.11
ID1	17.26	4.05	40.96	6.31	31.42
ID2	10.35	7.18	27.13	14.73	40.61
ID3	4.84	13.16	60.68	8.83	12.49

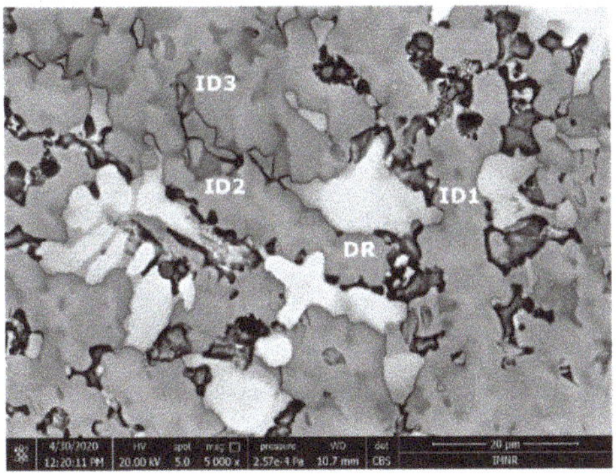

Figure 15. SEM-EDS images for the as-cast $FeTa_{0.5}Nb_{0.5}TiZr_{0.5}$ alloy sample with the selected points for EDS analysis.

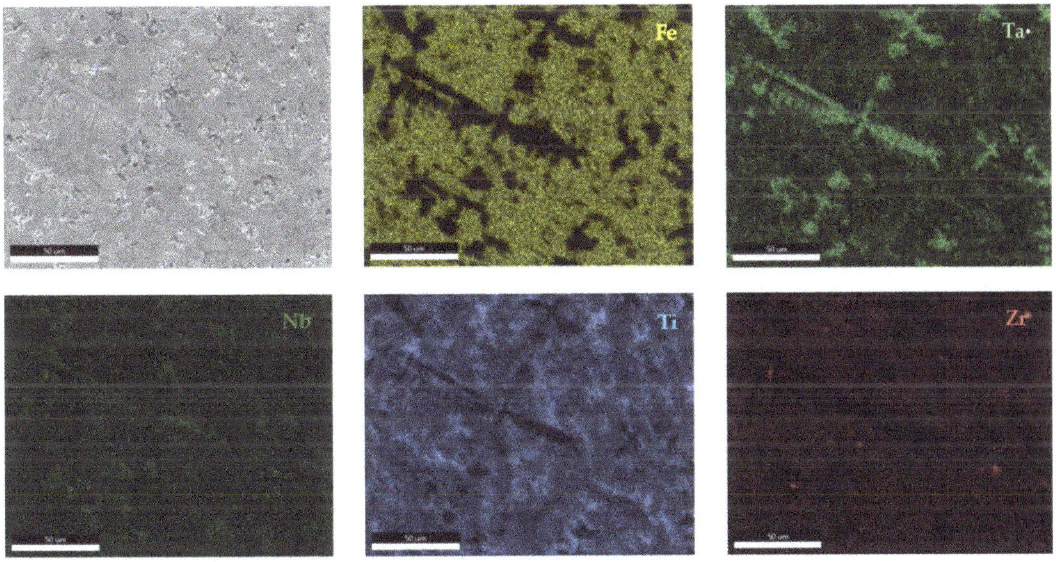

Figure 16. EDS mapping of $FeTa_{0.5}Nb_{0.5}TiZr_{0.5}$ alloy.

Compared to the alloy with a higher Ti concentration, the $FeTa_{0.5}Nb_{0.5}TiZr_{0.5}$ alloy showed a reduced phase composition in the XRD analysis report (Figure 17). Several structures, BCC–β1, BCC–β2, Laves phase, and the complex cubic phase, were detected. Due to the phase proportion calculations, it was observed that the laves–C14 phase (56 wt.%) was predominant, followed by BCC–β1–A2 (22 wt.%) phase, BCC–β2–A2 (12 wt.%), and complex cubic–D8a (10 wt.%). The XRD results for the $FeTa_{0.5}Nb_{0.5}TiZr_{0.5}$ alloy show fewer phases than the alloy with a higher Ti content, but unstable BCC–A2 variations are still present.

According to the SEM images (Figure 18), the $FeMnNb_{0.5}TiZr_{0.5}$ alloy sample had a large dendritic structure. The EDS analysis (Table 7 and Figure 19), shows that the dendritic area (DR) contained mostly Mn, Ti, and smaller quantities of Fe, Zr, and Nb. In the interdendritic area, the presence of two phases can be observed (ID1 and ID2). These phases had a high concentration of Ti, followed by Nb, Mn, Fe, and Zr.

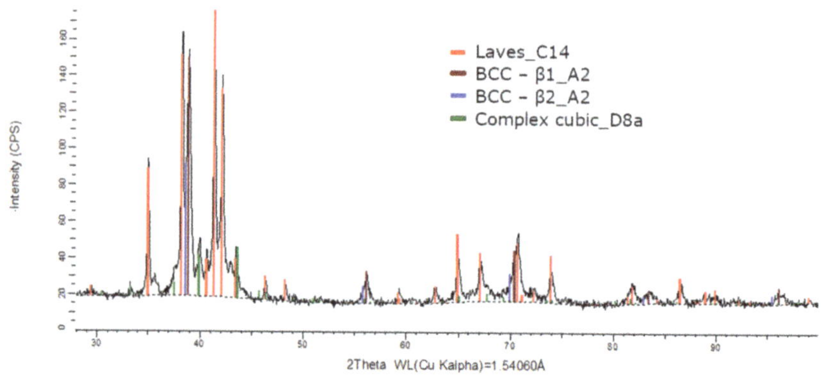

Figure 17. X-ray diffraction pattern for the $FeTa_{0.5}Nb_{0.5}TiZr_{0.5}$ alloy.

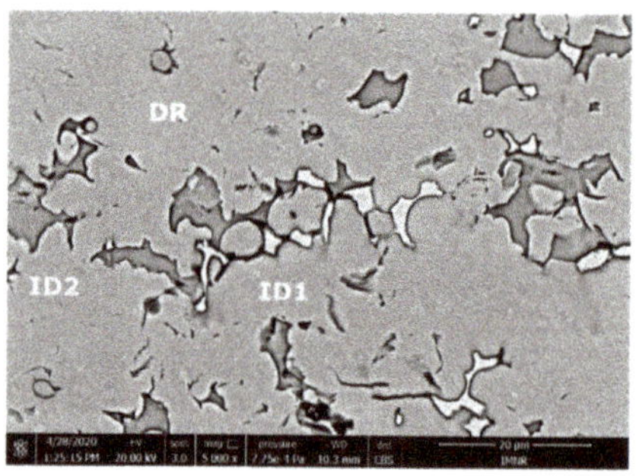

Figure 18. SEM image of the $FeMnNb_{0.5}TiZr_{0.5}$ alloy sample with the selected points for EDS analysis.

Table 7. EDS composition for the $FeTa_{0.5}Nb_{0.5}TiZr_{0.5}$ alloy, in wt.%.

Phase	Composition, wt.%				
	Zr	**Nb**	**Ti**	**Ta**	**Fe**
DR	7.33	16.20	30.65	32.41	13.41
ID1	6.82	12.70	38.86	25.72	12.70
ID2	7.00	13.26	39.33	20.78	17.21

The XRD phase analysis, presented in Figure 20, reveals several structures—the BCC–β1, Laves, and hexagonal compact phases. The Laves–C14 phases were present in a higher proportion (82 wt.%) than the BCC–β1–A2 (14 wt.%) and HCP–A3 (4 wt.%).

The samples were analyzed to determine the microhardness (Table 8) and it was observed that the FeTa0.5Nb0.5TiZr0.5 was the alloy with the highest microhardness (898.02 HV), followed closely by $FeTa_{0.5}Nb_{0.5}Ti_{1.5}Zr_{0.5}$ and $FeMnNb_{0.5}TiZr_{0.5}$. After the re-melting of the $FeTa_{0.5}Nb_{0.5}Ti_{1.5}Zr_{0.5}$ alloy, the microhardness decreased by approximately 13%.

The selection criteria applied to the $FeTa_{0.5}Nb_{0.5}TiZr_{0.5}$, $FeTa_{0.5}Nb_{0.5}Ti_{1.5}Zr_{0.5}$, and $FeMnNb_{0.5}TiZr_{0.5}$ alloys showed good potential for the formation of mixed structures containing either solid solutions (SS) or intermetallic compounds (IM). The experimental findings match the criteria calculations for the $FeTa_{0.5}Nb_{0.5}TiZr_{0.5}$ and $FeTa_{0.5}Nb_{0.5}Ti_{1.5}Zr_{0.5}$ alloys, where high proportions of SS and IM were distinguished. Nevertheless, for the $FeMnNb_{0.5}TiZr_{0.5}$ alloy, the experimental findings show much larger contents of intermetallic Laves-type phases than the criteria calculation results.

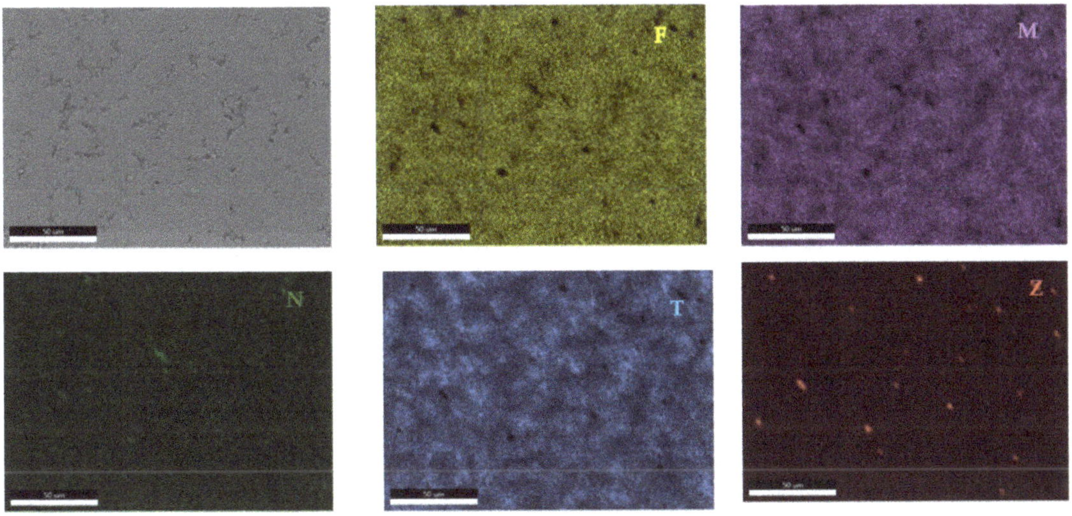

Figure 19. EDS mapping of the $FeMnNb_{0.5}TiZr_{0.5}$ alloy.

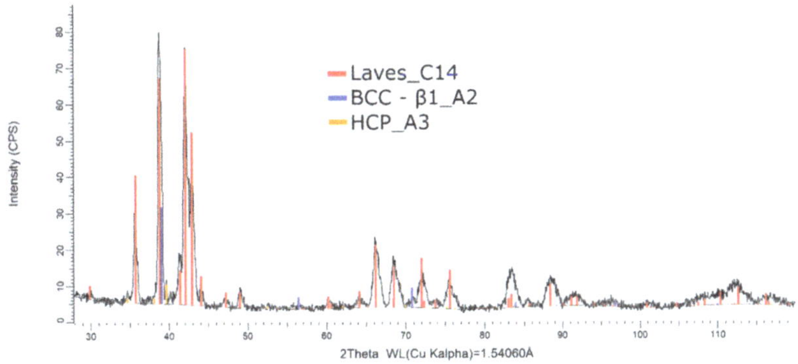

Figure 20. X-ray diffraction pattern for the $FeMnNb_{0.5}TiZr_{0.5}$ alloy.

Table 8. Vickers microhardness for the obtained alloys.

Alloy	Vickers Microhardness, HV
$FeTa_{0.5}Nb_{0.5}Ti_{1.5}Zr_{0.5}$	802.8
re-melted $FeTa_{0.5}Nb_{0.5}Ti_{1.5}Zr_{0.5}$	699.0
$FeTa_{0.5}Nb_{0.5}TiZr_{0.5}$	898.2
$FeMnNb_{0.5}TiZr_{0.5}$	802.9

The CALPHAD method produced similar phase configurations with those obtained in the experimental results for the $FeTa_{0.5}Nb_{0.5}TiZr_{0.5}$ and $FeTa_{0.5}Nb_{0.5}Ti_{1.5}Zr_{0.5}$ alloys, where BCC and Laves phases had similar proportions. The thermodynamic simulation for the $FeMnNb_{0.5}TiZr_{0.5}$ alloy showed a significantly lower proportion of the Laves phase in the experimental results.

The phases that were present in both representations for the FeTaNbTiZr system were BCC–A2, HCP–A3, and Laves. A complex $D8_a$ phase was identified in the as-cast alloy sample and was not indicated by the CALPHAD simulation. The simulation of FeMnNbTiZr solidification behavior showed a larger number of phases (BCC–A2, HCP–A3, β–Mn, and Laves) than in the experimental results (Laves, BCC–A2 and HCP–A3).

Overall, the results obtained in the experimental trials relate well to the preliminary alloy structural modeling results, and the differences encountered in the phase composition can be attributed to the nonequilibrium nature of the casting process.

4. Conclusions

This paper presents the selection process and characterization of complex concentrated alloys with reduced contents of critical raw materials. With the analysis of the properties of the main elements used for obtaining complex concentrated alloys with reduced contents of critical raw materials, two alloy systems are proposed—FeTaNbTiZr and FeMnNbTiZr. Considering the variations of the contained elements, the MatCalc simulation program was used to analyze the solid solutions' redistribution during the solidification process. To design the optimal composition for both selected alloy systems, calculations of the kinetic and thermodynamic criteria were performed, varying the proportions of each element. For the FeTaNbTiZr alloy system, it was identified that high percentages of Ta, Nb, and Zr stimulate the formation of structures based on solid solutions. Ti has good biocompatible properties, but it is important it is kept at low percentages because it is included in the critical raw materials list. Besides reducing the Ti content, the substitution of Ta with Mn was also considered because of its high availability and known potential for the formation of solid solution structures, thus leading to the second selection for the alloy system—FeMnNbTiZr.

From the simulation results performed by CALPHAD modeling, the semi-empirical criteria calculations, and the critical characteristics of the used elements, several compositions were selected for further analysis—$FeTa_{0.5}Nb_{0.5}Ti_{1.5}Zr_{0.5}$, $FeTa_{0.5}Nb_{0.5}TiZr_{0.5}$, and $FeMnNb_{0.5}TiZr_{0.5}$.

The selected alloys were prepared in an induction furnace for improved homogeneity and reduced processing stages. Significant structural changes were observed after the remelting process, shown for the $FeTa_{0.5}Nb_{0.5}Ti_{1.5}Zr_{0.5}$.

The empirical results provided by SEM-EDS and XRD analyses of the $FeTa_{0.5}Nb_{0.5}Ti_{1.5}Zr_{0.5}$ (melted and remelted), $FeMnNb_{0.5}TiZr_{0.5}$, and $FeTa_{0.5}Nb_{0.5}TiZr_{0.5}$ alloys indicate a significant content in solid solutions as well as in intermetallic phases. The as-cast $FeTa_{0.5}Nb_{0.5}Ti_{1.5}Zr_{0.5}$ alloy presented a refined structure, containing a preponderant BCC–β1–A2 solid solution phase and characterized by different types of dendritic structures, with a variable phase composition from the center to the borders. The structure of the remelted alloy can be described by larger and branched dendrites, in comparison to the small and well-dispersed dendrite fragments of the melted composition. The differences between the melted and remelted $FeTa_{0.5}Nb_{0.5}Ti_{1.5}Zr_{0.5}$ alloys, in terms of dendrite size and element distribution, were observed in the EDS analysis.

The as-cast $FeTa_{0.5}Nb_{0.5}Ti_{1.5}Zr_{0.5}$ alloy showed a dendritic structure with uniform element distribution, besides small size segregations of Zr, while the remelted one presented well-defined dendrites. The as-cast $FeTa_{0.5}Nb_{0.5}TiZr_{0.5}$ alloy sample, where the titanium content was reduced, had a small number of phases compared to the $FeTa_{0.5}Nb_{0.5}Ti_{1.5}Zr_{0.5}$ alloys (melted and remelted), and the dendritic structures have larger branches. The CCA composition based on manganese, $FeMnNb_{0.5}TiZr_{0.5}$, had a smaller number of phases and a large dendritic structure.

Considering the XRD results, it can be observed that the melted and remelted $FeTa_{0.5}Nb_{0.5}Ti_{1.5}Zr_{0.5}$ alloys showed three different BCC–A2-type phases—a Laves phase, a complex cubic phase, and a hexagonal compact phase, with similar phase proportions. With the decrease in Ti content, the number of phases decreased compared to the previous alloy, but unstable BCC-A2 variations were still present. The $FeMnNb_{0.5}TiZr_{0.5}$ alloy composition presented a structure based on BCC–β1–A2, Laves, and hexagonal compact phases.

By comparing the criteria and with the MatCalc simulation with the empirical results, it was observed that several variations of the BCC–A2 phase were present in the prepared samples, while more intermetallic phases were present in the simulation diagrams. However, the structural findings are similar in terms of phase constitution. BCC–A2, HCP–A3, and Laves appeared in both representations.

The studied alloys represent good options for further studies on the replacement and improvement of conventional alloys with biocompatibility properties.

Author Contributions: Conceptualization, B.-A.Ș., I.-C.B., and I.A.; methodology, M.T.O. and I.A.; software, D.M., I.-C.B., and I.A.; validation, N.C., M.B., and A.M.-J.P.; formal analysis, S.-E.B.; investigation, A.-N.G. and M.T.O.; resources, M.B. and A.M.-J.P.; data curation, B.-A.Ș., D.M., and N.C.; writing—original draft preparation, I.-C.B.; writing—review and editing, M.T.O. and S.-E.B.; visualization, D.M.; supervision, B.-A.Ș. and A.-N.G.; project administration, I.-C.B. and I.A.; funding acquisition, D.M. and M.B. All authors have read and agreed to the published version of the manuscript.

Funding: This work was supported by research grants of the Romanian National Authority for Scientific Research and Innovation, CNCS/CCCDI—UEFISCDI, project numbers: ERANET-M-COMⓟTRANS-1, within PNCDI III and PN-III-P2-2.1-PED-2019-2022, within PNCDI III.

Institutional Review Board Statement: Not applicable.

Informed Consent Statement: Not applicable.

Data Availability Statement: Not applicable.

Conflicts of Interest: The authors declare no conflict of interest. The funders had no role in the design of the study; in the collection, analyses, or interpretation of data; in the writing of the manuscript, or in the decision to publish the results.

References

1. Constantinescu, D.; Cârlan, B.A. A short history of the iron and steel industry in Central Europe during the Roman Iron Age. In Proceedings of the 25th International Conference on Metallurgy and Materials Metal 2016, Brno, Czech Republic, 25–27 May 2016. Available online: http://metal2016.tanger.cz/en/ (accessed on 30 January 2017).
2. Tylecote, R.F. *A History of Metallurgy*, 2nd ed.; The Institute of Materials: London, UK, 1992; ISBN 1-902653-79-3.
3. Santos, G.A. The importance of metallic materials as biomaterials. *J. Tissue Eng. Regen. Med.* **2017**, *3*, 300–302. [CrossRef]
4. Wintermantel, E.; Mayer, J.; Ruffieux, K.; Bruinink, A.; Eckert, K.L. Biomaterials, human tolerance and integration. *Chirurg* **1999**, *70*, 847–857. [CrossRef]
5. Chen, Q.; Thouas, G.A. Metallic implant biomaterials. *Mater. Sci. Eng. R Rep.* **2015**, *87*, 1–57. [CrossRef]
6. Silver, F.H.; Christiansen, D.L. *Biomaterials Science and Biocompatibility*; Springer: Berlin, Germany, 1999.
7. Love, B. *Metallic Biomaterials*; Academic Press: Cambridge, MA, USA, 2017. [CrossRef]
8. Gunnarsdóttir, I.; Davidsdottir, B.; Worrell, E.; Sigurgeirsdóttir, S. Sustainable energy development: History of the concept and emerging themes. *Renew. Sustain. Energy Rev.* **2021**, *141*, 110770. [CrossRef]
9. European Commission Report. Critical Raw Materials Resilience: Charting a Path towards Greater Security and Sustainability. Communication from the Commission to the European Parliament COM (2020) 474 Final, Brussels. Available online: https://www.europeansources.info/record/critical-raw-materials-resilience-charting-a-path-towards-greater-security-and-sustainability/ (accessed on 3 September 2021).
10. Hofmann, M.; Hofmann, H.; Hagelüken, C.; Hool, A. Critical raw materials: A perspective from the materials science community. *Sustain. Mater. Technol.* **2018**, *17*, e00074. [CrossRef]
11. Cantor, B.; Chang, I.T.H.; Knight, P.; Vincent, A.J.B. Microstructural development in equiatomic multicomponent alloys. *Mater. Sci. Eng.* **2004**, *375–377*, 213–218. [CrossRef]
12. Gorsse, S.; Couzinié, J.P.; Miracle, D.B. From high-entropy alloys to complex concentrated alloys. *Comptes Rendus Phys.* **2018**, *19*, 721–736. [CrossRef]
13. Gorsse, S.; Miracle, D.B.; Senkov, O.N. Mapping the world of complex concentrated alloys. *Acta Mater.* **2017**, *135*, 177–187. [CrossRef]
14. Mukherjee, S. Complex concentrated alloys (CCAs)—Current understanding and future opportunities. *Metals* **2020**, *10*, 1253. [CrossRef]
15. Zhang, Y. Solid-solution phase formation rules for multi-component alloys. *Adv. Eng. Mater.* **2008**, *10*, 534–538. [CrossRef]
16. Manzoni, A.M.; Glatzel, U. New multiphase compositionally complex alloys driven by the high entropy alloy approach. *Mater. Charact.* **2019**, *147*, 512–532. [CrossRef]
17. Syrovatka, M. On sustainability interpretations of the ecological footprint. *Ecol. Econom.* **2020**, *169*, 106543. [CrossRef]
18. Agren, J. CALPHAD—An approach to predict microstructural stability. *Mater. Sci. Eng.* **2020**. [CrossRef]
19. Zhang, C.; Zhang, F.; Chen, S.; Cao, W. Computational Thermodynamics Aided High-Entropy Alloy Design. *JOM* **2012**, *64*, 839–845. [CrossRef]

20. Senkov, O.N.; Miller, J.D.; Miracle, D.B.; Woodward, C. Accelerated exploration of multi-principal element alloys with solid solution phases. *Nat. Commun.* **2015**, *6*, 1–10. [CrossRef] [PubMed]
21. Chen, H.; Mao, H.; Chen, Q. Database development and Calphad calculations for high entropy alloys: Challenges, strategies, and tips. *Mater. Chem. Phys.* **2018**, *210*, 279–290. [CrossRef]
22. Lederer, Y.; Toher, C.; Vecchio, K.S.; Curtarolo, S. The search for high entropy alloys: A high–throughput ab–initio approach. *Acta Mater.* **2018**, *159*, 364–383. [CrossRef]
23. Troparevsky, M.C.; Morris, J.R.; Kent, P.R.; Lupini, A.R.; Stocks, G.M. Criteria for predicting the formation of single-phase high-entropy alloys. *Phys. Rev.* **2018**, *5*, 1–6. [CrossRef]
24. Takeuchi, A.; Inoue, A. Calculations of Mixing Enthalpy and Mismatch Entropy for Ternary Amorphous Alloys. *Mater. Trans. JIM* **2000**, *41*, 1372–1378. [CrossRef]
25. Yang, X.; Zhang, Y. Prediction of high-entropy stabilized solid-solution in multi-component alloys. *Mater. Chem. Phys.* **2012**, *132*, 233–238. [CrossRef]
26. Poletti, M.G.; Battezzati, L. Electronic and thermodynamic criteria for the occurrence of high entropy alloys in metallic systems. *Acta Mater.* **2014**, *75*, 297–306. [CrossRef]
27. Guo, S. Effect of valence electron concentration on stability of fcc or bcc phase in high entropy alloys. *J. Appl. Phys.* **2011**, *109*, 103505. [CrossRef]
28. Singh, A.K.; Kumar, N.; Dwivedi, A.; Subramaniam, A. A geometrical parameter for the formation of disordered solid solutions in multi-component alloys. *Intermetallics* **2014**, *53*, 112–119. [CrossRef]
29. Cardonne, S.M.; Kumar, P.; Michaluk, C.A.; Schwartz, H.D. Tantalum and its Alloys. *Int. J. Refract. Hard Met.* **1995**, *13*, 187–194. [CrossRef]
30. Ferro, P.; Bonollo, F. Materials selection in a critical raw materials perspective. *Mater. Des.* **2019**, *177*, 107848. [CrossRef]
31. Cârlan Șerban, B.A. Theoretical and Experimental Researches Regarding the Obtaining of New High Entropy Alloy with Biocompatible Properties. Ph.D. Thesis, Politehnica University of Bucharest, Bucharest, Romania, October 2020.

materials

Article

Unusual Force Constants Guided Distortion-Triggered Loss of Long-Range Order in Phase Change Materials

Jiong Wang [1], Dongyu Cui [1], Yi Kong [1,*] and Luming Shen [2]

[1] Powder Metallurgy Research Institute, Central South University, Changsha 410083, China; wangjionga@csu.edu.cn (J.W.); dongyucui@csu.edu.cn (D.C.)
[2] School of Civil Engineering, The University of Sydney, Sydney, NSW 2006, Australia; luming.shen@sydney.edu.au
* Correspondence: yikong@csu.edu.cn

Abstract: Unusual force constants originating from the local charge distribution in crystalline GeTe and Sb_2Te_3 are observed by using the first-principles calculations. The calculated stretching force constants of the second nearest-neighbor Sb-Te and Ge-Te bonds are 0.372 and −0.085 eV/Å^2, respectively, which are much lower than 1.933 eV/Å^2 of the first nearest-neighbor bonds although their lengths are only 0.17 Å and 0.33 Å longer as compared to the corresponding first nearest-neighbor bonds. Moreover, the bending force constants of the first and second nearest-neighbor Ge-Ge and Sb-Sb bonds exhibit large negative values. Our first-principles molecular dynamic simulations also reveal the possible amorphization of Sb_2Te_3 through local distortions of the bonds with weak and strong force constants, while the crystalline structure remains by the X-ray diffraction simulation. By identifying the low or negative force constants, these weak atomic interactions are found to be responsible for triggering the collapse of the long-range order. This finding can be utilized to guide the design of functional components and devices based on phase change materials with lower energy consumption.

Keywords: unusual force constant; phase change materials; GeTe; Sb_2Te_3

Citation: Wang, J.; Cui, D.; Kong, Y.; Shen, L. Unusual Force Constants Guided Distortion-Triggered Loss of Long-Range Order in Phase Change Materials. *Materials* **2021**, *14*, 3514. https://doi.org/10.3390/ma14133514

Academic Editor: Antonio Caggiano

Received: 11 May 2021
Accepted: 21 June 2021
Published: 24 June 2021

1. Introduction

Phase change materials can exist in at least two different phases, such as a crystalline phase and an amorphous phase, featuring rapid and reversible switching between phases with large property contrasts. The most widely used phase change materials for rewritable optical disks are GeTe-Sb_2Te_3 pseudobinary compounds [1–3]. The unique property of GeTe-Sb_2Te_3 also makes it an excellent candidate for various applications in computer science, especially in the field of non-volatile computer memory [4,5]. Important progress has been achieved in the past 20 years to develop new phase change materials and understand their phase change mechanisms [6–15]. It has been known that local interactions between atoms play an important role in phase changes [4,16]. For example, a recent study reveals that local distortions in crystalline can trigger the collapse of long-range order, leading to the formation of the amorphous phase without going through the liquid state [4]. Furthermore, local disorder can also induce localized amorphization even in crystalline phase [16]. According to the simple and early-established theoretical considerations, the transition between the amorphous and the crystalline states, such as the "umbrella flip" model [17], has been attributed to rapid crystallization from the intrinsic similarity in atomic arrangements. However, their atomic arrangements are not yet clear, which results in poor understanding of the phase exchange mechanism. The evidence observed experimentally also indicates that the local atomic arrangements in the amorphous and crystalline states differ considerably [18]. In addition, several recent theoretical reports strongly suggest that the "umbrella flip" model needs to be reevaluated [1,4,19,20].

To further fully understand the underlying physics behind phase change, a solid understanding of the bonding mechanism is a prerequisite. Thus, many studies have been conducted to reveal the chemical bonding nature of phase change materials [20–22]. For example, Wuttig et al. found that resonant bonding is a unique fingerprint and is responsible for the physical properties of the crystalline phase change materials [21]. Through electron localization function analysis, Ma et al. identified the chemically bonded atom pairs and found that the threefold p-type bonding is prominent in amorphous $Ge_2Sb_2Te_5$ [20]. Recent publications by Kolobov et al. also confirmed the proposed resonant bonding [7] and p-type bonding [23] in GeTe.

Directly knowing the force acting between atoms in the phase change material can help us understand the atomic origin of the rapid and reversible switch between the crystalline and amorphous phases, because the force constants (FCs) can quantitatively show us the binding strengths between specific atoms. These strengths are, however, very difficult to be determined experimentally, due to the extremely small distance between atoms in nature. When the eigenvectors are not known, things will get worse because there is no unique solution if one wants to extract the FCs from the measured phonon frequencies [24,25]. In the present study, to gain understanding of the origin of phase change mechanisms, we have extracted the FCs between the atoms in GeTe and Sb_2Te_3 through the approach of chemical bonding analysis by using the first-principles calculations.

2. Materials and Methods

Theoretically, the FCs between atoms can be calculated directly by combining the frozen-phonon method with the first-principles calculations [26,27]. The phonon frequencies can be reduced using the Fourier transformation of the FCs. The reliability of these obtained FCs can then be verified through the comparison between the deduced phonon frequencies from the calculated FCs and the available Raman, infrared, and neutron diffraction experimental data. Without loss of the generality, stable GeTe and Sb_2Te_3 phases are studied in the present work. The reason why the GeTe-Sb_2Te_3 pseudobinary compounds are not chosen is that the precise atomic structures of these pseudobinary compounds remain unknown, due to the randomly distributed Ge and Sb atoms and the intrinsic vacancies [28]. For example, a recently proposed atomic model for $Ge_2Sb_2Te_5$ contains 240 atoms in the unit cell [19], which will become a very challenging problem in terms of computing demands.

In the present first-principles calculations, the Vienna ab initio simulation package (VASP) [29,30] and ALKEMIE (an intelligent computational platform for accelerating materials discovery and design) [31] are used. The calculations are conducted in a plane-wave basis with cut-off energy of 400 eV. The projector-augmented-wave potentials are used to describe the electron-ion interaction. The electron configuration is $[Ar]3d^{10}4s^24p^2$ for Ge, $[Ar]5S^25P^3$ for Sb, and $[Ar]5S^25P^4$ for Te [32]. The exchange and correlation effects are described by the generalized gradient corrections proposed by Perdew-Burke-Ernzerhof [33]. The integration in the Brillouin zone is performed on the special *k* points determined from the Monkhorst-Pack [34] scheme over $17 \times 17 \times 17$ and $13 \times 13 \times 13$ meshes for GeTe and Sb_2Te_3, respectively. The unit cells are fully relaxed with respect to both their volume and shape, as well as to the atomic positions.

Furthermore, the FC matrix and the corresponding phonon frequencies are calculated with the frozen-phonon method, which is implemented in the alloy theoretic automated toolkit (ATAT) [35,36], combined with the VASP. The cutoff distance in constructing a supercell for all the FC calculations is 20 Å, resulting in 256 and 208 atoms for GeTe and Sb_2Te_3, respectively. Gamma point is used in the integration of the Brillouin zone. A test with $3 \times 3 \times 3$ Monkhorst-Pack mesh for Sb_2Te_3 with 208 atoms shows that there is almost no difference between the results integrated from the $3 \times 3 \times 3$ mesh and gamma point. In addition, opposite-sign perturbations are also applied to ensure that the effect of the third-order FCs cancels out exactly in the fit [36].

To study the distortion effect, a 4 × 4 × 1 supercell containing 96 Sb and 144 Te atoms constructed from the conventional hexagonal cell of Sb_2Te_3 is introduced in the present molecular dynamics (MD) simulations. Considering that the short simulation time can be approached with the current first-principles calculations, the simulation system is considered as in an isolated thermodynamic system, i.e., the microcanonical ensemble (NVE) is used in the present MD simulations. An initial temperature is set to 300 K. Each MD job is run to 3 ps with 3 fs time step. Results collected from two selected MD jobs running to 12 ps show that there are no qualitative differences between the results due to short running time.

3. Results

We have carried out detailed calculations of phonons in GeTe and Sb_2Te_3. The crystalline structures used in the calculations are shown in this section. The calculated phonon frequencies at the gamma point are listed for GeTe and Sb_2Te_3, respectively. The selected phonon spectra and phonon density of states along with the available experimental data are alsp shown. The FCs are then reported, while the electronic origins of the unusual FCs are demonstrated. Next, the important findings in these results are given in detail.

3.1. Crystalline Structures

GeTe is crystallized in the rocksalt structure at temperatures above ~700 K, and will distort into a rhombohedral structure at lower temperatures [28]. Sb_2Te_3 crystallizes in a rhombohedral structure as well [28]. The primitive cells of these two stable phases are shown in Figure 1, with the hexagonal conventional cells. The simulated lattice constant of the GeTe primitive rhombohedral cell containing two atoms is 4.3431 Å with the interaxial angle of 58.089°, which agrees well with the corresponding experimental values of 4.3061 Å and 57.942° [37], respectively. The simulated lattice constant of the Sb_2Te_3 primitive rhombohedral cell containing five atoms is 10.5759 Å with the interaxial angle of 23.706°, which is slightly larger than the corresponding experimental values of 10.4469 Å and 23.551° [38].

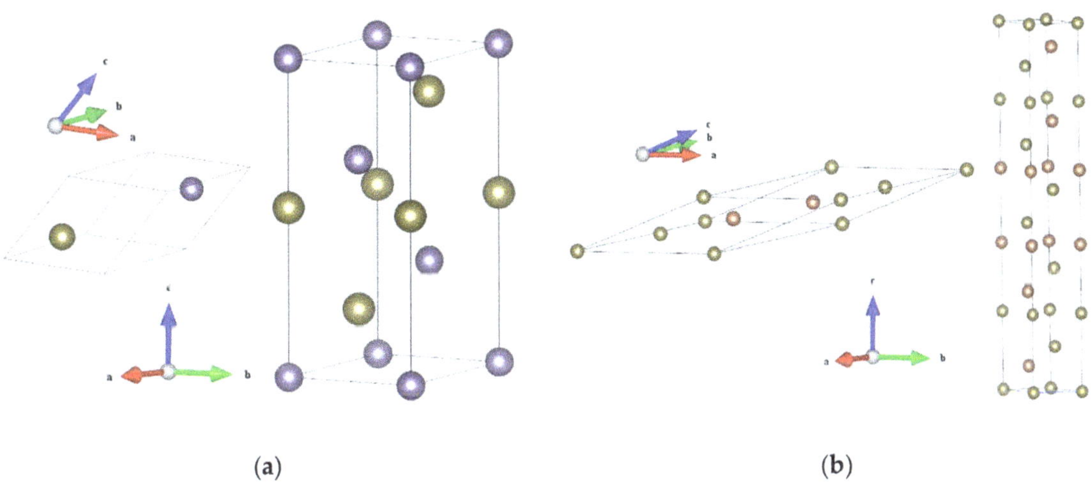

Figure 1. Primitive (left) and conventional (right) cells of (**a**) GeTe and (**b**) Sb_2Te_3. The purple, orange and brown balls represent Ge, Sb and Te atoms, respectively.

3.2. Phonon Frequencies

Tables 1 and 2 list the calculated phonon frequencies at the Brillouin zone center (in cm^{-1}) for GeTe and Sb_2Te_3, respectively, as well as the available experimental data and other theoretical results [39–44]. Figure 2a–c show the phonon density of states (DOS) for Sb_2Te_3 and GeTe, and phonon spectra for Sb_2Te_3 along Γ-Z directions in reciprocal space, respectively, as well as the available neutron scattering [45] and time-of-flight spectrometer [46] data. Here, group theoretical analyses show that in rhombohedral GeTe with two atoms in unit cell, with E mode it is double degenerate, two modes A1 and E are Raman and infrared active, respectively [41]. In addition, for Sb_2Te_3, there are five atoms in unit cell, thus total 12 optical modes exist. Due to the fact that E and A modes represent displacement in the a–b plane and along the c axis, respectively, all the four vibration E optical modes along Γ–Z direction are double degenerate; thus, only eight frequencies can be measured with Raman observation [47]. In short, the overall good agreement between the calculated phonon frequencies and the measured data imply that the FCs calculated from the supercells are reasonable.

Table 1. Calculated phonon frequencies (in cm^{-1}) from the present and previous studies at the zone center for GeTe, as well as the available Raman experimental data.

Method	VASP	Cal.	Cal. [a]	Cal. [b]	Exp.	Exp.	Exp.
E(TO)	90.63	73	73	73	98	80	88
A1(LO)	123.94	152	121	115	140	122	123
Ref.	Present	[39]	[39]	[40]	[41]	[48]	[40]

[a] Complete screening of the dipole-dipole interaction; [b] Consideration of hole (hole concentration is 2.1×10^{21} holes/cm^3).

Table 2. Calculated phonon frequencies (in cm^{-1}) from the present and previous studies at the zone center for Sb_2Te_3, as well as the available experimental data.

Eg1	A1g1	Eu1	Eu2	A1u1	Eg2	A1u2	A1g2	Method	Ref.
43.7	64.0	82.9	98.7	106.6	109.1	133.4	166.5	VASP	Present
46	62	72	99	108	113	145	166	PWSCF	[44]
	69				112		165	Exp.	[42]
43		56						Exp.	[43]
		67	91	110		157		Exp.	[49]

PWSCF: Plane-Wave Self-Consistent Field software.

3.3. Force Constants

Figure 3 demonstrates the calculated stretching and bending FCs in crystalline GeTe and Sb_2Te_3. The stretching action is defined as a change in the length of the bond between two atoms due to an axial force, while the bending action refers to a change in the lateral distance between two atoms due to a force normal to the bond direction. Technically, the stretching FCs can be calculated by:

$$Stretching\ FC = \vec{\mathbf{u}}^{T} \cdot (\mathbf{fc} : \vec{\mathbf{u}}) \quad (1)$$

where $\vec{\mathbf{u}}$ is the 3×1 unit vector defined as the direction from atom a to atom b, $\vec{\mathbf{u}}^{T}$ is the transposition vector of $\vec{\mathbf{u}}$ and $\mathbf{fc}$ is the 3×3 FC matrix between atom a and atom b. Bending FCs are the average of the two perpendicular bending FCs, which can be calculated by:

$$B1 = \vec{\mathbf{a}}^{T} \cdot (\mathbf{fc} : \vec{\mathbf{a}}) \quad (2)$$

and

$$B2 = \vec{\mathbf{b}}^{T} \cdot (\mathbf{fc} : \vec{\mathbf{b}}) \quad (3)$$

where unit vector $\vec{a}$ is perpendicular to vectors $\vec{u}$ and $\vec{b} = \vec{a} \otimes \vec{u}$, normal to the plane containing $\vec{u}$ and $\vec{a}$. *B1* and *B2* represent the separated bending FCs along the two perpendicular directions. It needs to be noted that this stretching-bending force constant model was first proposed by Ceder et al. [50]. In this model, the coordinate system of each force constant matrix is transformed following Equations (1), (2) or (3). Thus, the FCs shown in Figure 3 are not the original 3 × 3 FC matrix but a scalar.

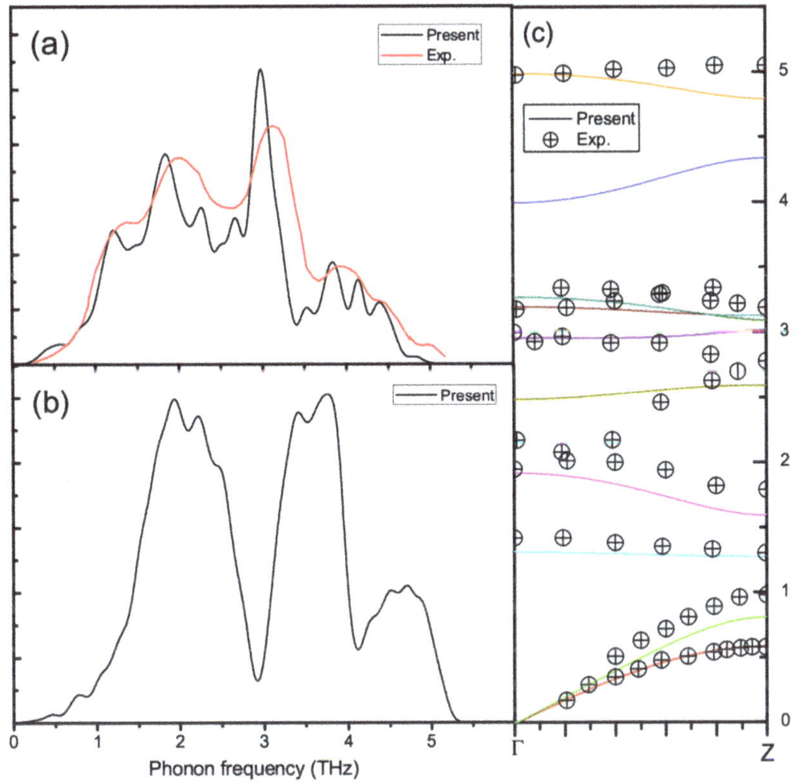

Figure 2. (**a**) Phonon DOS (black line) and the time-of-flight spectrometer data [46] (red line) for Sb_2Te_3, (**b**) phonon DOS for GeTe and (**c**) phonon spectra for Sb_2Te_3 with neutron scattering data [45] (cross-filling of circle).

Some general features can be found from Figure 3. The first and most important unusual feature is that the stretching FCs of the second nearest-neighbor (NN) Sb-Te and Ge-Te bonds are 0.372 and −0.085 eV/Å^2, respectively, much lower than those of the first NN bonds, which are equal to 1.933 eV/Å^2. Besides a rather small bending FC (<0.05 eV/Å^2), the stretching FC of the Ge-Te bond at the second NN distance even becomes negative. These dramatic changes of the FCs are really astonishing, as the differences between the first and second NN distances are only 0.17 Å and 0.33 Å for Sb_2Te_3 and GeTe, respectively.

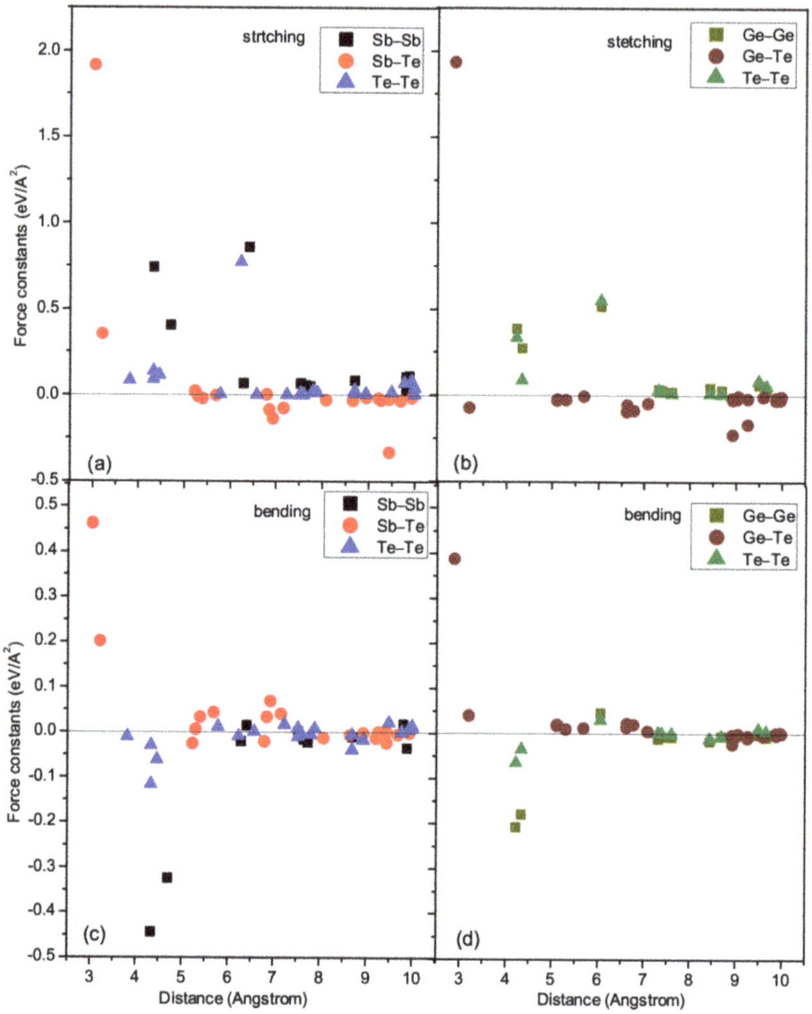

Figure 3. Stretching FCs of (**a**) Sb_2Te_3 and (**b**) GeTe, and bending FCs of (**c**) Sb_2Te_3 and (**d**) GeTe.

The second interesting feature is that the bending FCs of the Ge-Ge and Sb-Sb bonds at the first and second NN distances between like atoms show large negative values, suggesting that the Ge-Ge and Sb-Sb bonds would be ready to move under slight shear perturbations, in accordance with their layer structure nature. It needs to be noted that negative FC between two atoms along a certain direction does not mean that the two atoms will leave, for the stability of a structure depends on the collective behavior of atomic vibrations [51,52]. As the phonon DOS shown in Figure 2a,b, all phonon modes are really positive, which means the two structures are dynamically stable. More details can be found from the separated bending FCs. In GeTe, the bending FCs of Ge-Ge bonds at a distance of 4.2170 Å are $B1 = -0.170$ eV/Å^2 and $B2 = -0.242$ eV/Å^2, while these values become $B1 = -0.014$ eV/Å^2 and $B2 = -0.342$ eV/Å^2 at a distance of 4.3431 Å. The bending FCs of these slightly longer Ge-Ge bonds are rather asymmetric, suggesting that the shear deformation would be strongly direction-dependent. In Sb_2Te_3, the Se-Se bending FCs are $B1 = -0.309$ eV/Å^2 and $B2 = -0.577$ eV/Å^2 at a distance of 4.3447 Å, while these values become $B1 = -0.046\ \mathrm{eV/Å^2}$ and $B2 = -0.602\ \mathrm{eV/Å^2}$ at a distance of 4.7076 Å. More profound asymmetric shear behaviors are observed.

Thirdly, the second largest stretching FCs are found at the third NN distance between like atoms (around 6 Å), higher than most of the shorter bonds (except for the bonds between the first NN unlike atoms), making significant contribution to the stability of the phase. Much shorter bonds such as Ge-Ge and Sb-Sb at around 4 Å, and Sb-Te and Ge-Te at around 3 Å, are weaker than these longer bonds between like atoms.

Fourthly, the overall distributions of the stretching and bending FCs are similar for GeTe and Sb_2Te_3. The highest stretching and bending FCs are around $2.0\ \mathrm{eV/Å^2}$ and $0.4\ \mathrm{eV/Å^2}$, respectively, for both GeTe and Sb_2Te_3. This similarity is consistent with the similar rhombohedral symmetries of these two stable phases.

3.4. Origin of the Unusual Force Constants

It is of high scientific interest to find out the physical origin of these unusual FCs. Hence, in the next section, the spatial valence charge density is analyzed to reveal the bonding nature in the studied structures [53]. Figure 4 shows the crystal structures of Ge-Te and Ge-Ge bonds at the first and second NN distances between unlike atoms and between like atoms, respectively, as well as the isosurface with value of −0.035 number of electrons per bohr^3 in spatial valence charge. From the comparison between Figure 4a,b, one can clearly see that higher charge densities are distributed along the first NN Ge-Te bonds, but are much lower than those along the second NN Ge-Te bonds. These differences suggest that the binding of the first NN Ge-Te is covalent type and much stronger than the binding of the non-covalent second NN bonds, which leads to the large difference between the FCs of these two bonds. Figure 4c shows that the first NN Ge-Ge bond is entirely located in the Ge layer, hence the tensile or compressive movement is limited by the adjacent Ge atoms, resulting in positive stretching FCs. In the out-of-plane direction, however, there are three equally distributed strong covalent first NN Ge-Te bonds, which are oriented by 42.7° to the Ge layer, exerting the attractive force to draw Ge atoms out from the Ge layer, and thus resulting in large negative bending FCs. The situation is similar for the second NN Ge-Ge bonds as shown in Figure 4d. The only difference is that the covalent Ge-Te bonds are not equally distributed, resulting in large asymmetric bending FCs. The reason why the second largest FCs of Ge-Ge and Te-Te bonds occur at distances around 6 Å is that their characteristics are similar to those of the first and second NN Ge-Te bonds. Hence, they are covalent bonds to some extent.

Due to the similar rhombohedra symmetry, the situations for Sb_2Te_3 are not discussed in detail. A noteworthy difference as demonstrated in Figure 5b is that although the isosurfaces near the second NN Sb-Te are not as profound as those near the first NN Sb-Te bonds, they are not spherical but bulged, indicating some covalent binding characteristics. These characteristics cause the stretching and bending FCs of the second NN Sb-Te bonds to be positive, different from those of the second NN Ge-Te bonds (shown in Figure 5a), where the stretching FC is negative and the bending FC is close to zero.

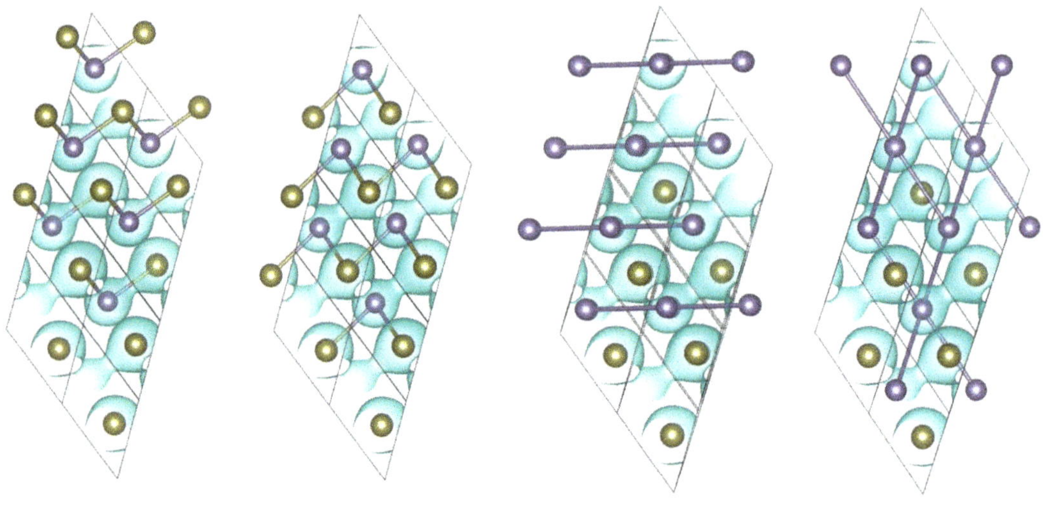

(**a**) Ge-Te first NN (**b**) Ge-Te second NN (**c**) Ge-Ge first NN (**d**) Ge-Ge second NN

Figure 4. Crystal structures shown with the isosurface at −0.035 electrons per $bohr^3$ in spatial valence charge density and the noteworthy bonds for (**a**) Ge-Te first NN, (**b**) Ge-Te second NN, (**c**) Ge-Ge first NN, and (**d**) Ge-Ge second NN. The bonds between atoms are shown with bicolor cylinder type.

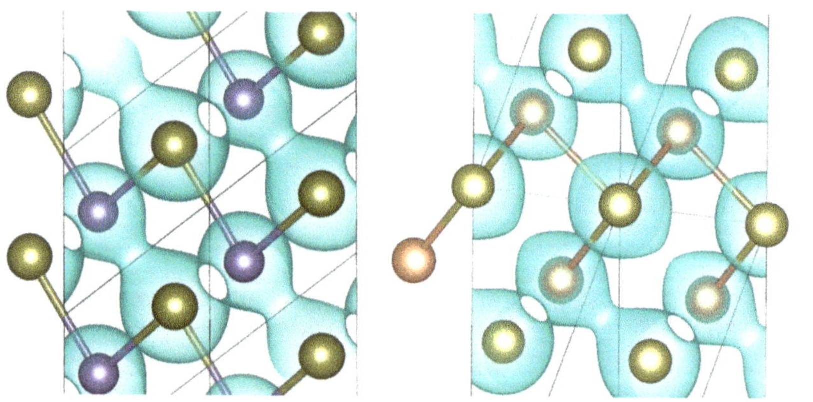

(**a**) GeTe (**b**) Sb_2Te_3

Figure 5. The bonds at the second NN between unlike atoms with the isosurface at −0.035 electrons per $bohr^3$ in the spatial valence charge density of (**a**) GeTe and (**b**) Sb_2Te_3.

4. Discussion

A thorough investigation on these unusual FCs will certainly help us understand the mechanism of phase change. From the calculated FCs, we can conclude that the first NN Ge-Te and Sb-Te bonds have much larger FCs than the other bonds. With this insight, we can explain a lot of experimental observations and simulation results for phase change materials. For example, through atomic simulations, Caravati et al. [54] found that the sharp peak of the bond angle distribution around Sb or Te atoms in amorphous Sb_2Te_3 is centered at 90.46°. According to our structure relaxation of Sb_2Te_3, the dihedral angle between the adjacent first NN Sb-Te bonds is 91.38°, while the dihedral angle between the adjacent second NN Sb-Te bonds is 85.36°. Since the first NN Sb-Te bonds have much larger FCs than the other bonds in Sb_2Te_3, the reported angle distribution by Caravati et al. is very close to that of the first NN Sb-Te bonds. This situation also holds true for GeTe-Sb_2Te_3 pseudobinary compounds such as $Ge_2Sb_2Te_5$. From the structure relaxation of this study, the dihedral angle between the adjacent first NN Ge-Te bonds is 94.6°, while that of the adjacent second NN Ge-Te bond is 82.39° in GeTe. Through atomic simulations, Sun et al. [55] found that the sharp peaks of the bond angle distributions around Ge, Sb and Te atoms in amorphous $Ge_2Sb_2Te_5$ are centered at ~97°, ~90° and ~89°, respectively. That is to say, although the total pair correlation functions for amorphous and crystalline states are rather different [18], microscopically the strongest Te-Ge and Te-Sb binding is preserved in phase change materials. Furthermore, other examples can also be seen in rather different situations. We all know that besides temperature, high pressure can also induce phase change. According to Krbal et al. [56], among the remaining bonds in $GeSb_2Te_4$ under pressure of 46 GPa, 59% of them are Ge-Te and Sb-Te bonds, while only 8% of them are Ge-Ge and Sb-Sb bonds.

We would like to further stress the importance of quantitatively knowing the FCs. Recently, Kolobov et al. [4] demonstrated that appropriate distortion in crystals can trigger the destruction of the subsystem with weaker bonds and the subsequent collapse of the long-range order, which will generate the amorphous phase without going through the liquid state, thus making it possible to significantly reduce the energy consumption through the use of shorter pulses or excitation of coherent optical phonons. As motivated by Kolobov et al. [4], we further demonstrate that through locally distorting the bonds, the phase change materials can be amorphized through MD simulations. From the results of FCs in Sb_2Te_3, we know that the FCs at the second NN distance are much lower than those at the first NN distance between unlike atoms, although the differences between the first and second distances are only 0.17 Å. Figure 6 shows the first and second Sb-Te interactions in hexagonal conventional cell with polyhedral view. From the figure, it is noted that each Sb atom is connected with three surrounding Te atoms to form a tetrahedron. However, it is not possible to entirely separate the first and second NN interactions between Sb-Te atoms. Nevertheless, we consider here one possible distortion method, which is to compress or stretch the two Sb atoms along the <001> direction as shown in Figure 6b,c, in order to weaken or strengthen, respectively, the first and second interactions simultaneously.

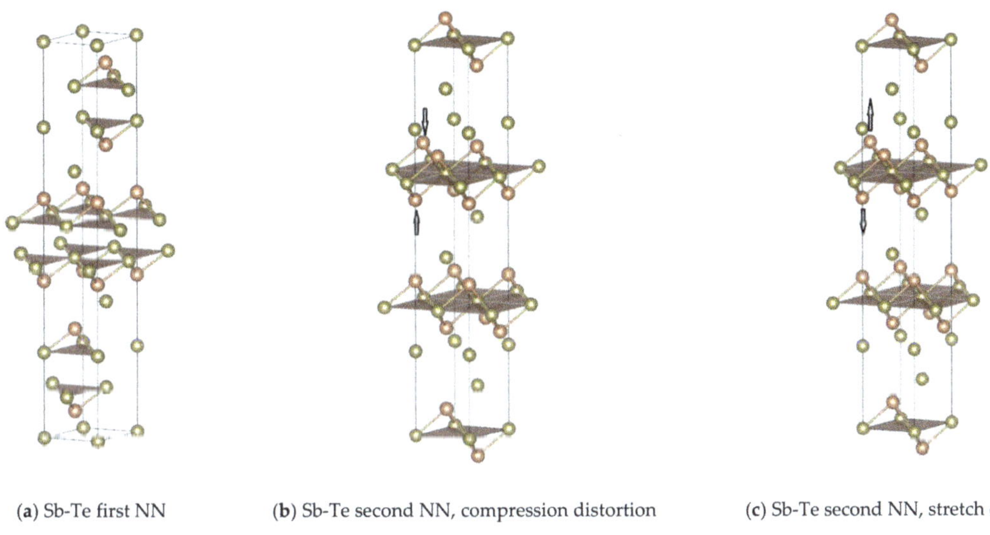

(**a**) Sb-Te first NN (**b**) Sb-Te second NN, compression distortion (**c**) Sb-Te second NN, stretch distortion

Figure 6. The first and second NN Sb-Te interactions shown in conventional hexagonal cell with polyhedral view: (**a**) first NN, (**b**) second NN, compression distortion, and (**c**) second NN, stretch distortion with arrows.

A 4 × 4 × 1 supercell containing 240 atoms constructed from the conventional hexagonal cell of Sb_2Te_3 is introduced to study the above mentioned distortion effect. The step size of the distortion along the <001> direction is 0.5% of the lattice constant of the conventional hexagonal cell in the direction normal to the basal plane. For the purpose of convenience, we use the negative value to represent the elongation distortion of the Sb atoms along the <001> direction in the tetrahedron, as well as the positive value to represent the compressive distortion. The distorted structures still retain the long-range order of the crystalline phase, as confirmed by the generated X-ray diffraction pattern shown in Figure 7. In the X-ray simulation, the wavelength λ = 1.540562 Å in Cu Kα1 emission lines is used as radiation source. The diffraction occurs in the range from 5 to 90° with an increment size of 0.05°. In all distorted structures, the spectra are dominated by the crystalline peaks, showing characteristics of the ideal Sb_2Te_3 model. These observations suggest that the introduced disturbances are local in nature, i.e., the structure remains crystalline evidenced by the Bragg X-ray diffraction experiment.

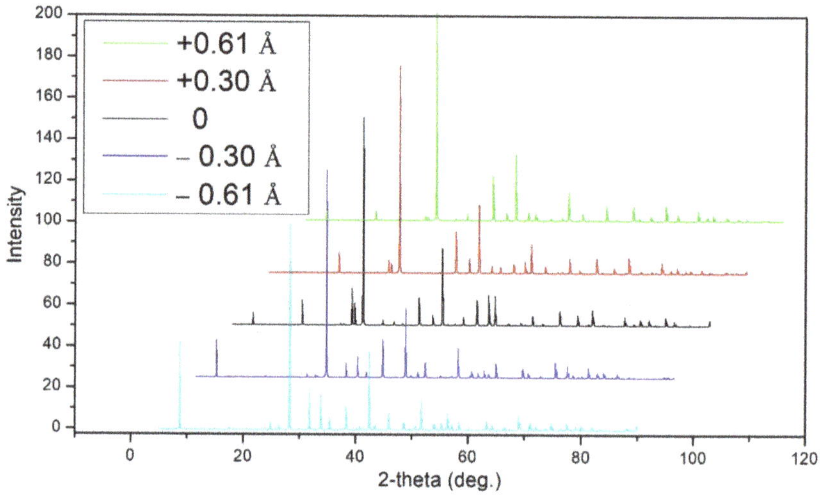

Figure 7. X-ray diffraction spectra of Sb_2Te_3 generated for the ideal and distorted structures.

From the system energy point of view, the distortion energy varies with the distortion distance along the <001> direction as shown in Figure 8. Bond elongation along the <001> direction of the tetrahedron results in the compression of the first NN Sb-Te bonds which features strong force constants, while the tension of the second NN Sb-Te bonds features weak force constants. From Figure 8 it is noted that when the magnitude of distortion is smaller than 0.17 Å, the distortion energy of stretching process is higher than that of compressing process, suggesting that the strong bonds are more difficult to be compressed and the weak bonds are easier to be stretched, consistent with the FCs results. When the magnitude of distortion is larger than 0.17 Å, the bond initially featuring strong FCs will become longer than the bond initially featuring weak FCs, leading to the exchange of the relative strong and weak characteristics of FCs, namely, the initial stronger bonds become weaker and the initial weaker bonds become stronger.

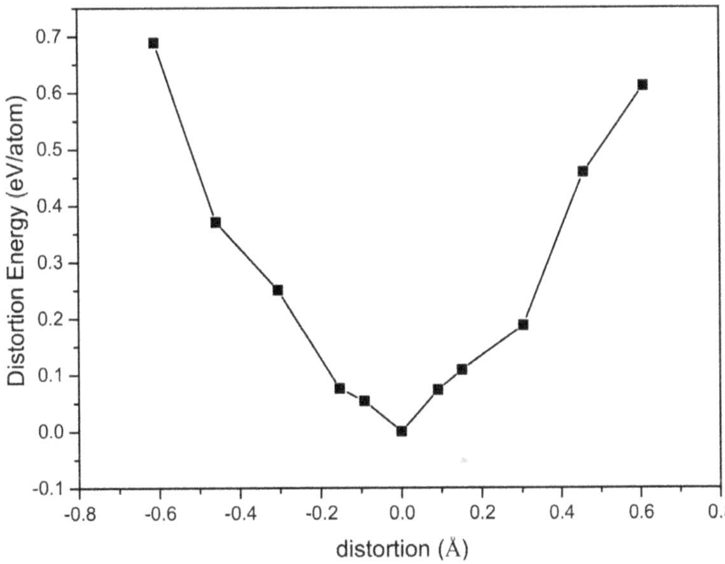

Figure 8. The distortion energies of Sb_2Te_3 vary with the distortion distances along the <001> direction.

The distorted structures are then deployed into MD simulations to see whether these locally disturbed crystalline structures can become amorphous without initially presetting very high temperature in MD simulations [57]. The obtained pair correlation function, or radial distribution function (RDF) g(r) for different distortions is shown in Figure 9. To simplify the figure, the average RDF values with the equal distance of the stretching and compressing distortion are shown here. From the figure, one can find that with increasing distortion, the initial crystalline structure will finally become amorphous during the simulated time. As different distortion will lead to different distortion energies, quantitatively we can know the different bond strengths between atoms. This finding is of importance for guiding us on where and how the appropriate distortion should be applied, and thus optimizing the design and fabrication of components/devices based on phase change materials with lower energy consumption.

Moreover, it is worthy to note that for GeTe and Sb2Te3, rhombohedral structure is closely related to rocksalt structure. For example, GeTe will distort into a rhombohedral structure at lower temperatures from rocksalt structure at temperatures below ~700 K [28]. Is the direction of the vector connected with the direction of the atomic displacement in the phase-shift from rhombohedral structure to rocksalt structure or vice versa? Do the calculated negative FCs of the rhombohedral GeTe reflect, in a relevant way, the distorted rocksalt cubic structure? The present work concentrates on the transition from the crystalline to the amorphous state at room temperature; such investigations are beyond the scope of this paper and will be the subject of future study. Transformation between rhombohedral structure and rocksalt structure from other perspectives can be found in literature as [28,47,58].

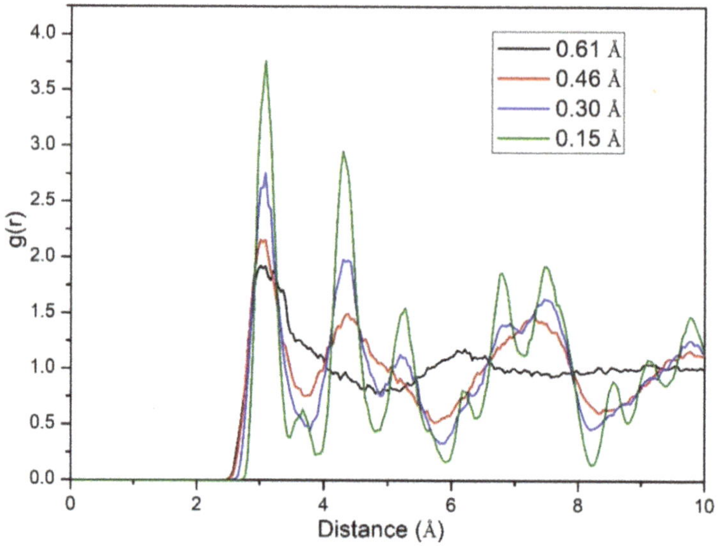

Figure 9. RDF of different distorted structures of Sb_2Te_3 after 3ps NVE MD simulations. It is clearly shown that when the distortion is large enough, the locally distorted structure will lose its long-range order and be finally amorphized.

5. Conclusions

Through analyzing the FCs of GeTe and Sb_2Te_3 by using the first-principles calculations, we find that the overall characteristics of the stretching and bending FCs are similar for GeTe and Sb_2Te_3. In particular, (1) the stretching FCs between unlike atoms at the first NN distance in GeTe and Sb_2Te_3 are equal to 1.933 eV/Å^2; (2) the stretching FC of Sb-Te bonds at the second NN distance is only 19% of that at the first NN distance, while the stretching FC of Ge-Te bond at the second NN distance even becomes negative; (3) the bending FCs of the first and second NN bonds between like atoms feature large negative values, and are strongly direction-dependent; and (4) the second NN bonds between like atoms in Sb_2Te_3 and GeTe are found to have the second largest stretching FCs.

The origin of these unusual FCs is explained through analyzing the spatial valence charge density. Our first-principles molecular dynamic simulations reveal the possible amorphization of phase change materials through local distortions of the bonds featuring weak or strong FCs, while the crystalline structure remains observed by the X-ray diffraction experiment. We believe that the obtained FCs are of a general nature and thus can be used to understand the phase change mechanism of the GeTe-Sb_2Te_3 pseudobinary compounds. This interesting result is of importance for understanding phase change materials as it might provide a recipe to generate two-phase-based devices with lower energy consumption.

Author Contributions: Conceptualization, J.W. and Y.K.; methodology, J.W. and Y.K.; investigation, J.W., D.C. and Y.K.; resources, J.W., Y.K. and L.S.; data curation, J.W. and Y.K.; writing—original draft preparation, J.W., D.C. and Y.K.; writing—review and editing, J.W., D.C., Y.K. and L.S.; visualization, J.W. and Y.K.; supervision, Y.K.; project administration, J.W.; funding acquisition, J.W. All authors have read and agreed to the published version of the manuscript.

Funding: This research was funded by the National Key Research and Development Program of China (Materials Genome Initiative), grant number 2017YFB0701700, and the National Natural Science Foundation of China, grant numbers 51771234 and 51601228.

Institutional Review Board Statement: Not applicable.

Informed Consent Statement: Not applicable.

Data Availability Statement: Data are contained within the article.

Acknowledgments: The financial support from the National Key Research and Development Program of China (Materials Genome Initiative) (grant number 2017YFB0701700), the National Natural Science Foundation of China (grant numbers 51771234 and 51601228) is greatly acknowledged. First-principles calculations were partially carried out at the High Performance Computing of Central South University.

Conflicts of Interest: The authors declare no conflict of interest.

References

1. Lencer, D.; Salinga, M.; Wuttig, M. Design Rules for Phase-Change Materials in Data Storage Applications. *Adv. Mater.* **2011**, *23*, 2030–2058. [CrossRef]
2. Gu, M.; Zhang, Q.; Lamon, S. Nanomaterials for optical data storage. *Nat. Rev. Mater.* **2016**, *1*, 16070. [CrossRef]
3. Bilovol, V.; Fontana, M.; Rocca, J.; Chanduví, H.M.; Navarro, A.M.; Gil Rebaza, A.; Errico, L.; Liang, A.; Errandonea, D.; Ureña, A. Structural, vibrational and electronic properties in the glass-crystal transition of thin films Sb70Te30 doped with Sn. *J. Alloys Compd.* **2020**, *845*, 11. [CrossRef]
4. Kolobov, A.; Krbal, M.; Fons, P.; Tominaga, J.; Uruga, T. Distortion-triggered loss of long-range order in solids with bonding energy hierarchy. *Nat. Chem.* **2011**, *3*, 311–316. [CrossRef] [PubMed]
5. Zhang, W.; Mazzarello, R.; Wuttig, M.; Ma, E. Designing crystallization in phase-change materials for universal memory and neuro-inspired computing. *Nat. Rev. Mater.* **2019**, *4*, 150–168. [CrossRef]
6. Sosso, G.C.; Salvalaglio, M.; Behler, J.; Bernasconi, M.; Parrinello, M. Heterogeneous Crystallization of the Phase Change Material GeTe via Atomistic Simulations. *J. Phys. Chem. C* **2015**, *119*, 6428–6434. [CrossRef]
7. Kolobov, A.V.; Fons, P.; Tominaga, J.; Hase, M. Excitation-Assisted Disordering of GeTe and Related Solids with Resonant Bonding. *J. Phys. Chem. C* **2014**, *118*, 10248–10253. [CrossRef]
8. Deringer, V.; Lumeij, M.; Stoffel, R.P.; Dronskowski, R. Mechanisms of Atomic Motion Through Crystalline GeTe. *Chem. Mater.* **2013**, *25*, 2220–2226. [CrossRef]
9. Krbal, M.; Kolobov, A.V.; Fons, P.; Tominaga, J.; Elliott, S.R.; Hegedus, J.; Giussani, A.; Perumal, K.; Calarco, R.; Matsunaga, T.; et al. Crystalline GeTe-based phase-change alloys: Disorder in order. *Phys. Rev. B* **2012**, *86*, 045212. [CrossRef]
10. Matsunaga, T.; Fons, P.; Kolobov, A.V.; Tominaga, J.; Yamada, N. The order-disorder transition in GeTe: Views from different length-scales. *Appl. Phys. Lett.* **2011**, *99*, 231907. [CrossRef]
11. Xu, Y.; Wang, X.; Zhang, W.; Schäfer, L.; Reindl, J.; Bruch, F.V.; Zhou, Y.; Evang, V.; Wang, J.; Deringer, V.L.; et al. Materials Screening for Disorder-Controlled Chalcogenide Crystals for Phase-Change Memory Applications. *Adv. Mater.* **2021**, *33*, e2006221. [CrossRef] [PubMed]
12. Sun, L.; Zhou, Y.-X.; Wang, X.-D.; Chen, Y.-H.; Deringer, V.L.; Mazzarello, R.; Zhang, W. Ab initio molecular dynamics and materials design for embedded phase-change memory. *NPJ Comput. Mater.* **2021**, *7*, 29. [CrossRef]
13. Kang, L.; Chen, L. Overview of the Role of Alloying Modifiers on the Performance of Phase Change Memory Materials. *J. Electron. Mater.* **2021**, *50*, 1–24. [CrossRef]
14. Yuan, H.; Wang, J.; Hu, B.; Zhao, R.; Du, Y.; Zhang, S.-Y. Thermodynamic assessment of the Te-X (X = As, Si, Co) systems. *Calphad* **2020**, *68*, 101743. [CrossRef]
15. Dong, C.C.; Wang, J.; Hu, B.; Zhu, L.P.; Qu, Q.; Du, Y. Thermodynamic modeling of the Te-X (X = Zr, Ce, Eu) systems. *Calphad* **2020**, *74*, 102281. [CrossRef]
16. Siegrist, T.; Jost, P.; Volker, H.; Woda, M.; Merkelbach, P.; Schlockermann, C.; Wuttig, M. Disorder-induced localization in crystalline phase-change materials. *Nat. Mater.* **2011**, *10*, 202–208. [CrossRef] [PubMed]
17. Kolobov, A.V.; Fons, P.; Frenkel, A.; Ankudinov, A.L.; Tominaga, J.; Uruga, T. Understanding the phase change mechanism of rewritable optical media. *Nat. Mater.* **2004**, *3*, 703–708. [CrossRef]
18. Wuttig, M.; Yamada, N. Phase-change materials for rewriteable data storage. *Nat. Mater.* **2007**, *6*, 824–832. [CrossRef]
19. Liu, X.Q.; Li, X.B.; Zhang, L.; Cheng, Y.Q.; Yan, Z.G.; Xu, M.; Han, X.D.; Zhang, S.B.; Zhang, Z.; Ma, E. New Structural Picture of the Ge2Sb2Te5 Phase-Change Alloy. *Phys. Rev. Lett.* **2011**, *106*, 025501. [CrossRef]
20. Xu, M.; Cheng, Y.Q.; Sheng, H.W.; Ma, E. Nature of Atomic Bonding and Atomic Structure in the Phase-Change Ge2Sb2Te5 Glass. *Phys. Rev. Lett.* **2009**, *103*, 195502. [CrossRef]
21. Shportko, K.V.; Kremers, S.; Woda, M.; Lencer, D.; Robertson, J.; Wuttig, M. Resonant bonding in crystalline phase-change materials. *Nat. Mater.* **2008**, *7*, 653–658. [CrossRef]
22. Wełnic, W.; Pamungkas, A.; Detemple, R.; Steimer, C.; Blügel, S.; Wuttig, M. Unravelling the interplay of local structure and physical properties in phase-change materials. *Nat. Mater.* **2006**, *5*, 56–62. [CrossRef]
23. Kolobov, A.V.; Fons, P.; Tominaga, J. Local instability of p-type bonding makes amorphous GeTe a lone-pair semiconductor. *Phys. Rev. B* **2013**, *87*, 155204. [CrossRef]
24. Cochran, W. The Relation between Phonon Frequeneies and Interatomie Force Constants. *Acta Cryst. A* **1971**, *27*, 556. [CrossRef]
25. Leigh, R.S.; Szigeti, B.; Tewary, V.K. Force Constants and Lattice Frequencies. *Proc. R. Soc. Lond. A* **1971**, *320*, 505–526.

26. Van de Walle, A.; Ceder, G. The effect of lattice vibrations on substitutional alloy thermodynamics. *Rev. Mod. Phys.* **2002**, *74*, 11–45. [CrossRef]
27. Kong, Y.; Zhao, D.; Zhou, L.; Guo, H.; Du, Y. Energetic, mechanical, and vibrational stability of metastable OsC phase. *J. Appl. Phys.* **2010**, *108*, 083523. [CrossRef]
28. Da Silva, J.L.F.; Walsh, A.; Lee, H. Insights into the structure of the stable and metastable (GeTe)m(Sb2Te3)n compounds. *Phys. Rev. B* **2008**, *78*, 224111. [CrossRef]
29. Kresse, G.; Furthmüller, J. Efficient iterative schemes for ab initio total-energy calculations using a plane-wave basis set. *Phys. Rev. B* **1996**, *54*, 11169–11186. [CrossRef]
30. Kresse, G.; Hafner, J. Ab initio molecular dynamics for liquid metals. *Phys. Rev. B* **1993**, *47*, 558–561. [CrossRef] [PubMed]
31. Wang, G.; Peng, L.; Li, K.; Zhu, L.; Zhou, J.; Miao, N.; Sun, Z. ALKEMIE: An intelligent computational platform for accelerating materials discovery and design. *Comput. Mater. Sci.* **2021**, *186*, 110064. [CrossRef]
32. Blöchl, P.E. Projector augmented-wave method. *Phys. Rev. B* **1994**, *50*, 17953–17979. [CrossRef] [PubMed]
33. Perdew, J.P.; Burke, K.; Ernzerhof, M. Generalized gradient approximation made simple. *Phys. Rev. Lett.* **1996**, *77*, 3865–3868. [CrossRef] [PubMed]
34. Monkhorst, H.J.; Pack, J.D. Special points for Brillouin-zone integrations. *Phys. Rev. B* **1976**, *13*, 5188–5192. [CrossRef]
35. Van de Walle, A.; Asta, M.; Ceder, G. The alloy theoretic automated toolkit: A user guide. *Calphad* **2002**, *26*, 539–553. [CrossRef]
36. Van de Walle, A. Multicomponent multisublattice alloys, nonconfigurational entropy and other additions to the Alloy Theoretic Automated Toolkit. *Calphad* **2009**, *33*, 266–278. [CrossRef]
37. Goldak, J.; Barrett, C.S.; Innes, D.; Youdelis, W. Structure of alpha GeTe. *J. Chem. Phys.* **1966**, *44*, 3323–3325. [CrossRef]
38. Anderson, T.L.; Krause, H.B. Refinement of the Sb_2Te_3 and Sb_2Te_3Se structures and their relationship to nonstoichiometric $Sb_2Te_{3\text{-}y}Se_y$ compounds. *Acta Cryst. B* **1974**, *30*, 1307–1310. [CrossRef]
39. Shaltaf, R.; Durgun, E.; Raty, J.-Y.; Ghosez, P.; Gonze, X. Dynamical, dielectric, and elastic properties of GeTe investigated with first-principles density functional theory. *Phys. Rev. B* **2008**, *78*, 78. [CrossRef]
40. Shaltaf, R.; Gonze, X.; Cardona, M.; Kremer, R.K.; Siegle, G. Lattice dynamics and specific heat of α-GeTe: Theoretical and experimental study. *Phys. Rev. B* **2009**, *79*, 075204. [CrossRef]
41. Steigmeier, E.; Harbeke, G. Soft phonon mode and ferroelectricity in GeTe. *Solid State Commun.* **1970**, *8*, 1275–1279. [CrossRef]
42. Richter, W.; Kohler, H.; Becker, C.R. A Raman and far-infrared investigation of phonons in the rhombohedral V2-VI3 compounds. *Phys. Stat. Sol. B* **1977**, *84*, 619–628. [CrossRef]
43. Richter, W.; Krost, A.; Nowak, U.; Anastassakis, E. Anisotropy and dispersion of coupled plasmon-LO-phonon modes in Sb_2Te_3-need modify. *J. Phys. B Condens. Matter* **1982**, *49*, 191–198. [CrossRef]
44. Sosso, G.C.; Caravati, S.; Bernasconi, M. Vibrational properties of crystalline Sb2Te3from first principles. *J. Phys. Condens. Matter* **2009**, *21*, 095410. [CrossRef]
45. Antimony telluride (Sb2Te3) phonon dispersion, phonon frequencies. In *Landolt-Börnstein Database: Antimony Telluride (Sb_2Te_3) Phonon Dispersion, Phonon Frequencies*; Madelung, O.; Rössler, U.; Schulz, M. (Eds.) Landolt-Börnstein—Group III Condensed Matter; Springer: New York, NY, USA, 1983; Volume 41C.
46. Rauh, H.; Geick, R.; Kohler, H.; Nucker, N.; Lehner, N. Generalized phonon density of states of the layer compounds Bi2Se3, Bi2Te3, Sb2Te3and Bi2(Te0.5Se0.5)3, (Bi0.5Sb0.5)2Te3. *J. Phys. C Solid State Phys.* **1981**, *14*, 2705–2712. [CrossRef]
47. Wang, B.-T.; Souvatzis, P.; Eriksson, O.; Zhang, P. Lattice dynamics and chemical bonding in Sb2Te3 from first-principles calculations. *J. Chem. Phys.* **2015**, *142*, 6. [CrossRef]
48. Andrikopoulos, K.S.; Yannopoulos, S.N.; Voyiatzis, G.A.; Kolobov, A.V.; Ribes, M.; Tominaga, J. Raman scattering study of the a-GeTe structure and possible mechanism for the amorphous to crystal transition. *J. Phys. Condens. Matter* **2006**, *18*, 965–979. [CrossRef]
49. Drope, R. *Zur Gitterdynamik von V2VI3-Verbindungen: Tunnelspektroskopische und infrarotspektroskopische Messungen an Sb2Te3-Einkristallen*; RWTH Aachen University: Aachen, Germany, 1975.
50. Wu, E.J.; Ceder, G.; van de Walle, A. Using bond-length-dependent transferable force constants to predict vibrational entropies in Au-Cu, Au-Pd, and Cu-Pd alloys. *Phys. Rev. B* **2003**, *67*, 7. [CrossRef]
51. Adjaoud, O.; Steinle-Neumann, G.; Burton, B.P.; van de Walle, A. First-principles phase diagram calculations for the HfC–TiC, ZrC–TiC, and HfC–ZrC solid solutions. *Phys. Rev. B* **2009**, *80*, 7. [CrossRef]
52. Liu, J.Z.; Ghosh, G.; van de Walle, A.; Asta, M. Transferable force-constant modeling of vibrational thermodynamic properties in fcc-basedAl−TM(TM=Ti, Zr, Hf) alloys. *Phys. Rev. B* **2007**, *75*, 15. [CrossRef]
53. Kong, Y.; Li, J.H.; Kong, L.T.; Liu, B.X. Role of spatial valence charge density on the metastability of an immiscible binary metal system at equilibrium. *Phys. Rev. B* **2005**, *72*, 024209. [CrossRef]
54. Caravati, S.; Bernasconi, M.; Parrinello, M. First-principles study of liquid and amorphousSb2Te3. *Phys. Rev. B* **2010**, *81*, 014201. [CrossRef]
55. Sun, Z.; Zhou, J.; Blomqvist, A.; Johansson, B.; Ahuja, R. Formation of Large Voids in the Amorphous Phase-Change Memory Ge2Sb2Te5 Alloy. *Phys. Rev. Lett.* **2009**, *102*, 075504. [CrossRef] [PubMed]

56. Krbal, M.; Kolobov, A.V.; Haines, J.; Fons, P.; Levelut, C.; Le Parc, R.; Hanfland, M.; Tominaga, J.; Pradel, A.; Ribes, M. Initial Structure Memory of Pressure-Induced Changes in the Phase-Change Memory Alloy Ge2Sb2Te5. *Phys. Rev. Lett.* **2009**, *103*, 115502. [CrossRef]
57. Anzellini, S.; Burakovsky, L.; Turnbull, R.; Bandiello, E.; Errandonea, D. P–V–T Equation of State of Iridium Up to 80 GPa and 3100 K. *Crystals* **2021**, *11*, 452. [CrossRef]
58. Jeong, K.; Park, S.; Park, D.; Ahn, M.; Han, J.; Yang, W.; Jeong, H.-S.; Cho, M.-H. Evolution of crystal structures in GeTe during phase transition. *Sci. Rep.* **2017**, *7*, 12. [CrossRef]

MDPI
St. Alban-Anlage 66
4052 Basel
Switzerland
Tel. +41 61 683 77 34
Fax +41 61 302 89 18
www.mdpi.com

Materials Editorial Office
E-mail: materials@mdpi.com
www.mdpi.com/journal/materials

www.ingramcontent.com/pod-product-compliance
Lightning Source LLC
Chambersburg PA
CBHW040312270926
51850CB00007BC/1133
9783036556604